THE SEI WHALE: Population Biology, Ecology and Management

111615

QL
737
.C424
H67
1987

THE SEI WHALE:
Population Biology, Ecology & Management

JOSEPH HORWOOD

QL 737 .C424 H67 1987
Horwood, J.
The sei whale : population
biology, ecology and...
111615 05010913 c.1

EMAR LIBRARY
FISHERIES AND OCEANS CANADA
501 UNIVERSITY CRESCENT
WINNIPEG, MB
R3T 2N6 CANADA

CROOM HELM
London • New York • Sydney

© 1987 Joseph W. Horwood
Croom Helm Ltd, Provident House, Burrell Row,
Beckenham, Kent, BR3 1AT

Croom Helm Australia, 44-50 Waterloo Road,
North Ryde, 2113, New South Wales

Published in the USA by
Croom Helm
in association with Methuen, Inc.
29 West 35th Street
New York, NY 10001

British Library Cataloguing in Publication Data

Horwood, Joseph
The sei whale: population biology,
ecology and management.
1. Sei whale 2. Mammal populations
I. Title
599.5′1 QL737.C424

ISBN 0-7099-4786-0

Library of Congress Cataloging-in-Publication Data

Horwood, Joseph.
The sei whale.

Bibliography: p.
Includes index.
1. Sei whale. I. Title.
QL737.C424H67 1987 599.5′1 87-13727
ISBN 0-7099-4786-0

Printed and bound in Great Britain by
Biddles Ltd, Guildford and King's Lynn

CONTENTS

TABLES AND FIGURES

Tables

Figures

PREFACE

This book is primarily about the population biology of the sei whale. Population biology is the study of those aspects of the life of an animal or plant that determine how and why numbers vary, and how the organism, as a component of a population, responds to fluctuations in its numbers and its environment. Knowledge of the population biology is essential for the medium and long term management of any organism.

The management of the great whales has attracted public concern for many years, with the UN Stockholm Conference on the Environment of 1971 calling for a moritorium on commercial whaling. This did not occur but management practices did respond to international concern, and quotas, agreed by the International Whaling Commission, were reduced, and many species and stocks were fully protected. This caused distress to the whaling industry with losses of income and jobs. At the same time many of the 'conservation' organizations argued that on moral grounds whaling should cease, and that the science of management was inadequate to set quotas without undue risk.

This book does not attempt to present a single answer for this issue; instead its approach is to present all the relevant information on one species of whale. It can be regarded as a comprehensive review for cetacean specialists, but for those more generally interested in the management of biological populations, or of whales in particular, the book shows the great wealth of knowledge achieved in some areas and the hiatuses in others. The reader is left, better informed one hopes, to decide whether we can manage the exploitation of the large whales if this is required.

The sei whale has been specifically chosen. The intensive exploitation of this 15 metre, 19 tonne, baleen whale occurred over a decade from about 1955, and followed the reductions in stock size and quotas of blue and fin whales. By this time scientists were regularly sampling from the catches, and they were able to obtain estimates of vital biological parameters over a period of substantial declines in population numbers; this was not so for the blue whales. Consequently one might anticipate that our knowledge of the population biology of the sei whale might be better than that of the other species of large whale. However nature has conspired against this, and the long periods of small catches and the effects of exploitation of other species have complicated the story. Given these and other problems I have resisted drawing speculative conclusions, considering that where we do not know it is of more use to stress this fact. Particularly I have avoided using information inferred from other related species; this is a necessary technique for management where decisions are required from incomplete information, but the sei whale is only a sei whale because it is different from the other species, and we risk being misled by superficial similarities.

ACKNOWLEDGEMENTS

I am indebted to many friends and colleagues for the assistance they have given to me in the preparation of this study. In particular I extend my thanks to the following for their help and patience; U. Arnason, P. Best, H. Braham, J. Breiwick, S. Brown, J. Croxall, G. Donovan, R. Gambell, C. Lockyer, Y. Masaki, S. Ohsumi, J. Sigurjonsson, and S. Wada, and to A. Preston, D. Garrod, A. Jamieson, J. Shepherd and other colleagues at the Fisheries Laboratory, Lowestoft for their valuable assistance. Access to important, unpublished manuscripts was generously granted by P. Best and C. Lockyer, and E. Mitchell and V. Kozicki; references to them are cited in the text.

The following organizations are gratefully recognised for permitting the publication of tables, figures and photographs, which are explicitly cited in the text: Far Seas Fisheries Research Laboratory, Japan; I.C.E.S.; Institute of Oceanographic Sciences, U.K.; Sea Mammal Research Unit, U.K.; Whales Research Institute, Japan. In addition, publication of Crown copyright materials are with permission of the Controller of Her Majesty's Stationery Office, and such tables and figures are identified by the copyright symbol and year of first publication.

I am indebted to the following organizations which provided essential financial support to allow the publication of this book:

Japan Whaling Association, Tokyo, Japan.

National Marine Mammal Laboratory, Seattle, USA.

North Slope Borough, Barrow, Alaska, USA.

Chapter One

GENERAL DESCRIPTION

INTRODUCTION

The "most graceful of all whales, as its proportions are so perfect, and wanting the clumsy strength of the two larger Balaenoptera (fin and blue), sperms and Megaptera (humpback)", so considered Haldane (1909) in his description of the sei whale from the Scottish whale fishery of 1908. He also added, ".. it is also far the best to eat, the flesh tasting of something between pork and veal, and quite tender". Andrews (1916a) likened the speed of the sei whale with that of the cheetah, as capable of great speed over short distances but soon tiring. The sei whale has been found in all oceans but tends to be more offshore than the other species of large whales, and Jenkins (1921) remarked that until catching started about the turn of the century it was considered to be the rarest of the European whales. From the USSR Red Book of threatened species Ivashin (1985) still describes the sei whales as rare, but since the turn of the century its numbers have been significantly reduced in all oceans.

Like most of the large whales the sei spends the winters in the warm sub-tropical waters, migrating to the temperate or polar regions to feed in the summer, but the sei whale does not penetrate the polar waters as far as do the blue, fin, humpback and minke whales. The coastal summer fisheries have always found the local abundance of sei whales erratic; perhaps associated with availability of the supply of planktonic food.

Dorsally the whales are coloured grey to dark grey, and this colour extends down the sides and under the flippers. Ventrally the colour is usually a greyish-white from the chin backwards, and this may cover the extent of the entire ventral grooves, but more usually the light colouring stops short of the end of the grooves. It does not cover the mandibles. In some individuals a whitish area may extend ventrally to the flippers, and a white streak may be seen going backwards from the ear. Matthews (1938) described the colour patterns of numerous sei whales and considered that the colour intensity, and extent of the light ventral area, was much more variable than in other whales. In one example the white region stopped only 1.5 m from the chin. The baleen plates are much finer than in the other rorquals and are usually dark to black in colour, with the bristles being white to greyish-white. At the anterior end white baleen plates are often found and these are sometimes associated with a white spot on the jaw. Rare white or albino individuals have been seen. No colour differences are found between the sexes.

The sizes of adults differ with sex with the females about 0.5 to 0.6 m larger than the males. Calves are born at a length of 4.5 m, and grow quickly to about 11 m at age 2 years. From 10 years of age their lengths change very little with the males typically 14.5 to 15.0 m and the females 15.2 to 15.8 m. In the Antarctic several whales of 18.3 m (60ft) or over have been caught with similar, but less reliable, records from the North Atlantic. The sei whale of the Southern Hemisphere is somewhat larger than that of the Northern Hemisphere.

The distribution and behaviour of the sei whale are characteristic, with its distribution in space and time being different from other species. It is known as a fast and graceful swimmer and feeds by skimming pelagic crustaceans from near the surface; it will feed at concentrations of food that are thought inadequate for other rorquals.

This is but a brief description to give some impression of the character of the sei whale, and the rest of the chapter focuses on those aspects that define and describe the sei whale as being different from the other whales; some aspects are covered in greater detail in subsequent chapters.

TAXONOMY

The order Cetacea has two sub-orders of living whales and one of extinct forms. The living whales are divided onto those with teeth, the odontocetes, and those with baleen plates, the mystecetes. In the sub-order Mysticeti are four families of which three have living representatives, these three groups are of the gray whales, the right whales and the rorquals. The rorquals are in the family Balaenopteridae which has two living genera and several extinct forms. The taxonomy given below is essentially that of Simpson (1945) and Scheffer and Rice (1963), but Simpson separates out the blue whale as a distinct genus.

Order
Cetacea, Brisson 1762

Sub-order
Archaeoceti, Flower 1883 - all extinct
Odontoceti, Flower 1867 - toothed whales
Mysticeti, Flower 1864 - baleen whales

Within the Mysticeti:

Family
Cetotheriidae, Cabrera 1926 - all extinct
Eschrichtiidae, Ellerman and Morrison-Scott 1951 - gray whales
Balaenidae, Gray 1825 - right whales
Balaenopteridae, Gray 1864 - rorquals, humpback

Within the Balaenopteridae:

Genus
Megaptera, Gray 1864 - humpbacks
Balaenoptera, Lacepede 1804 - rorquals

Species
M. novae-angliae, Borowski 1781 - humpback
B. musculus, Linnaeus 1758 - blue
B. musculus brevicaudata, Zemsky and Boronin 1964 - pygmy blue
B. physalus, Linnaeus 1758 - fin
B. acutorostrata, Lacepede 1804 - minke
B. borealis, Lesson 1828 - sei
B. edeni, Anderson 1878 - Bryde's
B. brydei, Olsen 1912 - Bryde's

In addition to the above some authors have recognised an additional southern hemisphere form of minke whale: *B. bonaerensis* Burmeister 1867. There is still some confusion about the status of the two species of Bryde's whales. Omura (1959) originally considered the two forms to be the same, but later speculated that they may be the onshore and offshore types described by Best (1977a); Best recommended that both forms be retained.

Separate populations exist within a species of whales, and some differences are probably sufficient to warrent sub-specific status, as in the different forms of Bryde's whales and minke whales (Best 1977a, 1982). Further Wada has identified with electrophoretic techniques what appears to be an intermediate form between sei and Bryde's whales (Wada, in preparation). Too little is known about the phylogeny of the sei whale but the balance of opinion is that only one species exists, but that this is composed of many populations; the concept of a population is discussed later. There is however little if any mixing between the populations of the two hemispheres, and Tomilin (1967) and Zemsky (1980) consider that the two groups form sub-species, the northern *B. borealis borealis* Lesson 1828, and the southern *B. borealis schlegeli* Flower 1884.

NOMENCLATURE

The sei whale is known as *Balaenoptera borealis* Lesson, this being the translation by Lesson (1828) of Cuvier's (1823) "Rorqual du Nord". Because of the comparative rarity of specimens the early nomenclature of the large whales was very confused and that of the sei perhaps the most mistaken. A review of the nomenclature of the large whales in general is given by Brinkmann (1967) with that of the sei whale in particular given by Andrews (1916a). The term "balaenoptera" is used to signify whales with a dorsal fin (or wing) and does not refer to the flippers as in Megaptera; its significance is to distinguish the whalebone whales with a dorsal fin, from the right whales which have none. The "borealis" indicates a northern form. The term "rorqual" is in common usage and is derived from the Norwegian word for whales having pleats or folds - these being the ventral grooves found in blue, fin, sei, Bryde, humpback and minke whales. Synonyms for *B. borealis* are given by Andrews (1916a) and Hershkovitz (1966), and include the following names:

Balaena rostrata, Rudolphi,1822
Rorqual du Nord, Cuvier,1823
Balaenoptera arctica, Temminck, 1841,
(=*B. artica*, Schlegel, 1842)
Balaenoptera laticeps, Gray, 1846

Balaenoptera Iwasi, Gray,1846
Sibbaldus laticeps, Gray,1864
Sibbaldus Schleglii, Gray,1864
Sibbaldius schlegelii, Flower,1865
Rudolphius laticeps, Gray,1866
Balaenoptera alba, Giglioli,1870
Pterobalaena alba, Giglioli,1874
Balaenoptera schleglii, Van Beneden and Gervais, 1880
Balaenoptera borealis, Cuvier,1881
Belaenoptera Schlegeli, Lahille,1899
Balaenoptera schlegellii, Dabbene,1902
Balaenoptera Schlegeli, Lahille,1914
Balaenoptera borealis borealis, Tomilin,1946
Balaenoptera borealis schlegelii, Tomilin,1946
Pterobalaena schlegeli alba, Tomilin,1957.

Rudolphi (1822) described a whale that had been stranded on the coast of Holstein, Germany in 1819 and whose skeleton had been removed to the Berlin Museum. He described the whale as *Balaena rostrata*, which was a misidentification of the specimen as a minke whale. Cuvier (1823) also studied the skeleton and recognised it as something other than a minke whale, and gave it the name "Rorqual du Nord"; Lesson (1828) translated this into latin, to give the present accepted designation. The name has survived much competition. Fischer (1881) used the name *B. borealis* Cuvier, but it was Gray (1866, 1871), in a series of articles, that created many new species from the few specimens of sei whale. Gray gave particular significance to the presence of a bicuspid first rib which is common in sei and Bryde's whales, but which also occurs occasionally in other large whales. The great scarcity of materials posed the major problem but as luck would have it one of the earliest specimens of sei whale, that described by Turner (1882) from a stranding in Firth of Forth, Scotland, did not have a bicuspid first rib.

A southern hemisphere form was identified as *S. schlegelii*, Flower, and later Tomilin created northern and southern sub-species of sei whale; although there is some current support for this separation, most authorities do not retain the sub-specific differentiation. By the turn of the century the sei whale had been exploited from several locations, and was clearly identified from the other rorquals, excepting the Bryde's whale. Although taxonomically distinct, much of the whaling industry did not distinguish between sei and Bryde's whales until a decade ago.

The common names of the sei whale differ across the world but the name "sei whale" is a derivitive of the Norwegian "sejhval", used by Norwegian whalers, and the international character of the industry, and the major role that Norwegians played in modern whaling, has meant that this name is recognised in many places. The name arose as the sei whales tended to arrive at the Norwegian north coast with the "seje" (coal-fish or saithe, *Pollachius virens* L.). They do not hunt this fish and any eaten were taken incidentally with their preferred food. Millais (1906) noted that the name sei whale was derived from the coal-fish because of its colour, but this does not seem to be true. In most of the general literature the sei whale is also referred to as Rudolphi's rorqual, but this name is seldom if ever used. Tomilin (1967) provided a list of common names for the sei whale although most of the English names are historical. These names are:

Rudolphi's rorqual, sei whale, Japan finner, Pollack whale, sardine whale, coalfish whale (English)
Saidyanoi kit, ivasevyi kit, ivasevyi polosatik, seival (Russian)
Seiwal (German)
Baleine noire (French)
Seihval, seichval, saaiwahl (Norwegian)
Seival (Swedish)
Seihval (Danish)
Noordische vinvisch, Noordse vinvis (Dutch)
Iwasi kujira, kazuo kujira (Japanese)
Ballena boba (Spanish)
Sandereydur (Icelandic)
Plejtvak severny (Czech) and
Baleia espardate (Brazilian/Portuguese, Paiva and Grangeiro 1965).

COLOUR

"I have at no time been more strongly impressed with the necessity for caution as to the degree of importance which should be placed upon colour when dealing with large cetaceans in an attempt to accurately define species than in the case of *B. borealis*. The range of individual colour is enormous - greater even than the blue or finback". This quotation from Andrews (1916a) should be remembered when attempting to identify all large whales. Collett (1886) observed many sei whales caught in the northeast North Atlantic and was aware of a large variety of colour. Although colour cannot be relied upon to distinguish the sei whale from other rorquals a general description can be given.

The dorsal surface is smooth and shiny of a dark steel to bluish-grey colour, and there is a tendency to have lighter pigmentation down the sides and on the posterior ventral surface. Omura (1950a) and Mitchell (1975) described the sei whale as lightly and variably spotted, but this is not to be confused with the sometimes numerous white scars that are caused by the parasitic copepod *Penella* (Andrews 1916a, Ivashin and Golubovsky 1978), the shark *Isistius* (Jones 1971, Shevchenko 1977), or lampreys (Nemoto 1955). The region of the ventral grooves almost always exhibits a white or light area. This extends back from the chin, most usually stopping just short of the ends of the grooves, but sometimes stopping after only one to two metres from the chin. In some individuals the whitish area extends up to the flippers. As in some fin whales a white streak is occasionally seen going downwards posterior to the ear. This description is consistent with the colours described from both southern (Matthews 1938) and northern (Andrews 1916a, Collett 1886) forms. Zemsky (1980) emphasised that the right lower lip and mouth cavity is grey whereas in many fin whales a white patch is seen; however sei whales do occasionally have anterior white baleen plates with a corresponding light patch on the upper jaw, next to the plates.

VENTRAL GROOVES

The extent of the ventral grooves, Figure 1.1, is an important feature in distinguishing sei whales from other rorquals. In both the sei and the minke whale the ventral grooves end well before the umbilicus, whereas in the other Balaenoptera, including the Bryde's whale, the grooves end at or posterior to the umbilicus. In sei whales the distance between the end of the grooves and the umbilicus is typically about 6.5% of the body length (measured horizontally from the snout to the notch of the flukes), are rarely is the distance less than 3%.

The number of grooves are fewer in sei whales than in blue, fin and minke whales, but similar to that in Bryde's whales. From South Georgia,

Figure 1.1: Sei whale, foreground, showing the ventral grooves ending anterior to the umbilicus. A fin whale is in the background.

Matthews (1938) reported an average count of 47 with a range of 40-62, compared with mean values of 86 and 89 in fin and blue whales respectively, and a range of 50-70 grooves in minke whales. Olsen (1913) reported that Bryde's whales had between 42 and 54 grooves. Matthews does not document how the grooves were counted, but I would assume he used the same technique as that employed in the pioneering study of Mackintosh and Wheeler (1929) on blue and fin whales. They counted the anterior terminations of the grooves starting from the middle of the jaw and working along the angle of the jaw, past the angle of the gape, to the last groove near the eye; the number was then multiplied by two. Omura et al (1952) counted the number of grooves from sei whales caught off Sanriku and Hokkaido, and found a range of 42-66 with a mean number of 52 or 53. They however counted from the base of the flipper to the mid-ventral line and a lesser number might have been expected. The range is large in both studies, and Omura et al did not consider the two sets of counts to be significantly different.

DORSAL FIN

The dorsal fin is situated just forward of a vertical line through the anus and is relatively larger than that of the fin and blue whale. The relative height is about 3 to 4.6% of the body length and this is about 1.5 times that of fin whales and 4 times that of the blue whale, (Gambell 1966). It is also in a more forward position than in these two species. Its shape is convex on the anterior edge, comes to a point and is deeply concave (or falcate) on the posterior edge. Its overall shape resembles that of the minke whale and Bryde's whale. Gambell (1966) considered the dorsal fin to be a valuable identification cue in about 80% of cases when comparing fin and sei whales. The fin is set at an angle of about 46° from the horizontal in sei whales compared with 33° in fin whales.

BEHAVIOUR AND DISTRIBUTION

The distribution of the sei whale does not provide a reliable means of identification, but a knowledge of its distribution and behaviour helps. The sei whale does not penetrate into the summer polar seas to the same extent as do the blue, fin, minke and humpback whales. In the Southern Ocean this leads to a large degree of segregation, with a large proportion of the sei whales staying north of the Antarctic Convergence. In both the North Atlantic and North Pacific the other whales feed in slightly warmer waters and the species are much more mixed, but nevertheless the sei whales seldom penetrate to the ice edge. Bryde's whales are seldom found in waters where the temperature is under 20°C and, although sei whales are usually found in waters of lower temperature, the two species can be found in the same localities.

The diving and feeding behaviour of sei whales is somewhat different from that of the other whales. Ingebrigtsen (1929) and others have reported sei whales feeding by swimming horizontally near the surface, skimming food. This is in contrast to blue, fin and humpback whales that frequently turn on their side to feed. The minke whale can utilise both methods. The sei whales tend to be shallow swimmers, with their head seldom emerging, and with no positive arching on diving. They tend to sink rather than dive, but again other whales may do this. Gambell (1966) described sei whales off

Durban surfacing almost horizontally whereas fin whales came up snout first and then wheeled over.

BALEEN

The baleen is a keratinous filter attached to the upper jaw, and its characteristics can be used to help identify species. The filter is composed of a series of parallel plates of approximately triangular shape, and fixed into the upper jaw perpendicular to the long axis of the body. Consequently water can be extruded from the mouth through the gaps in the plates. The plates vary in length, increasing posteriorly, but with a final reduction. At either end of the jaw the plates become very small and are known as hairs. Williamson (1973) arbitrarily defined hairs as having a width less than three times the thickness of the plate; however most people have started counting plates at a point they regard as "obvious", and thus small differences in counts are to be expected between observers. On the inside of the plates there is a fringe of bristles which helps to trap the food.

The colour of the sei whale's baleen is usually recorded as dark grey or greyish-black, with a light bristle fringe (Matthews 1938, Leatherwood et al 1976, Zemsky 1980), but Brown (1965) noted that often, on the inside, the baleen has sections of a yellowish-brown hue. Many authors have noted sei whales with white baleen at the anterior end, and from the South Atlantic, Brown (1965) found that 36% of those sampled had some white plates; Collett (1886) cited one specimen with over 50 white plates, and Brown (1965) one with 72 white plates.

The plates are rigid and short with the largest usually less than 80 cm in length. The total number on each side of the jaw is about 340-350, and the blue and fin whales also have about 350-360 plates and their jaws are much longer. The spacing between the plates is similar in the three species, and it can then be imagined that the plates of the sei whale must be much finer. Mead (1977) noted that the bristles are very fine, and at their base are about 0.1 mm in diameter compared with about 0.3 mm in the other species of Balaenoptera, including the Bryde's whale. The width of the sei whale baleen plate is between 30-50% of the length, and this also contrasts with 67-106% for blue whales and 60-73% for fin whales. Omura and Fujino (1954) and Best (1977a) demonstrated that the sei whale plates are typically larger than those of the Bryde's whale, the latter rarely exceeding 40 cm, and that the length to breadth ratio in sei whales is typically over 2.2, whereas for Bryde's whales it has been found to be always less than 2.2.

OSTEOLOGY

Vertebrae

The early records of the number of vertebrae of the sei whale are reviewed by Andrews (1916a), who provided additional material of his own from catches off Japan. He concluded that the vertebral count was:

C7 + D14 + L13 + Ca 22 or 23 = 56 or 57.
Nishiwaki and Kasuya (1971) described the complete osteology of a specimen that they prepared for the museum at Taiji and they reported a formula of:

C7 + D13 + L14 + Ca22 = 56.
Omura (1959) used a specimen of sei whale with the same formula as that given by Andrews as a basis of comparison with Bryde's whales. Some of the last caudal vertebrae are very small and care is needed not to overlook them during collection; consequently some small variation might be expected in the reported number as specimens were often collected in difficult conditions. The vertebral formulae of the other large whales are given below; they are taken from the review by Tomilin (1967) unless otherwise stated.

blue:	C7 + D14/15 + L13/16 + Ca26/30 =63/66 (Ichihara 1966a)
pygmy blue:	C7 + D14/15 + L15 + Ca28/29 = 64/66 (Ichihara 1966a)
Bryde, edeni:	C7 + D13 + L13 + Ca21 = 54 (Omura 1959)
brydei:	C7 + D10 + L14 + Ca21 = 52+ (Anderson 1878)
fin:	C7 + D15/16 + L13/16 + Ca24/27 =60/63
humpback:	C7 + D14 + L10/11 + Ca21/22 = 52/53
minke:	C7 + D11 + L12(+1) + Ca18(+2)= 48/50
gray:	C7 + D14 + L10/11 + Ca21/22 = 52/53
right, bowhead:	C7 + D13 + L12/13 + Ca22/23 = 54/56
southern:	C7 + D14 + L12 + Ca23 = 56

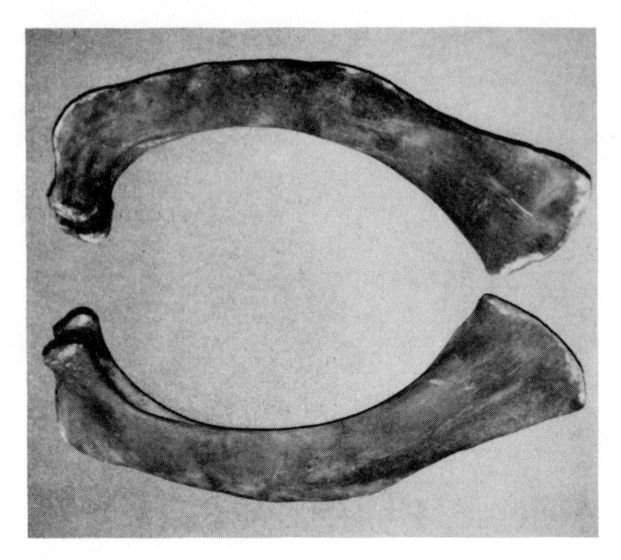

Figure 1.2: Bicuspid first ribs of a sei whale.

Compared with the sei whale the blue and fin whales have a significantly greater number of vertebrae, and the minke, humpback and gray whales significantly less.

The normal condition is that the seven cervical vertebrae are free in the sei whale, although Turner's specimen from the Firth of Forth had some fusion of the axis and atlas. They are also free in *B. edeni* and in *B. acutorostrata* (Omura 1959, Tomilin 1967), but in the other Balaenoptera two or more of the cervicals are ankylosed. Andrews (1916a) described the atlas bone as distinctive, being short, with the transverse processes much twisted, and a contracted neural canal. On cervicals 2-5 the transverse processes are united into a ring. The sixth has a short inferior process and the seventh has only a superior process. However Andrews reported a wide range of individual variation in the shape of the cervicals, and care is needed to use them diagnostically.

Omura (1959) and Omura et al (1970) compared the size of the vertebrae in sei, pygmy blue, and Bryde's whales, and they found that the relative length of the first five caudal vertebrae increased in sei whales but decreased in the other two species. In comparing the vertebrae of the sei and Bryde's whales, Omura (1959) reported that the most pronounced osteological difference between the two

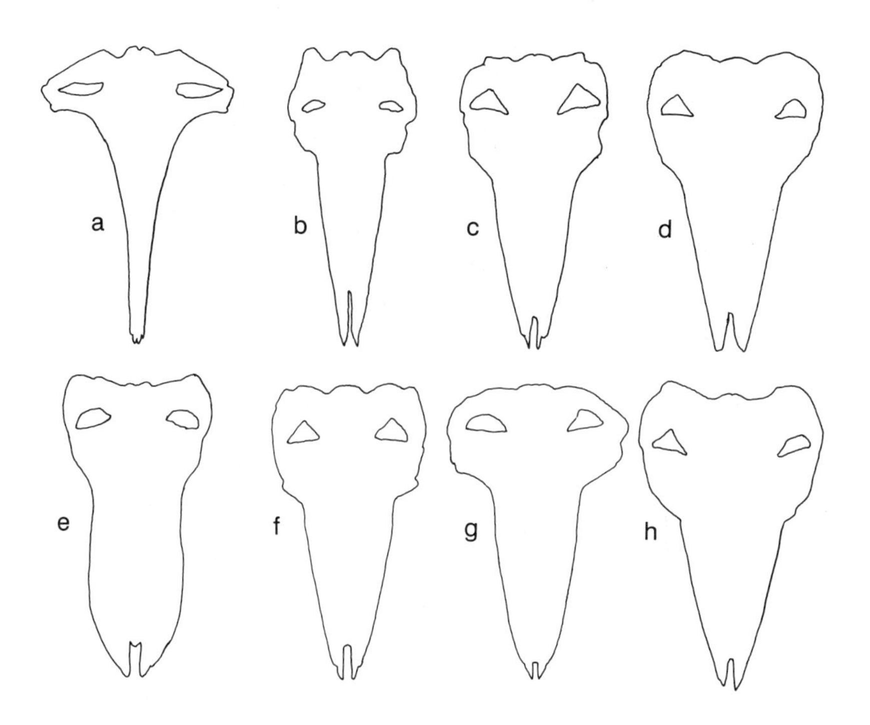

Figure 1.3: Dorsal profiles of the skulls of large whales, reduced to similar lengths. a: Northern right whale, b: gray whale, c: Bryde's whale, d: sei whale, e: blue whale, f: fin whale, g: humpback whale, h: minke whale.

species was the strong backward inclination of the spinous processes in the Bryde's whales compared with sei whales. The maximum inclination in Bryde's whales occurs between the 5-7th lumbar vertebrae.

The sei whale has 13 or 14 pairs of ribs but the most significant feature is that the first rib has a bicuspid proximal end, which is flattened distally, and this is shown in Figure 1.2. The rib is formed by a fusion of the true first dorsal rib and a cervical rib (Schulte 1916). It is a very useful taxonomic feature, but the bicuspid first rib is also general in Bryde's whales and occurs occasionally in other species; rarely a sei whale has a single first rib.

Other Bones

The skull provides a means of identifying the species of whales. Firstly the length of the skull is related to the size of the whale, and from the absolute size of the skull alone many species might be rejected as the source of the skull. For sei whales Tomilin (1967) described skulls that were 20.9-25.4% of the total body length, and from a range of species of large whales this percentage varied from 18-30%; Omura (1971) found a similar relationship. The gross shape of the skulls are very distinctive with, for instance, the rostrum of the right whale extremely narrow; the range of shapes of the skulls of large whales is illustrated in Figure 1.3 as outlines of the dorsal profiles.

Hosokawa (1951) reviewed the existence of a pelvis, tibia and femur in the Balaenoptera and concluded that the right whales possessed all three, the humpback, fin and possibly the blue had no trace of a tibia, and that the sei and minke whales had only a pelvis remaining; Hosokawa considered that pelvis shape could be used to segregate species. Subsequently Omura (1978) showed that in the Antarctic 25% of minke whales caught did have a femur attatched to the pelvis.

Omura (1964) demonstrated that the hyoid process can be used to distinguish species. The process consists of three bones, the basihyal, the thyrohyal and the stylohyal; usually the first two are completely fused. Unfortunately, as the hyoid process is not attached to the rest of the skeleton, it is often lost. Omura provided an identification key base on the morphometrics of the hyoid. This is given below.

1. The ratio of the height of the distal end of the combined basihyal and thyrohyal (a) to the overall length (b),(Figure 1.4):
<0.09 - Balaenoptera, go to 2
0.09-0.12 - humpback
>0.12 - gray, right

2. The ratio of the height at the centre of the fused base (c) to its length (b):
<0.18 - blue, fin
>0.18 - go to 3

3. The ratio of the greatest height of the combined bone (d) to the length (b):
<0.29 - minke
>0.29 - go to 4

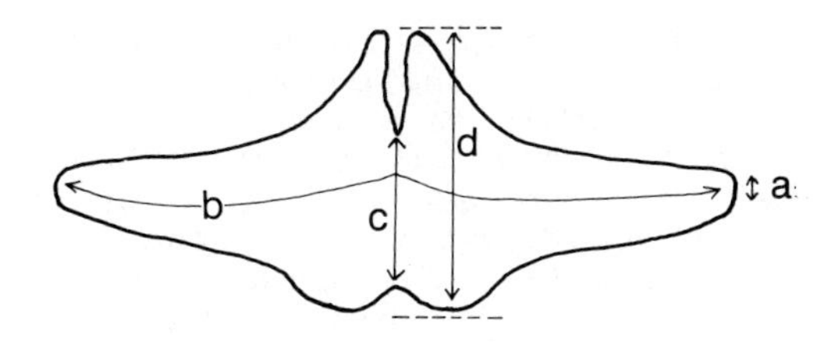

Figure 1.4: Fused basihyal and thyrohyal of the hyoid process. a-d are dimensions described in the text.

4. Stylohyals broad, flat, much curved forward: - sei
   less broad, thick and curved: - Bryde

The key does not include the bowhead or pygmy blue whales, and specimens of sei and Bryde's whales from the Natural History Museum, London could be identified with this key.

COMPARISON WITH BRYDE'S WHALES

The Bryde's whales have been recognised as different from the sei whales since the begining of the century, but the two types are sufficiently similar that they have often been recorded as the same species (sei) in catch statistics. For example catches from the Bonin Islands and off the coasts of Japan were called sei whales, and only distinguished as two species by Japanese scientists in the 1950s, (Omura et al 1952, Omura and Fujino 1954). Best (1977a) described similar problems in the records of catches from South Africa.

The two species are similar in colour, similar in size and have a similar, prominent, falcate dorsal fin; although that of the the Bryde's whale is slightly smaller. Bryde's whales are seldom found in waters less than 20°C and therefore tend to be restricted to between the latitudes 40°S and 40°N, (Omura 1959). The sei whale also inhabits these latitudes but in the summer is usually distributed more towards the poles. At sea the only safe way of distinguishing the two whales is by close inspection of the head. The Bryde's whale has three distinct head ridges, running along the length of the rostrum, whereas the sei whale has only a single central ridge. On surveys for whales such necessary and rigorous confirmation of species is very time consuming. Rice (1977) and Donovan (1983) described a range of characteristics and behaviours which help distinguish the two species; these include comments that the Bryde's whale has a smaller dorsal fin, which is at a steeper angle from dorsal tip to anterior base, the blow from the Bryde's whale is preceded by two small side blows, the body is more scarred, Bryde's whales swim less quickly and evade vessels by frequent changes in direction. However it is unlikely that identifications on this basis alone would be accepted unless made by very experienced observers.

If a whale has been caught or stranded then the features described earlier will easily distinguish the two species. The two most significant features are that in general the ventral grooves end at about the middle of the body in sei whales but in the Bryde's whales they extend to the umbilicus. Best (1977a) summarised the position by reference to the fact that, from 196 North Pacific Bryde's whales, in only 1.5% of the animals did the gap between the end of the grooves and the umbilicus exceed 3% of the total length, whereas in the Southern Hemisphere only 1.3% of 150 sei whales measured was this distance less than 3%. Secondly the bristles of the baleen plates are very coarse in Bryde's whales and other Balaenoptera, but very fine in sei whales. The width at the base is about 0.1 mm in sei whales and 0.3 mm or more in Bryde's whales. Although Omura and Fujino (1954) successfully used the length to breadth ratio of the plates to segregate North Pacific sei and Bryde's whales this ratio does not give a definitive answer, (Omura 1966, Best 1977a); typically however the ratio is over 2.2 in sei whales and less in Bryde's whales.

Omura (1966) presented a review of additional material on which to segregate the two species. In studying the lengths of various parts of the body he found that the greatest width of the skull is about 11% in sei whales compared with over 12% in Bryde's whales. (All body length measurements are given from the tip of the snout to the notch between the flukes in a straight line, after Mackintosh and Wheeler 1929). The relative lengths of the skull, and from the tip of the snout to the blowhole, are less in sei whales, but the skull width gives the best discrimination. In sei whales the relative distance from the notch in the flukes to the umbilicus is typically 49% but is 42% in Bryde's whales. In the sei whale the anus and dorsal fin are both found relatively more forward than in Bryde's whales. With all these latter morphometrics there is individual variation which can make their use in individual cases difficult.

Some of the skeletal differences have been described earlier, but Omura and Fujino (1954) showed that the palate of the Bryde's whale is very broad and concaved inwards at about the middle of its length, whereas the palate of the sei whale is much narrower and convexed outwards in the posterior region. In addition Nishiwaki (1974) emphasised that in sei whales the maxillary bones curve downwards at the tip whereas those of Bryde's whales are more horizontal, and, that from a dorsal view of the skull, the anterior edge of the nasal bones is convex in sei whales but concave in Bryde's whales.

GENETICS

Genetics is the study of biological variability and the composition of body proteins reflects this variability in individuals, lineages, species and phylogenies to above the species level of classification. Many different techniques are available with which to describe the variability seen in proteins but three generally used methods are cytogenetics, immunogenetics and the isolation of the allozymic forms of protein molecules. The genetics of the sei whale have received little attention, but some limited information shows that sei whales can be distinguished from other whales, and that different sub-populations could be identified should they exist.

When investigating the nature of myoglobins Kendrew et al (1954) found that crystals of myoglobin from sei and minke whales could be distinguished from those of blue, fin and humpback whales on the basis of their crystal lattice patterns. In addition they looked at the antibody responses of whale myoglobin to sperm whale antisera, and they found that myoglobin from sei and minke whales reacted strongly and again could be distinguished from the reactions given by blue, fin and humpback whales. Immunogenetics has identified the existence of different blood proteins within and between baleen whale species, and the techniques used are to identify antibody responses to serum antigens from known whales. Antibody responses are looked for in the natural immune system of whales, of a single species, or, between species, by the induced immune responses in animals such as rabbits or chickens inoculated with antigens; reviews are given by Fujino (1960) and Cushing (1964). Most of the work has been done with fin whales but two blood group systems (Bb and Ju) have been identified in fin, humpback, blue and Bryde's whales (Fujino 1953, 1956, 1960, 1964a). The Ju system is also present in sei whales (Cushing 1964), but Fujino (1963a) has shown that in Antarctic fin whales the three major Ju phenotypes can be split into eleven forms which react differently with Ju antisera, to give 66 groups.

Specifically for sei whales Fujino (1964b in Japanese) found that those caught off Vancouver were genetically different from those caught around the central Aleutian islands, and in addition that sei whale serum, used for natural antibody responses, showed a positive result with one of the antigens from a sperm whale. Fujino (1953) documented the natural response of human ABO blood to whale sera, but in this study his "sei" whales were taken from the Bonin Islands and were in fact Bryde's whales. Later (Fujino 1960) reported that both sei and fin whale sera showed some cross-reaction with a fraction of human B antigen. Of general significance is the finding of Fujino (1963a,b) that intra-uterine selection occurred in fin whales due to the incompatibility of maternal-foetal blood; this means that blood type is not freely hereditable, and may be presumed to apply also to sei whales.

The above immunological studies suffer from the disadvantage of being labour intensive and time consuming, although great resolution is possible. Another technique that is now more commonly used is that of zone electrophoresis in gel media (Pike 1964), and Tsuyuki et al (1966) successfully discriminated some whale species according to their distinctive pattern of muscle proteins. They distinguished sperm whales from blue, fin, humpback and sei whales, the patterns from the fin and humpback were both identifiable, but the blue and sei whales had a similar zonation. Borisov (1981) also demonstrated that haemoglobin fractions are different in the sperm whale, compared with sei, fin and minke whales, and that myoglobins can be used to distinguish fin whales from sei and minke whales; he also (Borisov 1980) confirmed the existence of polymorphisms in haptoglobins.

The introduction of zymogram methods, that is the combination of electrophoretic techniques with histochemical staining, gave more precision to the analysis of proteins. Interspecific comparisions were made by Wada et al (1973) of samples from sei, fin and sperm whales from the North Pacific, and from minke whales from the Antarctic. From serum esterases, sei and fin whales showed a similar banding that was different from the polymorphic forms of sperm and minke whales. From these four species, five major fractions were distinguished in lactate dehydrogenase (LDH I-V); some fractions had substantial subunits. The existence of polymorphisms indicates the possibility of using genes coding enzymes to identify populations within a species.

Electrophoretic techniques have also been applied to sei whales by Arnason and Sigurosson (1983). From sei, fin and sperm whales caught at Iceland they studied plasma proteins and esterase, red cell lysate, carbonic anhydrase esterase, acid citrate phosphatase, glucose phosphate transaminase, and glyoxalase; from sei whales they found polymorphisms of glucose phosphate transaminase.

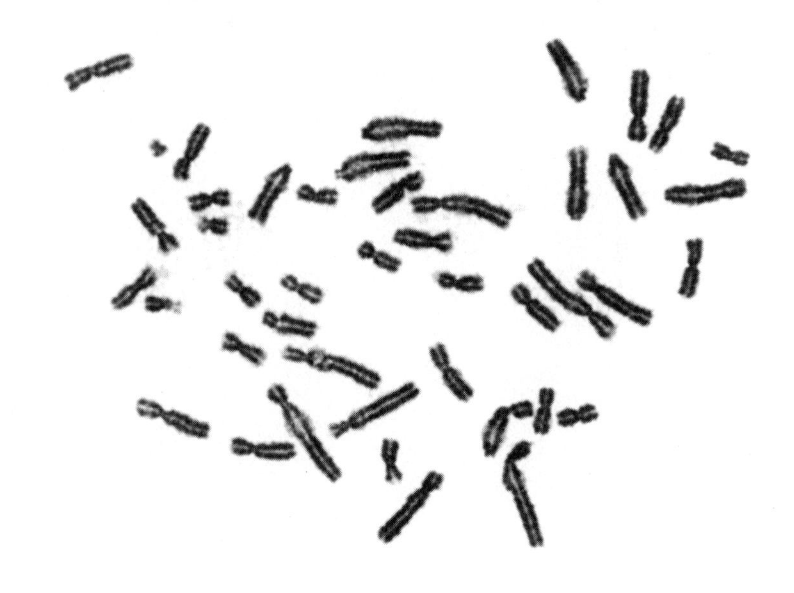

Figure 1.5: Plate of the karyotype of a male sei whale from Iceland.

The karyotype of a sei whale caught in the northern North Pacific was described by Kasuya (1966), and from 54 spermatogonial nuclei from a single whale he found a range of diploid chromosome number of 40 to 49; some 70% had 44 and this was determined as the actual number. Earlier studies on a range of Cetacea (Makino 1948, Nowosielski - Slepowron and Peacock 1955, Atwood and Razani 1965) have been clarified substantially by the studies of Arnason (1970, 1972, 1974); from 17 whale species, including the sei whale, all except two had a diploid number of 44 chromosomes. The two exceptions were *Physeter macrocephalus* and *Kogia breviceps*, the sperm whale and pygmy sperm whale, both of which had 42. The karyotype of a male sei whale is shown in Figure 1.5 and the dimensions of the chromosomes are given by Arnason (1974). Arnason classified the chromosomes into four groups on the basis of the ratio of their arms about the centromere, and sei and fin whales differ from gray and minke whales.

The above indicates that there is sufficient evidence that the chromosomes, blood groups and proteins could be used to separate the sei whale from other species, but the fragmentary observations have not yet been collated in such a way that the information could be used in an identification key.

Chapter Two

DISTRIBUTION

INTRODUCTION

It is known from general texts on whales that sei whales have been found in most ocean and sea areas, and this section reviews the information on which that knowledge is based. The main source of information comes from the whaling industry and records of catches have usually been preserved. The Secretariat of the International Whaling Commission (IWC) has succeeded the Norwegian Bureau of International Whaling Statistics (BIWS) as the repository of much of the catch data, which are published in summary form in the International Whaling Statistics (IWS). The catch data can usually give at least the location and season of the catches. In the polar areas the information on sei whales is good, but in the warmer waters there has been much confusion over the identification of sei and Bryde's whales. In many instances there have been attepts to resolve these problems and I have tried to make it clear whether sei whales have been unambiguously identified.

Data on sightings are often difficult to utilise. Sei whales can be easily recognised as rorquals, and although experience is needed to distinguish between sei and fin, and sei and minke whales, few problems should arise, however care, and much time, is needed to distinguish between sei and Bryde's whales in the warmer waters. One invaluble source of sightings information is the records of the Japanese whaling, scouting vessels; these vessels are used in a strategic role and have surveyed vast areas. The records have been reported to the Far Seas Fisheries Laboratory, Japan, and the techniques used are described by Ohsumi and Yamamura (1982). Whale catchers and experienced whalers are used and although random or pre-planned routes are not utilised the extensive effort employed, since 1965, has given very sound results.

Where possible strandings have been described. They are particularly useful in regions where there has been no catching, and they at least provide an opportunity to correctly identify the whale.

NORTH ATLANTIC

Catches

Norway, Murman coast and Northeast Atlantic. Collett (1886) described the sei whale from catches off the Finnmark coast of north Norway, and Jonsgard (1977) and Tonnessen and Johnsen (1982) show that sei whales were taken each year from 1885 to 1904. In 1885, 771 were caught from mid-May to early September, with the peak catches off Finnmark in late July. After July Collett thought the whales moved offshore and towards Murmansk. Cocks (cited extensively by Andrews (1916a)) described the early Finnmark fishery, and sei whales were often caught from May to August but with peak numbers in July. Ingebrigtsen (1929) caught sei whales occasionally off Bear Island and Spitzbergen, and one as far north as 79°N, and Jonsgard (1977) reported six sei whales taken from Spitzbergen over the period 1903-1927. In recent years sei whales have not been found off north Norway (Jonsgard 1974).

Hjort and Ruud (1929), Jonsgard (1974) and Jonsgard and Darling (1977) have described the sei whale fishery off western Norway. Off More the sei whales were found from May to August but occasionally catches were taken in April and September, and they were mainly taken from the edge of the continental shelf or over the deeper water.

Scotland and Ireland. Sei whales were caught from the islands of Orkney, Shetland and the Hebrides, and from off the west coast of Ireland, and Jonsgard (1977) records a peak catch in 1906 of 362. Millais (1906) reported that sei whales were common in the deep, fast moving waters around the three groups of islands in July to September. Details of the Scottish catches are given by Southwell (1904, 1905), and in a series of publications by Haldane (1905-1910), who described the largest catches as coming from about the island of St Kilda, off the Hebrides, in mid-June to early July. Thompson

(1928) presented catches by month and showed that sei whales were taken from April to October with the greatest numbers in June; however peak catches from off St Kilda were in June, and off the north of the Shetland Islands in July and August. In all locations catches were from along the shelf edge.

During whaling voyages Ingebrigtsen (1929) found only a few sei whales off Ireland and Fairley (1981) has summarised the catches from two Irish whaling stations at Innishkea and Blacksod, based mainly on the studies of Lillie (1910) and Burfield (1912, 1913, 1915). The largest number taken in any one year was 39, and all but one were caught by early or mid-June; one sei whale was caught in September.

<u>Spain, Portugal and Morocco</u>. Whaling has been carried out from shore stations in Portugal, Galicia in the northwest of Spain, Getares to the east of the Straits of Gibralter and Benzu, opposite, on the coast of former Spanish Morocco (Tonnessen and Johnsen 1982, Aguilar and Lens 1981, Sanpera and Aguilar 1984). Jonsgard (1977) reported a catch of 66 from the "coast of Spain and Portugal", and these were taken in the vicinity of the Portuguese land station just south of Lisbon, over the period 1925-1927 (Tonnessen and Johnsen). Although sei whales were not mentioned by Cabrera (1925) they were taken regularly from Benzu in a winter fishery (Aloncon 1964, Aguilar and Lens 1981). However Sanpera and Aguilar (1984) noted that when the Benzu whaling station was managed by Norwegians, in 1948, Bryde's whales and not sei whales were recorded. From 1950 to 1953, under Spanish control, only sei whales were recorded in the catch. Consequently it is clear that we cannot trust these identifications but Aguilar (pers. comm.) has inspected some baleen and found that at least some true sei whales were caught in the Straits.

Casinos and Vericad (1976) considered that sei whales were not common off Galicia, and Aguilar and Pelegri (1980) considered it useful to report on the catch of a single sei whale. Jonsgard (1977) recorded few catches of sei whales caught by Spain, but a reanalysis of catch records by Aguilar and Sanpera (1982) show a continued catch of sei whales off northwest Spain from 1957 onwards. The numbers are small but some of the records of fin whales may have been of sei, and it is possible that some Bryde's whales also may have been caught. The sei whales were caught from April to December, with a peak from mid-September to mid-November. The peak is associated with, but not necessarily caused by, the seasonal reduction in catches of fin and sperm whales.

Iceland. Sei whales have been a continuous but small part of the catch at Iceland since 1938 but, in recent years, over 100 per year have been taken (Jonsgard 1977). The whales are caught off the south or southwest of Iceland, mainly in the Denmark Strait, and Jonsson (1965) shows that they are caught mainly in August or September, but with some taken as early as June; a later description by Martin (1983) is similar. However Ingebrigtsen (1929) found sei whales were not common at Iceland and Saemundsson (1939) reported that few were found to the northwest, and that most catches were in the southeast of the island.

Faroe Islands. Sei whales have been caught regularly from around the Faroe Islands with occasional catches of over 100, but their numbers have been erratic, (Ingebrigtsen 1929, Jonsgard 1977). Catches occurred in the summer, and Dagerbol (1940) reported that the maximum catches were in July, with some taken in June and August.

West Greenland and Davis Strait. Berntsen, a Norwegian whaling captain, reported catches of two sei whales taken in July and August on the offshore banks, south of Disco Island (Hjort and Ruud 1929). Kapel (1979, 1985) reviewed the statistics of catches of sei whales from off West Greenland, and over the periods 1924-1939 and 1946-1958; catcher boats based in West Greenland took only eight sei whales. Kapel considered that they were caught only in years with warm water intrusions into the Davis Strait; the scarce data indicate catches in August and September (see also North Atlantic pelagic).

Canada. Whaling has taken place from Nova Scotia, Newfoundland and the coasts of Labrador. True (1903) reported the first evidence of the existence of sei whales from the western North Atlantic, from a catch from Placenta Bay, south Newfoundland, where, according to Millais (1906), they are found in August and September. From Newfoundland, and off Labrador Jonsgard (1977) reported a maximum annual catch of 39 sei whales, but greater numbers were taken from Nova Scotia, over the years 1966-1972, with a maximum annual catch of 235. The sei whales were caught from south to southeast Nova Scotia, in the deep water along the edge of the continental shelf, with their occurrence in two peaks in June-July and September-October (Mitchell 1974, 1975, Mitchell and Kozicki 1974). Sei whales were caught off the coasts of Labrador in August to November (Mitchell 1974a, Sergeant 1966). No sei whales have been taken from the Gulf of St Lawrence.

Pelagic whaling. From the pelagic whaling in the North Atlantic over the period 1929-1937 only 133 sei whales were caught (Jonsgard 1966b,1977). One hundred of these were taken in 1937 and the IWS of 1938 explains that three were caught south of Iceland, and 97 in the Davis Strait, however their exact position is uncertain.

Other southern areas. Jonsgard (1977) reported no catches of sei whales from operations about the Azores and Madeira. Ingebrigtsen (1929) claimed that 38 were caught off Capo Blanco, the western tip of Africa, in December to February, although it cannot be certain that these were not Bryde's whales. Varona (1965) described a young sei whale caught off southern Cuba in January but subsequent studies (Mead 1977) indicate that it was a Bryde's whale. Price (1985) described two Bryde's whales taken south of Grenada but no sei whales are mentioned.

Sightings

Northeastern North Atlantic. Although there is a long history of catches showing that sei whales were caught about the Faroe Islands, Scotland, Ireland, Norway and as far north as Spitzbergen and Nova Zemlya there are few records independent from catching operations. Christensen (1977, 1980) documented the records from observers on whale catchers from the most northern seas but no sei whales were seen. Tomilin (1967) says that sei whales were seen only rarely in the White Sea, and Jonsgard and Darling (1977) noted that sei whales have not appeared in recent years off the whaling grounds of northwest and north Norway.

Southeastern North Atlantic and Mediterranean Sea. Few sei whales have been recorded from these regions, and Allen (1916) thought them rare south of the Straits of Gibralter. Viale (1977, 1981), in her reviews of whales from the Mediterranean Sea and off Spain, does not mention the sei whale. Tomilin (1967), cited Mørch (1911) as having found groups of a considerable size off Portugal, and cites Buhigas (1917) as seeing sei whales off western Spain, but Mørch was referring to Portuguese West Africa (Angola). Kirpichnikov (1950) saw 10 sei whales in the Mediterranean Gulf of Genoa, and had several sightings off the west coast of Africa at 6°N, but it is of course possible that the latter animals were Bryde's whales. A recent sightings and markings cruise off Spain and Portugal in August covered the latitudes 33-48°N and found no sei whales (Sanpera et al 1984), and another in September from 38-45°N found three solitary sei whales between 40° and 43°N along the shelf edge (Aguilar et al 1983). A cruise from 42-54°N, in July and August, along the edge of the continental shelf found no sei whales (Sanpera and Jover 1985).

East Greenland and Iceland. Christensen (1977, 1980) reported on whales seen by observers on Norwegian catcher boats, and shows that a few sei whales were seen in the Denmark Strait in June to September and none elsewhere; none were reported from the results of the 1976-1978 seasons. Sigurjonsson (1983) reported on a whale sighting and marking cruise over June and July 1981, and 7 whales were seen in the southern Denmark Strait. Whales marked over the period 1979-1981 from Icelandic whalers were found along the shelf edge in waters of 600-2000 m depth. From surveys conducted under the International Decade of Cetacean Research (IDCR) Hiby et al (1984) described aerial searches in early June: one sei whale was seen in the Denmark Strait, and none were seen to the northeast and southeast of Iceland. A shipboard survey was undertaken about Iceland in June and July, but concentrated mainly in the Denmark Strait (Martin et al 1984), and again one sei whale was seen in the Denmark Strait and none between Iceland and Jan Mayen Island.

Davis Strait and Labrador Sea. Although sei whales were caught by Norwegian vessels in these waters no sei whales are reported from sightings from a Greenland catcher boat over a large range of months and years (Kapel 1979). Sightings from Norwegian vessels in the summer months, June to September, indicate no sei whales seen (Christensen 1977, 1980, Larsen 1981, Kapel and Larsen 1982, 1983, Kapel 1984). However Kapel (1985) says that catcher boats have seen sei whales off west Greenland and regularly off southwest Greenland. Observations have also been made from research cruises (Kapel 1979), and records of sei whales are rare. Larsen (1985) reported results from an aerial survey off West Greenland and one sei whale was found on the banks. Sei whales were reported by Perkins et al (1982a, b) in July and August, off the southwest of Greenland, and also by Mitchell (1974a) in deep water off southwest Greenland in May and June. Observations of whales have also been made by ornithologists on board support ships for oil platforms, and sei whales were recorded in July and August at about 65°N, however Kapel (1984) emphasised that these sightings were not made by experienced observers, and clearly the implication is that fin whales could have been mistaken for sei whales.

There have been several sightings of sei whales from off the coasts of Labrador, and the whales tend to be found in the deeper waters of the Davis Strait between Labrador and the southwest coast of Greenland. Millais (1906) says that sei whales were most numerous in summer off this coast and Christensen (1977) reported sei whales south of Hamilton Inlet Bank in June. Mitchell (1974a) and Mitchell and Kozicki (1974) described sightings from the Labrador Sea and Davis Strait. An aerial survey off the coast of Labrador in August 1980 found no sei whales (Hay 1982) and Mitchell (1974b) reported very few sei whales in the Labrador Sea regions. In Notre Dame Bay, on the north coast of Newfoundland, a cow and calf were seen in August (Whitehead 1979).

Southwestern North Atlantic. Canadian research cruises were undertaken for sighting and marking of whales over the period 1966 to 1971, and covered the regions from Nova Scotia to the coasts of Venezuela (Mitchell 1974a). The positions of the sei whales seen, which tend to be along the shelf edge, are illustrated by Mitchell and Kozicki (1974). Sei whales were seen in September east of Newfoundland, in August to the south of Newfoundland and in May to the southeast of Nova Scotia. To the south off Cape Cod sei whales were found in March and in July to October. Off Florida two sei whales were marked in February. In February and September whales were seen off Venezuela. Of the two whales marked off Florida, Mitchell and Kozicki make no comment as to any problems in identification, but of those seen and marked further south they say that both sei and

Bryde's whales could have been seen and marked. In 1978 and 1979 di Sciara (1983) undertook surveys along the coast of eastern Venezuela, using both ship and aeroplane. Although various authors were cited as identifying sei whales in this region (Erdman 1970, Erdman et al 1973, van Bree 1975) di Sciara saw none and claimed sei whales to be very rare in the area.

Strandings

Strandings of sei whales have been documented from Mexico to Massachusetts (Allen 1916, Miller 1924, 1927, 1928, Negus and Chipman 1956, Layne 1965, Villa-Ramirez 1969, Gunter et al 1973, 1974, Lowery 1974). Fortunately Mead (1977) conducted a pains-taking review which makes fascinating and admonitory reading; he was only able to confirm just over half of the identifications. Nevertheless strandings were validated from the Gulf of Mexico to Massachusetts. Sheldrick (1976, 1979) and Evans (1980) reviewed strandings from the European coasts. The system of reporting covering the UK and Ireland has been operating since 1913 and from 1913 to 1978 there have been nine strandings of sei whale. Two strandings are noted from the North Sea, a calf from Essex (Fraser 1974) and one from near Rotterdam (van Bree and Husson 1974, van Bree 1977), but of course much of the original taxonomic work was based on North Sea strandings (Turner 1882, Flower 1883, Lydekker 1895) and from the German Holstein coast (Rudolphi 1822). One stranding has been recorded from France from 1972 to date (Duguy and Aloncle 1974, Duguy 1975, 1978).

NORTH PACIFIC

Catches

Western North Pacific. Sei whales have been caught from several coastal stations of the western North Pacific. From the northern regions sei whales have been taken in summer from the Soviet land stations of Kamchatka, as shown in the IWS, and described by Tomilin (1967) and Klumov (1963). Tomilin described the sei whale as being more common off Kamchatka than further north off the Chukotka Peninsula, but still rare compared with the concentrations found off Japan. Sei whales were also taken from the Kuril Islands. Records of sei whales caught in the western North Pacific have been published in various places, but many records have not distinguished between sei and Bryde's whales, and these include the reports from the Bonin Islands of Mizue (1950), and from the coasts of Japan by Omura (1950a). Later Omura et al (1952) and Omura and Fujino (1954) demonstrated that "sei" whales caught in the 1950s about the Bonin Islands, Okochi and Korea were mainly Bryde's whales whereas those from Sanriku (N.E. Honshu) and Hokkaido were mainly sei; only in July to August were a few Bryde's whales caught from northern Japan. However catches from the whaling season of November 1935 to July 1936, from Okochi and Oshima, were all Bryde's whales but those from Sanriku, Hokkaido and also the Bonin Islands were all sei whales. They explained the difference by noting the fact that the 1935/36 catch from the Bonin Islands was mainly in March and April whereas in 1952 it was in May and June; a seasonal movement of both sei and Bryde's whales is implied. The timing of peak catches of sei whales from Japan is described by Nasu (1966) as September and October off Hokkaido, May to June from Sanriku and July and August from the Okhotsk Sea coast of north Hokkaido. Because of the mixing of the major water currents of the southern Kuroshio and northern Oyashio there is a hiatus of sei whale catches between the whaling grounds of northeast Honshu and east-southeast Hokkaido (Uda and Nasu 1956, Uda and Suzuki 1958). The IWS report sei whale catches from both Japan and Korea but there is no good evidence of sei whales having been taken from Korean land stations (Brownell 1981).

Eastern North Pacific. Catches from sei whales have been recorded from Alaska to the equator. Catches from the Alaskan whaling stations of Port Hobron and Akutan are described by Reeves et al (1985) but from 1924 to 1939 only three sei whales were caught, those from Port Hobron. The IWS show a sustained catch from British Columbia which Pike and MacAskie (1969) note were taken from June to August. Catches from California have been reviewed by Rice (1974, 1977) and never more than 100 were taken in one season. These were caught between May and October but peak catches were in August and September.

The IWS report that six sei whales were taken from off the coast of Mexico in 1935. In 1914 four sei whales were reported taken from off the coast of Mexico and from off the island of Gorgona (4°N). Bryde's whales are rare as far north as California but are found off Mexico so these catches cannot be confirmed.

Pelagic catches. The IWS show large numbers of sei whales caught between 1952 and 1975, although factory ships were hunting other species in the coastal regions much before this (Tomilin 1967, Ohsumi 1980a). The positions of catches of sei whales taken in Japanese operations have been given by Nishiwaki (1966), Nemoto (1963) and Masaki (1976a, 1977a). Nishiwaki and Nemoto showed that sei whale catches in the northern North Pacific were restricted to a narrow band to the south of the Aleutian Islands, from Kamchatka to the south coast of Alaska. These distributions were based on records before 1962 and over this period the main interests of the industry were blue and fin whales, which are found, in summer, about and to the north of the Aleutians, and sei whales were taken if they were encountered. With reductions in quotas of the larger whales the whaling operations moved further

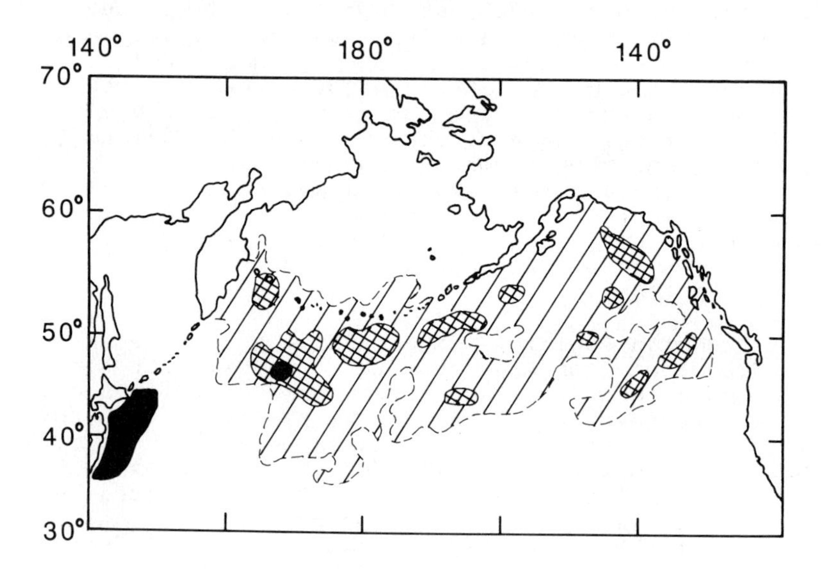

Figure 2.1: Catches of sei whales by Japan, 1952-1972 (no catching was allowed between 150-159°E). Bold: >200 per one degree square, cross hatched: 50-200, striped: 1-50, blank <1.

south (Ohsumi 1980, Horwood 1981a), and Masaki (1976a) gives the distribution of sei whales caught over the period 1952 to 1972 as far south as 35°N; this is illustrated in Figure 2.1. Under IWC regulations pelagic whaling for baleen whales was allowed north of 35°N from the coast of America to 150°W and from north of 20°N for the rest of the North Pacific, but very little effort has been expended in the most southern regions. Distributions of catches over the slightly extended period of 1952 to 1974 are given by Nemoto and Kawamura (1977) and they show that in the last few years sei whales were taken as far south as 25°N; a similar distribution is given by Privalikhin and Berzin (1978). Ohsumi (1977) showed that the large majority of sei whales were taken in waters of between 8 and 18°C. Details of the positions of individual catches are shown in illustration by Nemoto (1959) and Nemoto and Kawamura (1977). Nemoto (1959) reported that sei whales were found along the edge of the Bering Sea, and in 1958 61 sei whales were found in the Anadyre Gulf at 62°N in August. Nemoto explained their occurrence as exceptional and the whales were found in an intrusion of warm water of 8°C.

Sightings

Japanese scouting boats have provided extensive data on sightings from the North Pacific (Ohsumi and Yamamura 1982), and these data have been reported annually (Ohsumi and Wada 1974, Wada 1975-80) and summarised by month over the months April to September by Masaki (1976a, 1977a). Masaki's information comes from the period 1956 to 1972 (corrected in litt) when the vast majority of the sighting effort was expended to the north of 40°N but it can be seen that sei whales can be found in most places north of 30°N. In May concentrations are seen all over the North Pacific but particularly in the Gulf of Alaska and to the east of Honshu. In June and July high densities are found south of the Aleutian Islands, and off the coasts of Canada and the USA. In August sei whales are still seen in the Gulf of Alaska and some are north of 60°N, while in September they are still abundant around the Aleutian chain. Tomilin (1967) confirms that sei whales do go as far north as the Bering Strait and that they are sufficiently recognised by eskimos to be given a name - "komvokhaaks".

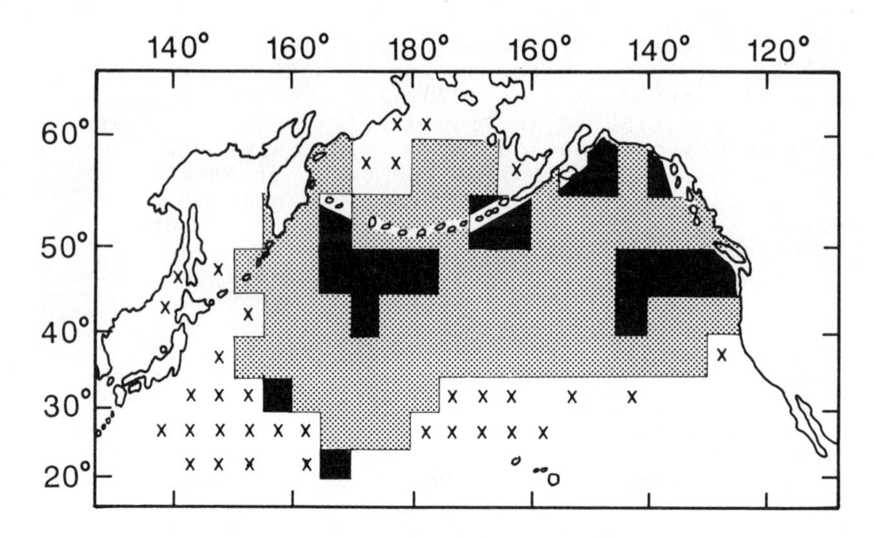

Figure 2.2: Density of sightings of sei whales from Japanese scouting vessels, 1965-1978. Bold: >100 per 10000 miles, stippled: 1-100, x: effort expended but none seen.

In the latter years the surveys extended further south and Wada (1977, 1980) shows that sei whales are seen at least as far south as 20°N. No dates are provided for these southern sightings, but Wada (in litt) has said that those found between 25 to 35°N were seen in May to early August. (There is a discrepancy between the distributions given by Masaki (1977a) and those of Wada (1980) in respect of the densities north of 60°N; it is shown as high by Masaki but nil by Wada. This appears to be due to Masaki using material from 1956 onwards and Wada from 1965 on; in fact sei whales rarely penetrate deep into the Bering Sea.) More recent reports are given in the Japanese research progress report (Japan 1984, 1985, Miyashita 1985) and few sei whales have been seen south of 30°S in summer. The sightings by month are shown on Figure 2.2.

Other sightings cruises have been undertaken in the North Pacific. Gray whales are caught for the native peoples of Chukotka by a Soviet catcher boat operating in June to October near to the coast. Records of whales sighted have been kept since 1969, but Ivashin and Votrogov (1982) reported that no sei were seen in this area. A cruise within 200 n. miles of the coast of the Gulf of Alaska in June to

August was described by Rice and Wolman (1982) but again no sei whales were seen. Similarly during an aerial survey of the traditional whaling areas of Port Hobron and Akutan in July and August no sei whales were seen (Stewart et al 1985). In the southwest North Pacific a cruise was undertaken in January to March from 35°N to 11°S and covering the longitudes 137-168°E; 50 Bryde's but no sei whales were found (Miyazaki and Wada 1978).

Ivashin and Veinger (1979) described the results from four cruises mainly in the equatorial regions of the eastern Pacific, and they reported seeing only three sei whales between the Sangar Strait and the Hawaiian Islands, and 10 about the Galapagos Islands. A co-operative Soviet-American expedition was undertaken in 1975, and has been reported on by Berzin (1978), Privalikhin and Berzin (1978) and Rice (1979). The cruise covered the region 10°S-20°N and from the coast of America to about 110°W. Over March to June Berzin reported seeing 63 sei whales, at least two of these were marked. This marking is reported by Privalikhin and Berzin as at 4°N, and they say the marking process, for which it is necessary to be very close to the whale, allowed the species to be "determined exactly". This opinion is not shared by Rice (1979). Rice (1977) described the distribution of sei whales off the coasts of America as being sparse but widely scattered from 18-35°N in December to March, and as far south as 32°N in May to October; he reported a school off Baja, California in June at 27°N. From the 1975 cruise Rice (1979) says "I saw no whales during the cruise of the VNUSHITEL'NYI that I identified as, or even suspected of being, either sei or fin whales". He considered all the sei whales noted above to be Bryde's whales. Although there may be many subtle differences of shape and behaviour between the two species the only reliable way of distinguishing these species is by the rostral ridges, and Rice claimed to have frequently seen these, and Berzin does suggest that some of the whales seen may have been Bryde's whales. Rice presents a well argued case from the field, but he also claims support from the lack of previous observations of sei whales in the area, and he claims that sei and Bryde's whales are mutually exclusive in the winter. This is not strictly correct as just south of the equator both catches and sightings of sei and Bryde's whales have been made in the same months, in virtually all months (Valdivia et al 1982, 1984). Recent observations have shown that sei whales are to be found well into the Sea of Cortez in January and March (Leatherwood pers comm).

SOUTH ATLANTIC

Catches

Brazil. Sei whales have been caught from the land stations at Costinha (7°S) and Cabo Frio (23°S) from at least 1947, and since then over 5000 sei whales have been taken from the two regions, with catches varying from a few hundred to almost 1000 per year. The catch from Costinha is given by Williamson (1975) and by da Rocha (1983), and Ohsumi and Yamamura (1978a) give the total from the two stations; unfortunately they are a combination of both sei and Bryde's whales. Ohsumi and Yamamura note "a few Bryde's were caught annually in recent years" and from Costinha, Williamson recorded that over the period 1967-1974 the sei to Bryde's whale ratio was 10:1. At Cabo Frio in 1961 453 sei and Bryde's whales were caught and Omura (1962) reported that by mid-September 14 Bryde's had been caught. The information confirms that from off Brazil catches of sei whale were much greater than those of Bryde's whale and, with the intensive exploitation of sei whales in the Antarctic during the early 1960s, the proportion of sei whales would have been greater in the preceding years.

Paiva and Grangeiro (1965, 1970) presented the catch by month from Costinha, over the period 1960-1967, and they show that the peak catch of sei whales occurred from July to September with very few taken in June and November; catches of other whales were negligible and thus the distribution of the catch with month will be a good index of true abundance. Da Rocha (1983 table 7) shows that after 1966 about one third of the sei whales caught were outside the months August to November. Paiva and Grangeiro illustrated the positions of the catches of the whales and it can be recognised that these are on the edge of the continental shelf.

Southwestern South Atlantic. From the islands of South Georgia, Ohsumi and Yamamura (1978) report an accumulated catch of sei whales since 1913 of over 15000, but the earliest record of sei whales taken from about these islands was in 1906 from New Island, to the westward of the Falkland Islands

(Horwood 1986). The early expedition of the Norwegian vessel ADMIRALEN in 1905 and 1906 is described by Tonnessen and Johnsen (1982), and although the ship visited the South Shetland Islands and took 58 whales, 24 of them blue, the other species cannot be identified. Similarly a British Government report (Misc. 1915) does not identify these whales and makes no mention of sei whales from South Shetland in 12 years of whaling. However Zenkovich (1969) in his table 3 refers to 97 sei whales taken at South Shetland in 1905/06. This is in error as this figure is the total catch of sei whales from the ADMIRALEN and was caught mainly, if not in total, from the Falkland Islands. Nevertheless a few sei whales were caught from the South Shetlands in later years (Zenkovich 1969, IWS vol 1). There are no records of catches from South Orkney and Andrews (1916a) quotes from a letter from a Captain Melsom saying that sei whales are not to be found in the South Orkney or South Shetland Islands. In a later account Andrews (1916b) does confirm that sei whales were caught from South Shetland.

There is an extensive record of catches from South Georgia and early catches over the period 1913 to 1931 were mainly taken in March with a few in February and April (Harmer 1931, Matthews 1938). By 1963-1965 peak catches were in January with a reduction in February and March (Gambell 1968). Matthews stressed that the whalers gave greater preference to fin and blue whales as they were much more valuable; consequently the timing of catches includes a component of selection. Over some seasons whaling was carried out all through the year (Misc. 1915) and although sei whales were taken their catch was restricted to the months of April to June (Zenkovich 1969).

West Coast of Africa. Sei whales have been caught from the Cape Province of South Africa to Gabon on the equator, and the IWS show sei whales caught from the French Congo, now Gabon, in small numbers from 1923 to 1937, with two years of greater activity in 1951 and 1952, (Anon 1951). Off Angola whales were taken from 1924 to 1928 and Olsen (1913) reported them taken in November in 1912. These catches were all from austral winter fisheries. Except for that from South Africa there is no information with which to judge if, or in what part, Bryde's whales were included in these catches.

Occasionally a sei whale was taken by the pelagic fleets on their way to and from the Antarctic, but over the period 1969 to 1976 the vessel SIERRA caught several hundred sei and Bryde's whales per year. The vessel operated out of the jurisdiction of the IWC and few details as to times and positions of the catches have been published. In the last year of operation, 1976, a catch of 23 sei and Bryde's whales was taken in January and February. Catches almost certainly came from both sides of the equator and the SIERRA was observed flensing off Nigeria in that year (IUCN 1979).

Further south details of catches are more precise and reviews of the catches of sei whales have been given by Matthews (1938), Best (1967a, 1974a) and Best and Lockyer (1977ms). Ohsumi and Yamamura (1978) show that over the period 1910-1976 over 9000 sei whales were caught from the Cape Province. Best (1967a) and Best and Lockyer explained that these statistics included Bryde's whales. From a total catch of sei and Bryde's whales of 721 in 1963, Best (1967a) identified 89 as Bryde's whales. In this year a research catch of 50 whales taken in the months of March to April, before the usual catching season, only seven were sei and none were caught in March. If these are deducted from the total for the season it leaves 46 Bryde's whales from a catch in the usual season of 671.

Matthews (1938) and Best (1967a) present similar distributions of catch by month with a peak in August to October and a second but smaller peak in May or June. Both authors refer to the annual occurrence of sei whales as erratic. Best (1967a) explained that sei whales are mainly found in waters of 16-18°C, 60-100 n. miles offshore, and not in the colder waters of the Benguela Current.

Sightings

Most of the information from sightings comes from operations associated with whaling. On the western side, as the Soviet fleets moved south, Budylenko (1977, 1978a) reported groups of 3 to 5 sei whales along the east coast of South America in October from 10-30°S. In 1914 "thousands" of sei whales were reported as seen between Cabo Frio and the Falkland Islands (Tomilin 1967). Lichter and Hooper (1983) summarised previous information as showing that sei whales are distributed from the coasts of Uruguay to Tierra del Fuego. Further south Liouville (1913) described sei whales in the

Bransfield Strait and Martha Bay of the South Shetland Islands, but the reliability of these sightings can be questioned. Liouville considered that other authors has misidentified sei whales as minke whales, but the southern positions of these whales makes it highly likely they were minke. That such apparent confusion existed means that whales were probably not approached at all closely, and a suspicion is cast on Liouville's identifications and on his ability to recognise sei whales.

On the eastern side, Mørch (1911) reported large groups of sei whales off Angola, and Budylenko (1970, 1978a) and Golubovsky et al (1972) found considerable numbers of sei whales between southwest Africa and Tristan da Cunha. The spotter plane used by whalers from Donkergat, Cape Province, provided detailed information on sei whale distributions (Best 1967a), and the sightings of sei whales in 1963, over the months March to October, give a bimodal pattern, with a peak in May and June and a second larger peak in August to October. A study of the direction of undisturbed whales showed that in April to July most (64-100%) were travelling northwards, and in August to October most (51-65%) were going to the south. Best (1977a) comments on the difficulty of identifying sei and Bryde's whales from the air, but the relatively low numbers of Bryde's caught in the whaling season means that the spotter plane information on sei whales would not be seriously distorted. In the months of March and April a research cruise found few sei whales, and Best (1967a) considered that whales previously reported as being numerous at that time of year were Bryde's whales. Johnson (1915) in his evidence to a British Colonial Office investigation on whaling said that "thousands" of sei whales came into Table Bay and False Bay, however no dates are given and the information appears exaggerated. Using the data from Japanese scouting vessels Masaki (1980) has shown that sei whales can be found from October to April from 30-40°S.

SOUTH PACIFIC

Catches

<u>South America</u>. Peruvian whaling began in 1951 operating for sperm whales but from 1968 baleen whales were taken from a land station at Paita at 5°S. From 1968 to 1972 records of catches of sei

and Bryde's whales are combined, but following a visit by Nishiwaki the two species are identified separately from 1973 (Valdivia et al 1981). Valdivia et al (1984) give catches of sei whales by month; normally land stations have operated for a six month period, a regulation of the IWC, and this period is coincident with the arrival or presence of the whales, but Peru has operated in most months. Sei whales were taken in all months over the period 1973-1978, except March, April and July, with peak catches in August to November. The IWS show sei whales taken off the coasts of Chile and Peru combined from 1914, and from Peru separately from 1947. These early catches are from pelagic operations (Clarke 1980). Tomilin (1967) mentions 15 caught off Ecuador in 1925 and Clarke (1980) also notes sei whales caught off Ecuador in 1926, but these could have been Bryde's whales.

Catches from Chile, and off its coasts, are recorded by Aguayo (1974), Clarke (1980) and Arriga (1981) and records of sei whales go back to 1914. Clarke thought that any proportion of these could be Bryde's whales and to date no approximate proportions have been estimated (Gallardo and Pasterne 1983). Pasterne et al (1983) documented a Bryde's whale caught as far south as 33°S.

Australia and New Zealand. Since 1909 a total of 32 sei whales are recorded as having been caught from the coasts of New Zealand, and eastern and western Australia (IWS, Ohsumi and Yamamura 1978). Best (1977a) was sent some information by R G Chittleborough on nine Bryde's whales landed in Australia. Two from the east coast did appear to be Bryde's whales but the seven from the west coast (Indian Ocean) would have been identified as sei whales on the extent of the ventral grooves. Off New Zealand in the Tasman Sea, south of 40°S, large catches have been taken by the pelagic fleets (Gaskin 1977). Gaskin (1968) confirmed that catches in 1956 and 1964 from the Tory Channel were sei whales.

Sightings and Strandings

Sightings have been documented from the Peruvian catcher boats (Valdivia and Ramirez 1981, Valdivia et al 1982) and sei whales, distinguished from Bryde's whales in their tables, were seen in the months October, November, January, May and June. In addition Peruvian marking cruises have been carried out in the region 2-9°S, 81-86°W in February 1980 and March 1981, but no sei whales were encountered (Valdivia et al 1981, 1982). An IWC cruise was conducted for Bryde's whales in the area 8°N-9°S, 100°W to the coast of America, in November and December 1982, but no sei whales were encountered (Landa et al 1983, Donovan 1984). Clarke (1962) claimed to have seen sei whales about the Galapagos Islands but later Clarke and Aguayo (1965) considered that they were probably Bryde's whales. Sei whales have been found from several cruises off Chile (Aguayo 1974), and Gallardo et al (1983) found, between 32-39°S and 75°W to the coast, two sei whales between 33-36°S and 25 Bryde's whales between 32-36°S. Clarke et al (1978) reported seeing a single sei whale on each of two cruises in 1958 and 1964 in November and December. Mikhaliev (1978) documents sei whales seen off Chile in November or December.

Other than the extensive Japanese scouting vessel surveys there are few reports of sightings in the South Pacific. Budylenko (1978a) says that between 40-50°S concentrations are seen in December to January, and that in the sub-tropical zone sei whales form large concentrations in November and December. Gaskin (1976) shows large aggregations off the west coast of New Zealand between 40-50°S in the austral summer, and Gaskin (1977) also noted that Bryde's whales are to be found off the north coast of the north island of New Zealand. A survey for Bryde's whales was reported by Ohsumi (1979a) which was undertaken from late October to late November, 1977, which mainly covered the region 0-30°S and 170°E-140°W, with a small leg extending to 40°S; from 16000 n. miles steamed no sei whales were seen. A similar cruise in November and December 1978 found seven sei whales north of 40°S (Ohsumi 1980b). Kawamura (1974, his figure 6.3) shows the whales seen in November to January from 0-70°S along the meridian of 170°W; sei whales were seen at 2°S, 34°S and south of 40°S. Masaki (1980), from Japanese scouting boat data, shows that in the IWC whaling Area V, (130°E-170°W), over the months October to April, sei whales are at their greatest densities from 30-40°S, but are rare in these months in the eastern South Pacific.

Strandings in New Zealand were reviewed by Gaskin (1968), but of six recorded specimens he could only confirm that one was definitely a sei whale; no strandings of sei whales occurred in more recent years Cawthorn (1984, 1985). One stranding of a sei whale was reported from New South Wales in October 1984 (Australia 1985).

INDIAN OCEAN

Catches

A long history of catches of sei whales is recorded from land stations on the coast of Natal (IWS, Ohsumi and Yamamura 1978, Best 1974b, 1975, 1977b), and a small number of Bryde's whales have also been landed at Durban (Best 1977a). Matthews (1938) described the seasonal catches over the period 1918-1926, and the peak catches were in May and June with a slow decline through to November. The density of whales off Durban has been described with an index of catch per unit of effort (CPUE) by Gambell (1968), for the period 1964-1966, and the index shows a slow rise from March to June and then a large rise to a maximum in September. The situation is a little different if CPUE data are used from the greater time period of 1954-1963 (Bannister and Gambell 1965, Gambell 1974). A bimodal distribution is seen similar to that found on the west coast of South Africa, with a small peak in May and June, a small drop in July before a large peak in September. A slight reduction occurs in October.

A few sei whales are recorded in the IWS as coming from south of Madagascar, but large numbers of Bryde's whales are found all around the island (Ohsumi 1979a), so the identification is uncertain. (See also Australia and New Zealand above).

Sightings

Records were kept of whales seen by the spotter plane at Durban, and these are summarised by Best (1974b, 1975, 1977b). The information also has been presented by months, from the period 1954-1963, by Bannister and Gambell (1965) as numbers seen per mile searched. As with the CPUE data a bimodal distribution is found with a small peak in May and June, a slight drop in July, and a large peak in September; the data were extended by Gambell (1968). The data were also analysed to investigate the direction of travel of whales off Durban, and it was found that they were going northeastwards in May and June, and mainly southwest in August and September.

Ohsumi (1979a, 1980b) described sightings surveys, mainly for Bryde's whales, north of 40°N. In 1977 surveys across the entire Indian Ocean found three sei whales in October and November; in contrast numerous Bryde's whales were seen about Madagascar. The 1978 survey covered the central and eastern Indian Ocean in the austral summer and again three sei whales were found; the positions of the sightings are not given.

Budylenko (1978a) described sei whales seen in the northern Indian Ocean in March, May and July, and also off India and Sri Lanka. Slijper et al (1964) reported solitary sei whales in January, May and October in the Aden Gulf, and one sei whale is recorded as stranded at the top of the Red Sea (Tomilin 1967). Rorvik (1980) reported a fin or sei whale off Mozambique in September. These records all suffer from possible confusion with Bryde's whales. Keller et al (1982) described aerial and ship surveys about the Seychelles Islands in April and June 1980. No sei whales were seen but they site a report of 1913 which refers to sei whales migrating to the Seychelles. A vessel survey off the coasts of Sri Lanka in May 1985 found Bryde's whales but no sei whales (Gunaratna et al 1985).

Gambell et al (1975) conducted a whale marking cruise over the period 24.11-2.2 that covered the region 20-42°S and 30-67°E. Sei whales were found in January at about 33°S 65°E. Budylenko (1978a) says that large concentrations of sei whales are to be found south of 30°S, and Masaki (1980), from the records of Japanese scouting vessels, shows that from October to April sei whales are at their highest densities between 30-40°S.

ANTARCTIC (South of 40°S)

Catches

The pelagic fleets have been restricted almost exclusively to the waters south of 40°S by whaling regulations. Since pelagic operations started in 1925/26 they have taken over 130,000 sei whales, although intensive exploitation did nót begin until 1957 because of the whalers preference for the larger species (Ohsumi and Yamamura 1978a, Horwood 1978). Catches by geographical regions from 1931/32-1971/72 are given by Omura (1973), and from 1956/57-1977/78 by Horwood (1980a). The distributions are very similar, and the cumulative catches since 1956/57 are given by ten degree squares in Figure 2.3.

Latitudinally the Antarctic Convergence acts as a barrier and the smaller sei whales, less than

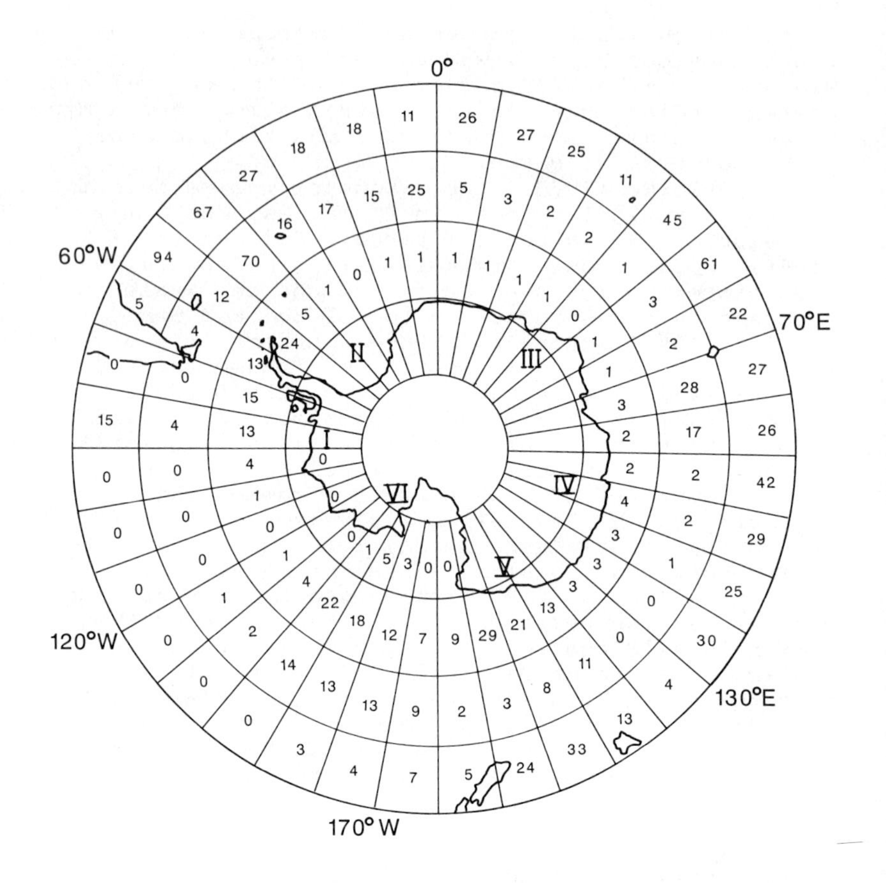

Figure 2.3: Cumulative catch of sei whales, in hundreds, by 10° squares, south of 40°S, 1956/57-1977/78.

about 14.6 m (48 ft), seldom go further south (Doi, Ohsumi and Nemoto 1967, Horwood 1986). Since 1931 catches have been taken from December to April, with peak catches in January and February, but this has varied with time. Before 1950 the peak catch was in March, from 1950 to 1960 it was in February, and after 1960 it was in January, and this is almost certainly associated with the early preference for blue and fin whales, although there is some argument that sei whales may be penetrating further south, following the reduction of the numbers of blue and fin whales. Kawamura (1974) has shown the preference of southern sei whales for the calanoid copepods, and with increases in temperature and changes in prey species, the sei whales move further south later in the whaling season (Nemoto 1959). Nasu and Masaki (1970) show that most of the sei whales were caught between the Subtropical Convergence and the Antarctic Convergence, and Figure 2.3 illustrates the northern concentration of catches between 60°W eastwards to 130°E associated with the more northern position of the Antarctic Convergence. In the other sector the Convergence and catches are much more to the south.

Within the Antarctic particularly favorable whaling grounds can be recognised and large catches have been obtained from the Tasman Sea and around the islands of Marion, Crozet and Kerguelen.

Sightings

There have been various surveys in these southern waters (e.g. Mikhaliev 1978, Horwood et al 1981) but the Japanese scouting boat data provide extensive information from a wide area (Ohsumi and Yamamura 1982). For the Antarctic these records have been reported by Ohsumi and Masaki (1974), Masaki and Fukuda (1975), Masaki (1976b, 1977b, 1979a, 1980) and Masaki and Yamamura (1978).

Masaki (1980) reviewed sightings from over the period 1965/1966 to 1977/1978 for the months of October to April, and by ten degree latitudinal bands, and his data have been rearranged in Table 2.1 to show the density of whales by whaling Area and by latitude. Over 48,000 sei whales seen

Table 2.1: Sighting density of sei whales seen per 1000 n. miles steamed by whaling Area and latitude, 1965 to 1978, mainly from November to March.

Area	30-40°S	40-50°S	50-60°S	60-70°S
I (60-120°W)	0	135	50	13
II (0-60°W)	123	375	24	120
III (0-70°E)	190	1619	148	1
IV (70-130°E)	205	258	121	3
V (130°E-170°W)	207	323	55	31
VI (120-170°W)	0	304	214	142

contribute to this table and over two million miles searched. (In passing it is interesting to note that to provide two Japanese vessels for the IWC/IDCR minke whale sighting and marking expedition of 1981/82 was estimated to cost $1.2 million). The data show that in the austral summer the highest densities are to be found between 40-50°S and as the catch data indicate the whales are found further south from 60-170°W.

Masaki (1980) does not give the data by month but the original material provided by the Far Seas Research Laboratory, Japan, does give that breakdown. The catch and sightings data show distributions in whaling Areas III and IV to be rather similar and combined data from them from the season 1969/70 are shown in Table 4.1. Although there are gaps the data reveal a movement of sei whales south through the season, with few further south than 50°S before January or February. In November the whales are spread between 30-50°S but by December most have moved south of 40°S. From the entire Southern Hemisphere Ohsumi and Yamamura (1978a) show that in January the vast majority of the sei whales are to be found from 30-50°S.

SUMMARY

The majority of information on the distribution of sei whales comes from whaling operations, both pelagic and from land stations, and the timing and location of these operations reflect the distribution of the large whales; however, until the 1950s, sei whales were not the main targets of the industry and so interpretation needs to be cautious. Nevertheless whaling has shown that sei whales are found in all oceans from the equator to near the ice edge, but within this broad description the distribution is strongly influenced by season, water depth and temperature.

In all oceans the sei whales are found widely dispersed in warm waters, near to the equator during the winter, and towards the polar regions in the summer, and catches from land stations and sightings reflect the seasonal migratory movements. In summer the sei whales do not penetrate into the polar seas as far as do the other rorquals, and catches tend to have been made in waters of 8 to 18°C. In the Southern Hemisphere the Antarctic Convergence appears to act as a barrier with only the larger sei

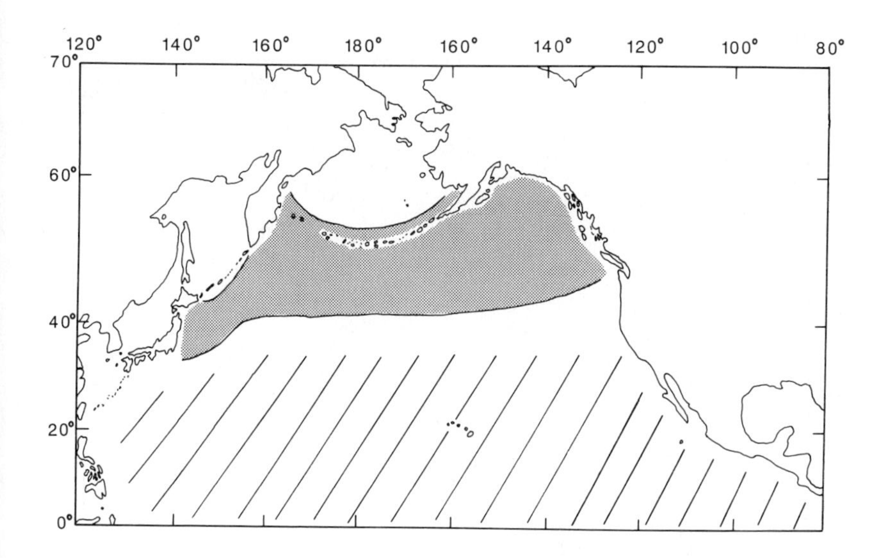

Figure 2.4: Schematic seasonal distribution of North Pacific sei whales. Between the bold lines is the summer distribution and the striped area is the presumed winter distribution.

whales going further south, and in the Northern Hemisphere the summer whaling grounds are in more temperate latitudes, where there are major frontal systems. Polar excursions are made only when there are favourable temperatures, either as eddies or late in the summer when temperatures are maximal. Summaries of the distributions are illustrated in Figures 2.4 to 2.6.

The sei whale is oceanic in habit and is rarely found in the marginal sea areas. Catches are usually from deep water, and from land stations catches were usually taken from along, or just off the edges of the continental shelves.

Local and annual occurrences of sei whales are erratic. Feeding concentrations can quickly disperse, as described in the next chapter, and land stations are much affected by the annual presence or absence of sei whales. Longer term changes are also observed with sei whales, once common off North Norway, having not been found there in forty years.

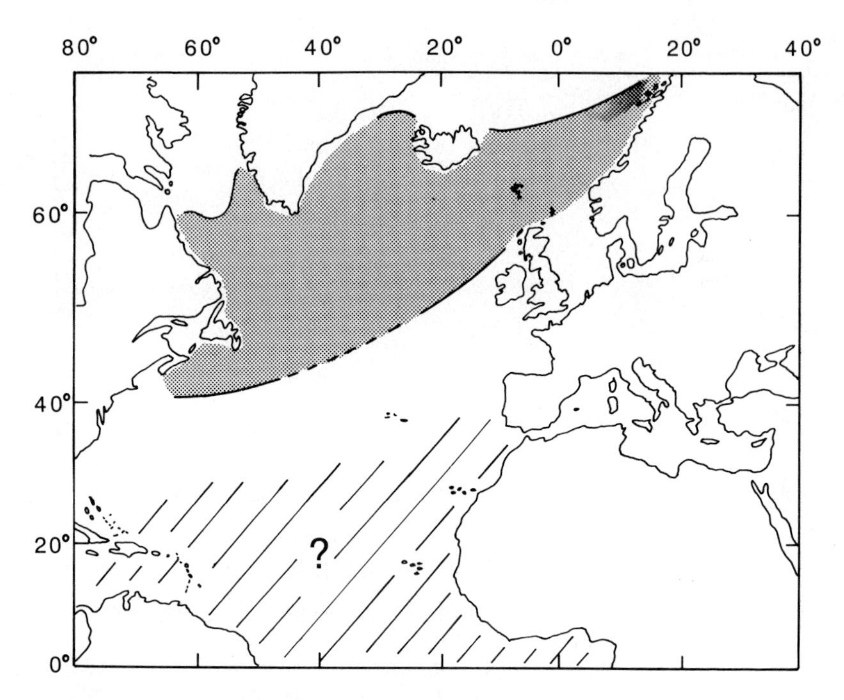

Figure 2.5: Schematic seasonal distribution of North Atlantic sei whales, as in Figure 2.4.

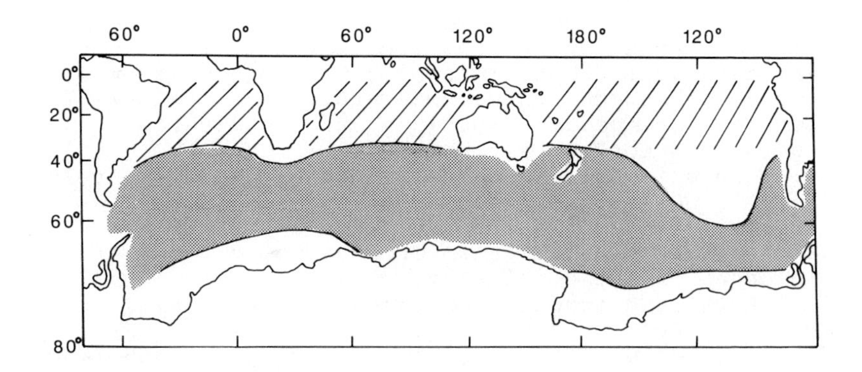

Figure 2.6: Schematic seasonal distribution of Southern Hemisphere sei whales, as in Figure 2.4.

Chapter Three

MARKING AND RECOVERY

INTRODUCTION

The recognition of individual whales over periods of time provides three sorts of information. The first indicates the movement of individual whales, and this can assist in the interpretation of general indices of population abundance, which in turn help to describe seasonal migrations. A consequence of this is that these data have often been used to assist in the delineation of stock boundaries. The second type of information is that of stock size. Information on the relative number of whales recognised to those not recognised can be used to obtain estimates of absolute abundance. Later sections describe how this information is used specifically and the assumptions and problems associated with the methods. Thirdly the known time between marking and recapture, especially if calves have been marked, allow ageing techniques to be validated.

Some species of whales possess patterns which lend themselves to recognition of individual whales. This seems to be so for right and killer whales (Payne et al 1983, Best 1985), and this has been demonstrated for humpback whales of the North Atlantic, from tail fluke patterns (Katona and Krauss 1979). In addition some individual whales can be identified by natural abnormalities, such as albinoism or disfigured dorsal fins, or by man made characters, such as harpoon scars or bullet holes in fins. For sei whales such characters have contributed little or no information; their natural patterns are subtle, and the sei whale is an oceanic species and is not as easy to approach as some others.

Hand harpoons used to be stamped with ship's marks, and whales lost by one ship have been recovered by others years later (Mitchell 1983). Similar incidents still occur (Martin 1982). This led to the idea of deliberately planting a mark in a whale and waiting for it to be recovered when the whale was killed. Omura and Ohsumi (1964) described deliberate marking of whales by whaling Captain T Amano. In 1910 a "marking rod" was shot into a blue whale which was recovered two years later when the whale was killed. However Brown (1977b) credits Johan Hjort with the idea of large scale marking, and in 1914 a "Hjort" mark was fired into a blue whale. Except for Amano's mark no other early records of returns are known, and experiments were conducted on different types of marks. Initial designs included a protruding part and again none of this design were recovered. By the 1932/33 Antarctic season the present Discovery Mark had been developed and this has proved relatively successful. The mark used for the larger whales, including sei whales, is a 23 cm long metal tube, fitted with a blunt lead head and fired into the back of the whale with a 12-bore gun. On each mark is a unique number and instructions for its return. If successful the mark buries itself in the blubber or muscle of the whale. For descriptions of movements little more is needed than accurate dates and positions of marking and recovery, but problems with marking are discussed in more detail when they are considered in relation to stock estimates.

From the 1945/46 Antarctic whaling season most countries co-operated in the marking of large whales, and this was encouraged by the IWC. Marking under this co-operation was referred to as the International Scheme. Not all marking was done through this Scheme and in particular the USSR retained its own marking programme, known as the Soviet Scheme, as did other countries for marking in the Northern Hemisphere. Nevertheless there has been substantial co-operation amongst all countries in the exchange of information on markings and recaptures. A review of the co-ordination is given by Brown (1974). One introduction of the International Scheme was of a standard notation for the fate of marks fired. These were the five categories of "Hit", "Possible Hit", "Hit Protruding", "No Verdict", "Ricochet" and "Miss". If a mark had not fully penetrated into the whale it was recorded as "Hit Protruding" and if it bounced off the whale it was a "Ricochet". With the movements of whale and vessel, spray and speed of the mark it is often not easy to mark whales, or to accurately judge the fate of a mark. For sei whales in the North Pacific Ohsumi and Masaki (1975) give a 59% hit rate, but from those sei whales where the marks were recorded as having missed they reported the recovery of marks as especially high (compared with other species).

In principle, and in general, a whale is marked and it is known that a sei whale, for instance, had mark number 12345 inside it. On being caught and flensed the mark would be found and it would be reported that mark 12345 was recovered from a sei whale on a given date and from a given location. However, in some instances the marks were not seen during flensing and were sometimes recovered much later from the traps in the blubber and meat rendering boilers. In these cases only an approximate date and general location could be given to the recovery. In addition, Ohsumi and Masaki (1975) reviewed the identification, by species, at marking and at recovery. In general the identification is consistent but instances such as a mark put into a sperm whale and found in a sei whale (No. 1271) show that errors can occur in recording, independent of problems in identification. Ohsumi and Masaki also show that of 82 whales recorded as sei whales at marking 13 were Bryde's whales, as identified on recapture. The proportion misidentified will depend upon the location of marking but again the serious problem of identifying sei and Bryde's whales at sea can be appreciated. Since the late 1950s this has not been a problem with Japanese records. In the next sections the numbers of sei whales marked and recovered are documented by ocean regions and by countries involved.

NUMBERS MARKED

North Pacific

The numbers of sei whales marked in the North Pacific are summarised in Table 3.1. The reports of Japanese marking over the period 1949-1962 are from Omura and Ohsumi (1964). As described in the earlier chapters identification of sei and Bryde's whales was not clarified until the middle 1950s, and from 1949 to 1955 sei and Bryde's whales are not distinguished in the marking records. Some whales were marked sufficiently to the north to be sure they were not Bryde's whales, but most were marked in regions where both species are found. From 1956 to 1962 sei whales were identified at the time of marking and 41 were marked. One small problem in

Table 3.1: Numbers of sei whales marked in the North Pacific

Country	Period	Number	Reference
Japan	1949-1955	192+	Omura & Ohsumi (1964)
	1956-1962	41	Omura & Ohsumi (1964)
	1963-1972	282+	Ohsumi & Masaki (1975)
	1973-1985	102	IWC marking reports
USSR	1954-1966	43+	Ivashin & Rovnin (1967)
	1967-1985	44	IWC marking reports
Canada	1955-1985	5	Brown (1977b), IWC reps.
USA	1962-1985	29	Brown (1977b), IWC reps.

+: sei and Bryde's combined

Table 3.2: Numbers of sei whales marked in the North Atlantic

Country	Period	Number	Reference
Canada	1950-1985	30	Brown (1979), IWC reports
Iceland	1950-1985	59	IWC reports

Table 3.3: Numbers of sei whales marked in the Southern Hemisphere.

Scheme	Period	Number	Reference
International	1955/56-1975/76	377	Brown (1978)
	1976/77-1984/85	160	IWC marking reports
Soviet	1954/55-1977/78	294	Ivashin (1980)
	1978/79-1984/85	17	IWC marking reports

these statistics is that, from 1949 to 1953, the fate of marks was recorded as only "Hit" or "Miss", with the Hit category including those marks that protruded out of the whale. Subsequent recordings followed the categories of the International Scheme.

From the period 1963-1972 numbers marked, under the Japanese programme, were given by Ohsumi and Masaki (1975). In this summary the 282 marked whales are of sei and Bryde's whale combined. Of these at least eight were Bryde's whales. After 1972, information, extracted from the whale marking reports and national progress reports, which are published annually in the Reports of the International Whaling Commission, shows that 102 sei whales were marked under the Japanese scheme. These data are given in Table 3.1.

Results of marking under the Soviet Scheme, over the period 1954-1966, were given by Ivashin and Rovnin (1967). Marking was with a Discovery-type mark and the first sei whale was marked in 1958. Over the period 43 sei whale were marked, mainly in the northern waters of the North Pacific. Ivashin (in litt) confirmed that those marked south of 40°N could be Bryde's whales. From the period 1967-1985 the IWC reports show 44 sei whale marked under the Soviet programme. A small amount of marking has been done by Canada and the USA, and this is reviewed by Rice (1974) and Brown (1977b), with subsequent information taken from the reports of the IWC.

North Atlantic

Marking of whales in the North Atlantic has not been intensive and was reviewed by Brown (1977b,c 1979). Since 1950, 30 sei whales were marked under the Canadian programme (Mitchell 1974a, Brown 1979), but of those a few marked to the north of Venezuela could have been Bryde's whales (Mitchell and Kozicki 1974). From Iceland the national progress reports to the IWC and Sigurjonsson (1983) show 59 sei whales marked; these were marked between 1979 and 1982. Details of whales marked are given in Table 3.2.

Southern Hemisphere

Marking in the Southern Hemisphere has mainly been carried out under the International Marking Scheme and the Soviet Scheme. Marking under the International Scheme was reviewed by Brown (1976a, 1977a,b, 1978) and Garrod and Brown (1980), and the

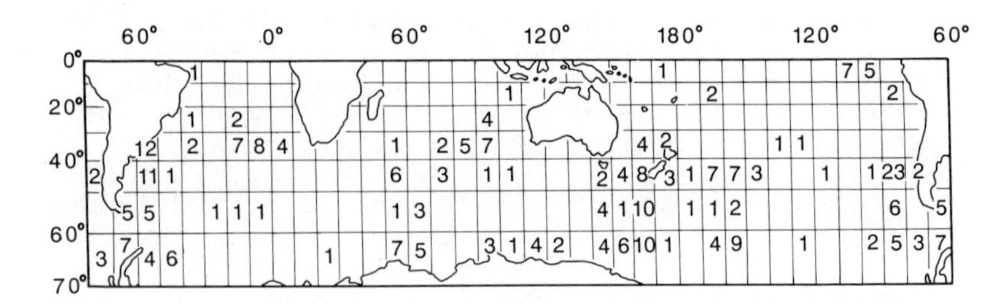

Figure 3.1: Numbers of marked sei whales in the Southern Hemisphere by 10° squares, from the Soviet Scheme, 1954/55 to 1977/78.

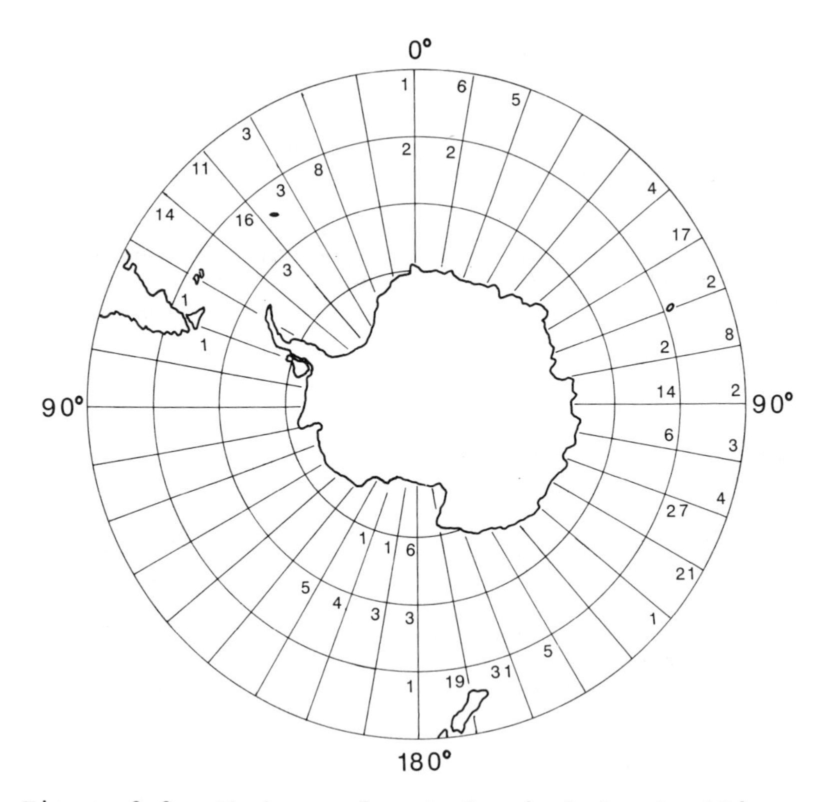

Figure 3.2: Numbers of marked sei whales by 10° squares, south of 40°S, from the International Scheme, 1955/56 to 1973/74.

number of sei whales marked under this scheme, since the 1955/56 Antarctic season is given in Table 3.3. Over the period 1934/33-1938/39 12 sei whales were marked (Brown 1977b), although Rayner (1940) recorded 28 sei whales marked in the Antarctic. The 377 sei whales recorded as marked by Brown (1978) come from 81 marked north of 40°S and 296 to the south. This may seem at variance with the figure of 282 sei whales marked over the period 1945/46-1975/76, being composed of 25 marked in 1975/76 (IWC reports) and 271 for the period 1945/46-1974/75 (Brown 1977b). However the 296 includes whales marked in the waters of Chile and New Zealand, south of 40°S, whereas the 271 is of whales marked in the open Antarctic. Marking reports and national progress reports of the IWC show that from 1976/77 160 sei whales were marked, mainly by Japanese expeditions.

Marking of sei whales in the Southern Hemisphere under the Soviet Scheme has been described by Ivashin (1971, 1980), and the numbers are given in Table 3.3. Of the 294 marked between 1954/55 and 1977/78, 213 were south of 40°S and 81 to the north. Subsequent reports to the IWC show another 17 were marked.

The significance of the 40°S boundary is that it is the northern limit of operations for catching of baleen whales by the pelagic whaling fleets. However, if all the marking and catching were done to the south, little information would be gleaned about the seasonal movements of the whales. Ivashin (1980) gives the distribution of the marked whales (north and south of 40°S) by the six whaling Areas, used by the IWC, and gives the distribution of Soviet marks by 10°squares; it can be seen that 18 (6%) were placed between 0°-20°S. Brown (1977a) gives a similar distribution of all marked whales, but for only south of 40°S. Figures 3.1 and 3.2 show the distribution of marks from the Soviet marking and from the International Scheme south of 40°S.

MARK RECOVERIES

North Pacific

The information on marks recovered from the North Pacific is given in Tables 3.4 to 3.7. From the beginning of the 1976 whaling season sei whales were protected by the IWC, and since then there have been no returns. The "approximate positions" referred to

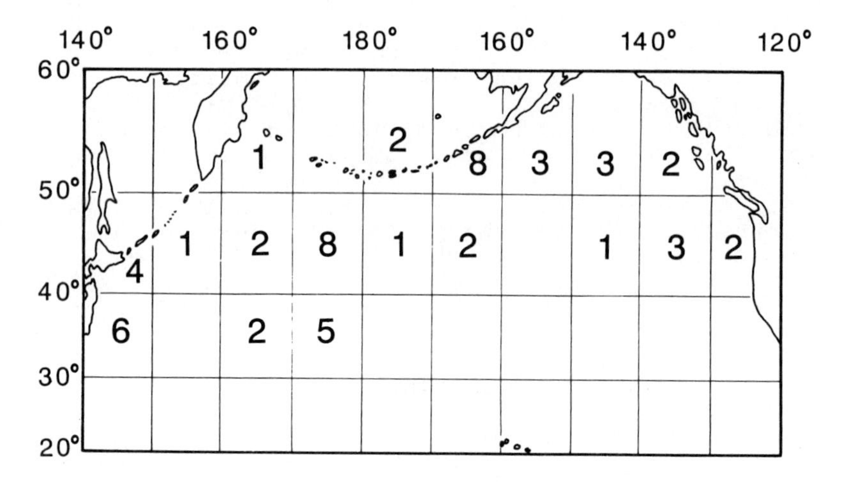

Figure 3.3: Numbers of recoveries of marked sei whales from the North Pacific, by 10° squares. Only known sei whales from known positions are included.

in the tables are usually due to the marks being found at stages of processing, rather than flensing, and if found in the cooking vats they could have been there for some time. Later on traps were fitted to the cookers, which were inspected regularly, but cookers without such traps were inspected sometimes only at the end of the whaling season.

The returns of Japanese marks were given by Omura and Ohsumi (1964) for the period 1949-1962 (Table 3.4). For the earlier dates, some sei whales were not distinguished from Bryde's whales, and these whales are identified in the table. Ten can be recognised as sei whales. The series of recoveries was extended by Ohsumi amd Masaki (1975) for the period 1963-1972 (Table 3.5); 61 marks were returned from sei whales. No further summary exists and recoveries of sei whales, for the post 1973 period, have been extracted from the Japanese progress reports to the IWC (IWC: 25-27). From these, Table 3.6 shows another 24 returns.

Table 3.4: Japanese marks recovered from the North Pacific, 1949 to 1962. Positions given to the nearest degree.

Date Marked	Date Recovered	Position Marked	Position Recovered	
21.07.49	25.06.53	39N 150E	39N 144E	*
8.05.50	15.07.50	27N 143E	37N 143E+	*
21.05.50	22.05.50	27N 144E	27N 144E	*
21.05.50	21.08.52	27N 144E	38N 143E	*
21.05.50	22.05.50	27N 144E	27N 144E	*
3.09.50	29.09.55	40N 146E	41N 145E	
15.05.51	27.05.51	26N 144E	26N 144E	*
23.08.51	6.07.52	42N 148E	39N 144E	*
29.08.51	9.06.61	41N 151E	37N 145E	
2.09.52	21.10.52	39N 149E	38N 142E	*
1.09.52	4.10.59	39N 149E	43N 147E	*
3.09.52	25.09.55	40N 153E	41N 148E	
7.09.52	8.08.55	40N 150E	41N 146E	
11.09.52	15.09.55	43N 148E	43N 146E	*
4.09.52	4.06.58	39N 154E	38N 146E	*
4.09.52	29.08.61	39N 154E	42N 147E	*
5.09.52	5.08.57	39N 156E	40N 144E	*
12.09.52	17.09.55	44N 149E	43N 147E	*
12.09.52	17.09.57	44N 149E	42N 150E	*
3.07.53	12.07.62	38N 162E	38N 144E	
7.07.53	12.09.53	42N 149E	39N 146E	*
7.07.53	29.09.58	42N 149E	41N 145E	*
9.07.53	10.07.55	42N 149E	40N 148E	
9.07.53	10.07.53	42N 149E	43N 148E	*
13.06.54	22.07.58	53N 163W	50N 178W	
13.06.54	22.07.58	53N 163W	50N 178W+	
29.06.60	18.07.62	57N 152W	49N 130W+	
28.07.60	20.08.62	56N 167E	Near Paramushir Is	

*: sei or Bryde's whales
+: approximate position

Another seven marks have been recovered from marking by other nations and are given in Table 3.7. The first two recoveries are from Ivashin and Rovnin (1967), and the last five from Ohsumi and Masaki (1975).

In total 102 marks have been recovered; the positions of recovery of known sei whales is shown in Figure 3.3. This can be compared with 738 marked but some of these were Bryde's whales and some returns were from whales not considered to have been "hit".

Table 3.5: Japanese marks recovered from the North Pacific, 1963 to 1972

Date Marked	Date Recovered	Position Marked	Position Recovered
12.06.54	19.07.64	54N 158W	52N 167W
3.07.54	27.06.65	56N 152W	56N 144W+
21.07.56	28.07.64	54N 158W	54N 158W
13.08.57	20.05.67	52N 168E	46N 171E+
13.08.57	21.05.67	53N 168E	46N 172E+
19.06.58	27.06.67	50N 176E	50N 172E+
29.06.59	14.07.66	54N 164W	50N 173W
4.06.59	22.06.65	53N 165W	54N 147W+
12.07.61	7.08.63	51N 152W	55N 141W
30.06.61	30.07 64	49N 167W	53N 157W+
28.07.62	29.05.69	50N 168W	37N 145E
15.06.62	19.06.64	51N 167W	52N 166W
22.05.63	2.09.66	48N 179W	50N 169E
17.06.63	12.06.64	55N 155W	54N 149W
16.06.63	13.07.66	53N 159W	51N 174W+
16.06.63	11.09.65	53N 158W	51N 168W+
16.06.63	13.07.66	53N 159W	51N 174W+
20.06.64	31.07.66	46N 130W	53N 143W+
29.05.64	12.07.64	55N 137W	49N 128W
18.06.64	20.06.64	56N 139W	55N 139W
10.07.64	14.07.64	55N 154W	55N 153W
23.05.65	3.06.67	49N 173W	50N 167W
23.05.65	7.06.67	49N 173W	51N 162W
27.05.65	14.07.65	50N 167W	51N 163W
2.06.65	14.06.66	55N 150W	53N 151W
2.06.65	21.07.70	55N 150W	48N 139W
18.07.66	20.07.66	51N 162W	51N 162W
15.07.66	29.07.71	56N 150W	51N 153W+
19.07.66	12.08.68	57N 141W	49N 147W+
17.08.66	8.07.69	51N 164W	46N 167E+
11.08.66	13.08.66	54N 144W	53N 145W
13.08.66	1.08.68	51N 177E	50N 175W+
17.05.67	21.06.67	45N 174E	49N 178W
17.05.67	15.06.67	45N 174E	50N 175W
24.05.67	31.05.71	47N 167E	48N 171E
24.05.67	18.06.68	47N 167E	49N 174E
28.05.67	25.07.67	52N 172E	53N 167E
30.06.67	31.07.68	50N 168W	51N 168W
3.08.68	13.07.70	54N 146W	49N 134W
16.05.68	28.05.68	43N 175E	45N 175E
9.08.69	14.06.71	41N 159E	44N 174E
23.08.69	15.06.72	40N 158E	35N 177E
23.08.69	10.06.71	40N 157E	42N 171E
25.08.69	26.05.71	40N 150E	36N 146E
27.08.69	10.06.70	53N 151W	45N 164W

Table 3.5 (cont'd)

Date Marked	Date Recovered	Position Marked	Position Recovered
7.10.70	20.05.71	41N 155E	36N 145E
7.10.70	25.06.72	41N 155E	48N 171E
17.10.70	12.08.71	39N 152E	44N 173E
16.10.70	7.06.72	38N 149E	38N 167E
9.07.70	7.07.71	49N 136W	47N 137W
22.05.71	29.07.72	44N 161W	45N 164W
26.01.72	12.06.72	15N 143E	47N 172E
5.02.72	15.06.72	21N 152E	35N 177E
5.02.72	7.09.72	21N 152E	pelagic grounds
6.02.72	23.07.72	22N 151E	45N 167E
6.02.72	14.06.72	22N 151E	37N 178E
6.02.72	3.07.72	22N 152E	39N 175E
7.02.72	3.06.72	23N 152E	42N 168E
9.02.72	24.05.72	24N 147E	39N 170E
9.02.72	8.06.72	24N 147E	39N 168E
9.02.72	26.06.72	24N 147E	41N 176E

+: approximate position

North Atlantic

In his review of 1979 Brown described recoveries from only three sei whales and these are described by Mitchell (1974c). Subsequently marking off Iceland has resulted in another eleven returns (Icelandic progress reports to the IWC, Sigurjonsson 1983, pers. comm.). The information is summarised in Table 3.8.

Southern Hemisphere

Although a few sei whales were marked in the 1930s, the first recoveries were made in the Antarctic whaling season of 1963/64. Brown (1968a) reviewed the recoveries from the International Scheme up to the 1966/67 season and the 21 returns are given in Table 3.9. No other reviews exist from this Scheme in the Southern Hemisphere, but Mr S G Brown has extracted the returns for me from 1967/68. These are given in Table 3.10 as the additional 30 returns up to 1977/78; the end of whaling for Southern Hemisphere sei whales under IWC jurisdiction. Recoveries from the Soviet Scheme are taken from Ivashin (1980) and the 30 are given in Table 3.11; slight amendements have been made by Ivashin (in litt). The positions of recoveries are illustrated in Figure 3.4.

Table 3.6: Japanese marks recovered from the North Pacific from 1973.

Date Marked	Date Recovered	Position Marked	Position Recovered
9.07.64	8.08.73	56N 153W	46N 149W
22.05.67	16.06.74	48N 174W	37N 169W
28.06.69	27.05.73	40N 149E	39N 172E
9.08.69	14.07.75	41N 159E	41N 169E
9.08.69	9.06.73	42N 158E	36N 169E+
9.10.70	30.06.73	41N 155E	40N 166E
17.10.70	29.06.74	39N 151E	39N 171E
17.10.70	24.05.73	39N 152E	38N 167E
26.01.72	10.07.74	15N 143E	49N 172E
6.02.72	24.05.73	22N 151E	38N 167E
6.02.72	27.06.73	22N 151E	42N 170E
6.02.72	7.08.73	22N 151E	48N 141W
6.02.72	17.05.75	22N 152E	35N 143E
7.02.72	4.07.74	23N 152E	39N 170E
7.02.72	21.07.74	23N 152E	48N 169E+
9.02.72	11.07.74	24N 147E	49N 176E
9.02.72	4.06.73	23N 147E	39N 167E+
9.02.72	4.06.73	23N 147E	40N 167E
6.02.73	4.07.74	24N 155E	39N 171E+
16.02.73	8.07.74	21N 179E	42N 172E
2.03.73	21.08.73	24N 166W	45N 156W
2.03.73	15.08.74	24N 166W	43N 163W
20.05.74	10.07.74	31N 157E	48N 172E+
22.05.74	29.06.74	33N 163E	39N 169E

+: approximate position

Table 3.7: USSR and other marks recovered from the North Pacific.

Date Marked	Date Recovered	Position Marked	Position Recovered
c18.06.62	25.07.63	54-55N 156-157W	55N 139W
22.04.64	14.08.64	42N 151W	48N 129W
-	12.06.67	-	37N 143E
25.03.66	4.09.69	30N 152E	42N 150E
4.07.66	-.08.68	42N 165E	pelagic grounds
16.04.66	-.09.68	42N 180	pelagic grounds
-	27.07.69	-	48N 140W

Table 3.8: Marks recovered from the North Atlantic

Date Marked	Date Recovered	Position Marked	Position Recovered
5.08.66	26.09.71	44N 55W	43N 65W
8.08.67	17.09.71	42N 65W	43N 65W
8.08.67	11.06.72	42N 65W	42N 66W
29.08.80	13.08.81	65N 28W	64N 28W
29.08.80	11.09.82	64N 28W	66N 28W
29.08.80	30.08.85	64N 28W	63N 23W
30.08.80	14.09.83	65N 29W	63N 20W
30.08.80	30.08.80	65N 28W	65N 28W
30.08.80	15.08.81	65N 28W	63N 27W
30.08.80	15.08.81	65N 28W	63N 27W
10.09.80	26.07.81	65N 28W	65N 29W
14.09.80	22.09.83	64N 28W	64N 21W
24.09.81	22.09.83	64N 27W	63N 21W
24.06.82	23.08.84	64N 27W	62N 21W

Table 3.9: International Scheme. Marks recovered from the Southern Hemisphere to the seasons of 1966/67.

Date Marked	Date Recovered	Position Marked	Position Recovered
10.02.62	15.02.67	58S 91E	57S 82E
18.01.64	20.01.64	54S 23W	53S 24W
18.01.64	19.02.66	54S 23W	44S 17W
6.03.64	19.01.65	48S 57W	47S 57W
7.03.64	13.01.65	51S 62W	46S 56W+
6.01.65	6.01.65	53S 48W	52S 47W
7.01.65	10.01.65	55S 48W	55S 49W
14.01.65	15.01.65	46S 57W	46S 56W
15.01.65	25.01.66	49S 54W	50S 50W+
17.01.65	19.12.66	46S 56W	42S 53W
17.01.65	18.01.65	47S 56W	48S 56W
17.01.65	21.01.65	46S 56W	45S 58W
21.01.65	3.02.65	52S 44W	53S 46W
26.01.65	18.01.66	49S 48W	44S 50W
18.02.65	19.02.66	50S 40W	44S 11E
18.02.65	19.02.65	50S 40W	50S 40W
19.02.65	27.02.65	57S 46W	55S 46W
19.02.65	20.02.65	49S 40W	50S 40W+
26.02.65	4.03.65	59S 49W	56S 45W
30.11.66	3.01.67	39S 0	41S 24E
15.02.67	19.02.67	59S 84E	59S 84E

+: approximate position

Table 3.10: International Scheme. Marks recovered from the Southern Hemisphere from the seasons of 1967/68 to 1977/78.

Date Marked	Date Recovered	Position Marked	Position Recovered
16.11.62	15.02.68	36S 81E	51S 73E
16.01.63	2.01.68	57S 83E	47S 81E
17.11.63	22.12.74	42S 169E	42S 160E+
18.01.64	14.12.68	54S 23W	41S 6E
17.02.64	20.03.68	47S 174E	45S 176E
12.02.64	6.01.69	43S 174E	42S 160E
17.12.66	26.01.76	47S 76W	61S 56W
17.02.67	9.02.68	45S 53E	47S 55E
18.03.68	9.12.75	44S 178E	42S 156E
post 4.68 *	4.01.73	*	43S 129E
27.11.68	3.01.72	42S 170E	48S 150E
27.11.68	3.01.72	42S 170E	48S 150E+
22.08.69	11.02.70	30S 33E	44S 51E
22.08.69	24.02.70	30S 32E	46S 71E+
22.08.69	28.03.73	30S 32E	46S 51E
22.08.69	1.04.73	30S 32E	43S 82E+
22.11.69	17.12.69	41S 112E	41S 115E+
27.11.69	5.01.78	41S 122E	44S 127E
28.12.69	6.01.70	40S 39W	42S 13E+
21.11.71	15.12.74	41S 123E	41S 152E
21.11.71	21.12.71	41S 123E	44S 118E
6.11.72	13.12.73	38S 157E	42S 124E+
1.12.72	18.12.73	43S 112E	42S 132E
9.02.73	15.02.76	64S 171W	65S 174W
15.11.73	13.12.74	40S 153E	41S 152E
2.12.73	19.01.75	42S 111E	42S 101E
6.12.73	19.12.74	41S 111E	41S 122E
6.12.73	23.01.75	41S 111E	42S 93E
25.12.73	29.12.73	46S 160E	46S 160E
22.01.74	26.01.76	33S 67E	43S 98E

+: approximate position

*: S. Brown (in litt) reported that this whale was marked off Albany, Western Australia. The exact position is unknown. It was probably marked after April 1972, but it may have been marked in the previous austral winter.

Table 3.11: Soviet Scheme. Marks recovered from the Southern Hemisphere.

Date Marked	Date Recovered	Position Marked	Position Recovered
1.04.55	9.03.65	61S 43W	52S 49W
14.12.57	20.02.65	56S 24W	50S 39W
22.01.61	28.12.71	62S 119E	43S 110E
7.02.62	21.12.62	54S 58W	55S 60W
14.02.62	22.01.65	46S 58W	46S 57W
14.02.62	14.05.63	46S 58W	24S 41W
16.01.63	26.12.70	40S 109E	42S 96E
16.03.63	19.03.63	60S 54W	60S 51W
6.01.64	30.12.74	53S 81W	46S 82W
26.01.64	27.12.64	45S 57E	41S 50E
15.01.65	10.03.69	51S 166E	67S 165W
17.11.65	21.01.66	36S 13W	44S 33W
17.11.65	10.09.66	36S 13W	35S 18E
17.11.65	9.02.66	36S 13W	46S 19W
30.10.66	26.01.74	36S 98E	41S 70E
19.11.66	13.01.68	36S 79E	51S 74E
31.01.67	22.03.67	55S 174W	0 east to 170W
22.02.67	6.03.67	65S 165W	64S 166W
3.02.68	13.02.68	56S 7W	59S 8W
29.03.68	16.12.70	51S 11W	41S 10E
14.11.69	10.01.71	33S 87E	43S 101E
22.11.70	9.01.72	32S 85E	42S 97E
5.01.71	29.11.71	39S 51W	44S 52W
29.11.71	--.03.77	38S 163E	0 east to 170W
26.11.72	27.01.76	41S 50W	51S 44W+
12.01.73	3.02.77	48S 81W	62S 62W
12.01.73	7.01.75	49S 81W	48S 83W
6.01.74	3.02.74	55S 81W	66S 68W
8.01.74	27.01.74	47S 81W	50S 81W
5.04.74	24.12.76	44S 155W	44S 152E

+: approximate position

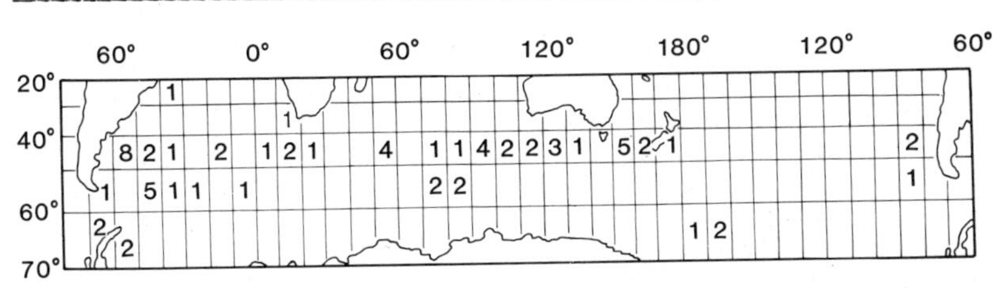

Figure 3.4: Numbers of recoveries of marked sei whales from the Southern Hemisphere, by 10° squares. Only known sei whales from known positions are included.

One additional mark is described by Brown (1976a) but not included in the above tables. This was of mark no. 16204 which was used to mark a solitary whale in February 1963 by P Best off South Africa. This mark was returned from catches from the SIERRA, in July 1973, at 00°32'S 05°10'E, from a 50 ft female sei or Bryde's whale. Best noted that the SIERRA had taken Bryde's whales, and the time and location of the recovery strongly suggest that this was a Bryde's whale. Brown (1978) shows one further mark returned from the marking season of 1955/56 but the details are not included in Brown (1968a). This mark was one fired into a whale recognised as a blue whale but returned six months later as coming from a sei whale, caught at Hauraki Bay, New Zealand. However, this could also have been a Bryde's whale as they inhabit the region.

SUMMARY

Initially, this chapter was not planned for it was envisaged that reference could be made to convenient summaries of marks and recaptures. This proved not to be the case with information on numbers marked and recoveries made scattered in various publications and with much unpublished. The information shows some 1658 sei whales marked from all oceans. A few of these would be Bryde's whales. The intensity of marking has varied with area, and few sei whales have been marked in the North Atlantic. In addition to the 1658 sei whale "hit" more would have been marked but not recorded as hit.

General aspects of whale marking are reviewed by Brown (1977b), but more specific aspects are described by Ohsumi and Masaki (1975). The latter study includes information on recoveries from multiply marked whales, position of recovery within the whale, condition of the wound caused by the mark, validity of identification of species at time of marking, and the proportion of mark returns from different categories of the fate of the marks.

The information on numbers marked and recoveries will be used in later sections associated with stock identification, migrations and estimates of abundance. However, it can be seen from Tables 3.5 and 3.10 that the longest period that a mark has been placed in a sei whale, and subsequently recovered, is 11 years, from both the North Pacific and Southern Hemisphere. Recoveries from Bryde's, sperm and fin whales have been made after 18 years.

Chapter Four

MIGRATION AND LOCAL MOVEMENTS

INTRODUCTION

Migrations are recognised by a movement of animals from one habitat to another, and with a return or circularity of movement. The large whales are renowned for their extensive annual migrations. Much is known of the migration of the larger whales and one must be wary of immediately considering the sei whale as behaving identically to the others, particularly the blue and fin whales, and information on the movement of sei whales needs to be evaluated on its own. The data fall into two categories. Firstly, there is information on the gross movements of populations; catches are found in one place in one season and in another place a few months later. Assumptions need to be made to combine such information. Secondly, data are available on the movement of individual whales and, although these are too few to describe population movements, they assist in interpreting the grosser indices of abundance.

Specific information on the distribution of populations with time is given by catch and sightings data. These are affected by the abundance of effort, and catch of sei whales, as an index to abundance, is particularly affected by the presence of other species, consequently preferred indices are catch and sightings per unit of effort (CPUE and SPUE). Nevertheless the location of whaling stations and the timing of land and pelagic operations are obviously related to the presence of the whales. Additional information is given by direction of swimming of the whales, presence of ectoparasites and diatom films. Finally, the marking schemes are an exceptional source of information.

In addition to the main migrations, sei whales are involved in movements which are often associated with environmental conditions partiularly on the feeding grounds, and these are described in the last section.

SEASONAL MOVEMENTS

North Pacific

Catches and CPUE. In the coastal western North Pacific Andrews (1916a) reported that catches were made in Japanese waters from June to August, and Nasu (1966) described the timing of sei whale catches as May to June off Sanriku, September to October off Hokkaido, and July to August from the Okhotsk Sea of Hokkaido. From Kamchatka and the Kuril Islands sei whale catches increased from late May to early June (Sleptov 1955 - cited in Masaki 1976a). Off the eastern coast, Nasu (1966) described catches of sei whales from May to June from British Columbia, and Pike and MacAskie (1969) gave the timing of the Canadian fishery as June to August. From California, Rice (1977) gave the time of catches as May to October, and Nasu (1966) as July to August. The pelagic fleets have operated over the six month summer of May to October. Kawamura (1973) explained that domestic legislation stopped the Japanese pelagic fleets operating south of 40°S until 1972 and Masaki (1976a) shows that from 1952 to 1972 the whales were caught between 40°N and the Aleutian Islands. No information is published on monthly pelagic catches and on CPUE.

Sightings. Relative densities of sightings of sei whales by Japanese scouting vessels are reviewed by Masaki (1976a, 1977a) and Wada (1980) for the periods from 1956 to 1979 and these are shown in Figure 4.1. Because of the incomplete coverage the data are not easy to interpret. From May to June densities increase north of 45°N and west of 160°W. Into July the densities continue to increase along and to the north of the Aleutian Islands and in the west there are fewer whales between 40-50°N. In August few are seen south of 45°N and substantial densities are still found to the east of Kamchatka in September. The distribution in the eastern North Pacific shows high densities of sei whales in the

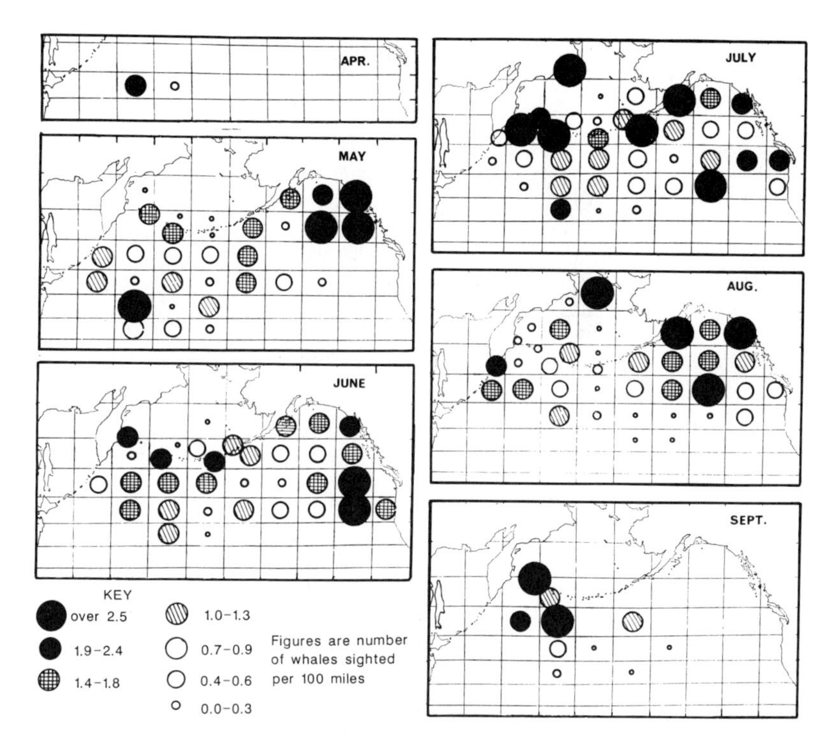

Figure 4.1: Monthly densities of sei whales from Japanese sightings up to 1972. Units are whales sighted per 100 miles.

Gulf of Alaska as early as May, with lower densities off the northwest coast of the USA found in August.

Marking. The mark recoveries from Tables 3.4-3.7 can be used to show the movement of whales within and between seasons. Masaki (1976a) presented information from mark recoveries to 1973 to show the north-south movement in relation to month of marking and recapture, and Figures 4.2a and 4.2b are of the same form as his but they include information on whales caught in 1974 and 1975. Figure 4.2a presents the data from whales caught after one year from marking and Figure 4.2b the data from whales recaptured within one year. It can be seen that the character of the movements is similar from both data sets.

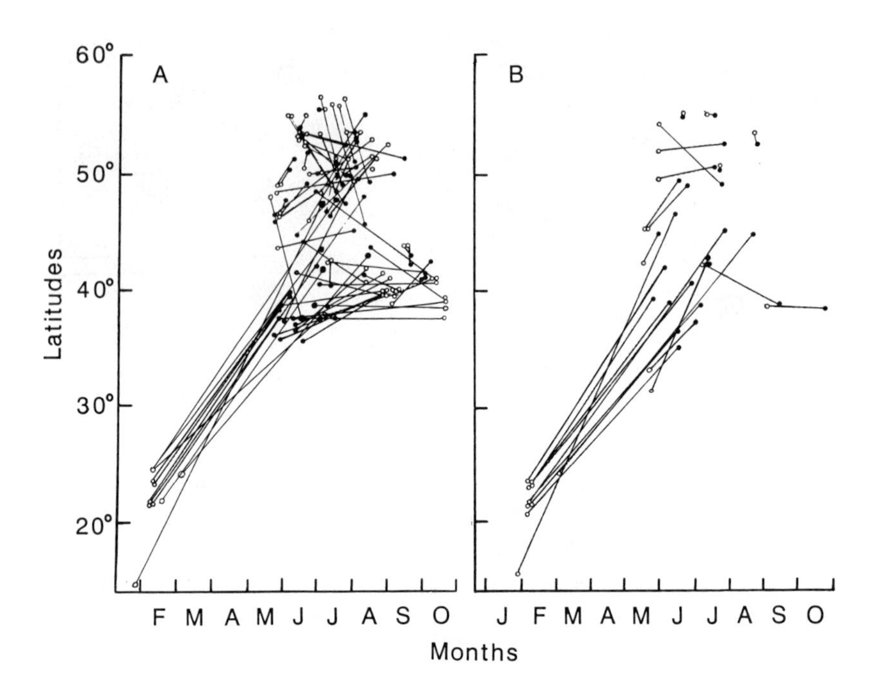

Figure 4.2: Seasonal and latitudinal movement of marked sei whales from the North Pacific. Circle: marking, dot: recapture. A: recaptures after one year from marking. B: recaptures within one year.

Sei whales were marked from January to March in latitudes below 25°N and these whales were all recovered from May to August in latitudes north of 35°N. The northward movement can be seen, from other marked whales, until the end of June. From July to September movements do not appear to show any preferred direction.

Oshumi (1980a) gave the latitudinal distribution of whaling effort by year and before 1966 pelagic effort was north of 40°N, only after 1973 did the pelagic fleets operate south of 30°N. Except for catches from the coast of Japan few recoveries could be expected south of 40°S. This then presents the problem that whales marked from 20-30°N could have remained there undetected. All we can conclude from these data are that some whales marked early in the year are to be found in northern waters

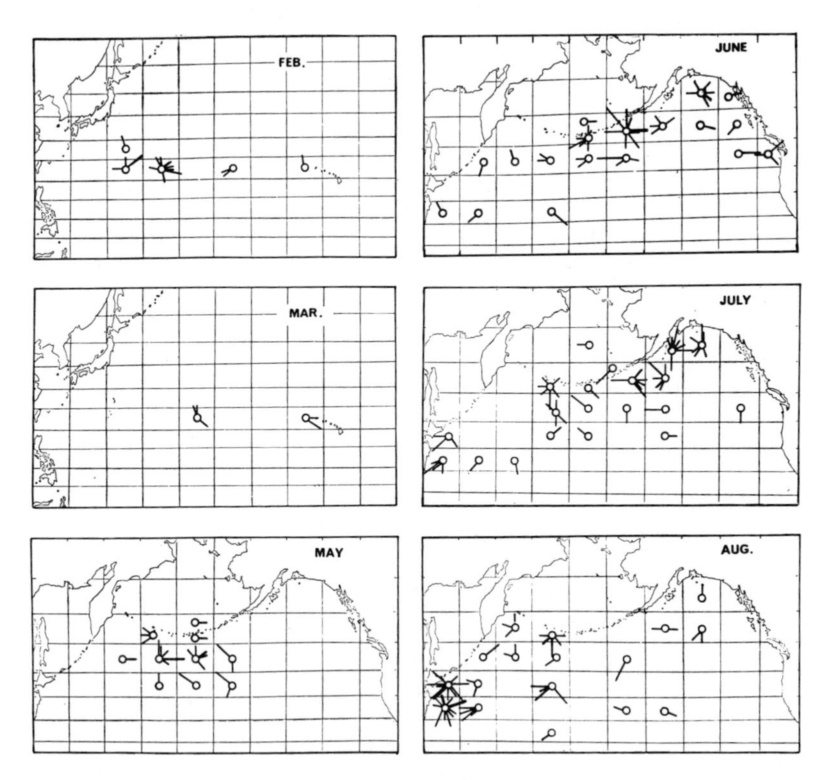

Figure 4.3: Directions of swimming of North Pacific sei whales by month.

during the summer, and that during the summer there is some limited latitudinal movement. In addition the figures do indicate that whales marked about 40°N tend to be found further north in mid-summer and that some whales marked in the summer in northern waters can be found there in subsequent years.

Direction of movements. The direction of swimming of sei whales by month has been presented by Masaki (1976a) and these data are illustrated in Figure 4.3. In February and March no preferred direction is observed. In May there is a tendency for the most southern whales to go north but the others show an east-west tendency. In June and July there is little directional movement except towards the SW from those whales off Japan. In the months of August to October few whales were found swimming to the north.

North Atlantic

Catches and CPUE. In the year of large sei whale catches from Finnmark, 1885, the sei whales were taken from mid-May to early September with the peak in July. More recently the sei whales have not been found in these northern waters, but off western Norway sei whales were caught mainly from May to August. Further south peak catches from the Faroe Islands were in July, from the Hebrides in June and from the Shetland Islands in July and August. A few sei whales were taken from Ireland in early June. From north-west Spain sei whales were caught from April to December with a small peak from mid-September to mid-November.

From Iceland sei whales are caught from August and September with a few taken in June. The Nova Scotia fishery took place in June to July and September to October, but off Labrador they were found in August to November (Mitchell 1974a). Mitchell and Kozicki (1974) show that sei whales caught off Labrador and Newfoundland were taken from early August to a peak in mid-October and then had a rapid decline. From the Canadian catches availability of sei whales may have been distorted by a preference for blue and humpback whales. No monthly CPUE data have been published for North Atlantic sei whales.

Sightings. No monthly data on sightings have been published.

Marking. Table 3.8 gives all the sei whale recoveries from the North Atlantic. The only information on migration that it yields is that sei whales marked off Nova Scotia in summer or autumn can be found in the same place at the same time years later, and similarly of sei whales in the Denmark Strait.

Southern Hemisphere

Catches and CPUE. From the South Atlantic the Brazilian catches from Costinha were taken from June to November with most taken from July to September and Paiva and Grangeiro (1965) show a single peak with more males caught than females in all months. They considered that this feature is not only due to protection of female with calves. From South Africa, Matthews (1938) and Best (1967a) found a bimodal distribution of catches from the Cape Province, with a peak seen in August to October and a second smaller peak in May or June. In the South Pacific peak catches from Paita, Peru, were taken in August to November with a hiatus from March to July. No seasonal data are available from Chile. From Natal a single peak of catches is seen in June, from Matthews (1938), and August from Gambell (1968).

Substantial catches were taken from South Georgia and prior to 1930 the peak catch of sei whales from the land station was in March, but by 1965 it was in January. The spatial distribution of pelagic catches of sei whale is tabulated by Omura (1973) and is illustrated in Figure 2.3. From 60°W eastwards to 130°E catches were mainly from 40° to 50° or 60°S, in the rest of the Antarctic they were caught mainly south of 60°S. Over the combined period 1931-1978 the peak catches were in January. However in the early years catches peaked later in both pelagic and land operations due to a preferential selection for fin and blue whales, (Horwood 1986).

Indices of abundance by month of catch per catcher-ton-days have been described for the Durban operations by Bannister and Gambell (1965) for the two periods 1954-1958 and 1959-1963, by Gambell (1968) for the period 1964-1966 and by Gambell (1974) for the period 1954-1963. This information is summarised in Figure 4.4. Most of the data sets show a bimodal distribution with very low abundance in March and April, a low peak in June with a small decline in July, and a large peak in September, declining in October. Bannister and Gambell displayed the information from the two periods separately since in the earlier years fin whales were of greater importance than later and were at their most abundant in June to August. Consequently there is the concern that the bimodal distribution might be an artefact of selection. However the later set of Bannister and Gambell shows an early shoulder to the distribution and that of Gambell (1968) for 1964-1966 still shows a bimodal distribution at a time of small catches of fin whales.

From the South Atlantic Best (1967a) described a similar distribution of CPUE from Donkergat in

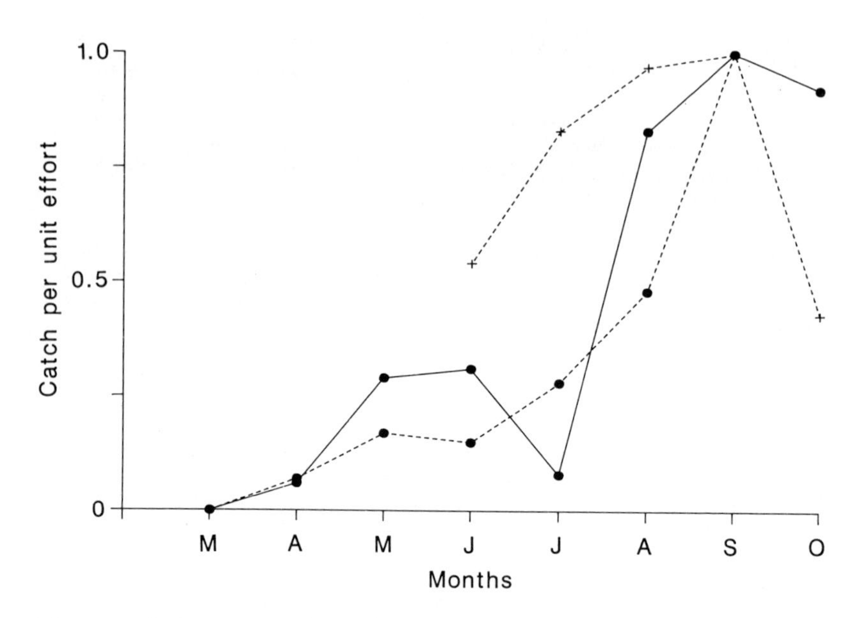

Figure 4.4: Relative indices of CPUE from Donkergat (dots + bold line), Durban (dots + dashed line) and Costinha (crosses).

1963. A small peak is seen in May and June, a hiatus in July and a larger peak in August to October. Additional information from Donkergat for the period 1962 to 1967 is given by Best and Lockyer (1977 ms) which substantially assist in interpreting these distributions and are described later. From Costinha, Paiva and Grangeiro (1965, 1970) gave catch per voyage for 1960 to 1967. Over the months June to October this index is low in June and October with a peak in August and September. These data do not appear to be subject to much confusion from the selection of other species. Data on CPUE from Durban, Donkergat and Costinha are illustrated in Figure 4.4.

Data on CPUE by month from the Antarctic are largely unpublished but are also difficult to interpret because of selection for other preferred species. Early data from South Georgia were subject to the same problems but Gambell (1968) considered that his information from the 1963/64 and 1964/65 seasons provided a valid index of abundance as then whaling for sei whales dominated. His CPUE show few sei whales from October to December and peak CPUE from January to March.

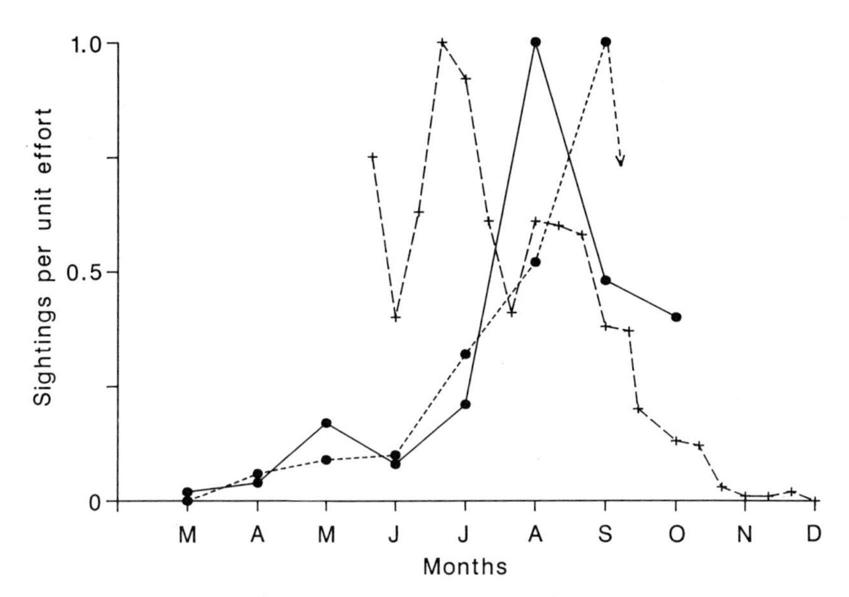

Figure 4.5: Relative sightings per unit effort. Symbols as in Figure 4.4.

Sightings. From Durban the number of whales sighted per miles flown by the whaling spotter aeroplane are described by Bannister and Gambell (1965) and Gambell (1968, 1974) for the same periods given for CPUE above. The sightings data are free of the whale selection problem that affects the interpretation of CPUE. Relative sightings are low in March, April and October and between these months show for most years a similar bimodal distribution to that found in the Durban CPUE. From Donkergat in 1963 Best (1967a) also found a bimodal distribution of spotter aeroplane sightings per miles flown. A small peak is seen in May with a larger one in August; again the relative sightings reflect the CPUE distribution.

Number of whales seen per whaling voyage, from the Brazilian operations at Costinha, were recorded by Paiva and Grangeiro (1965, 1970) for the period 1960 to 1967, but were not identified by species; however data presented to the IWC (IWC/38/Annex E) did include sightings of sei whales per day, and covered the period 1966 to 1983 for the months June to December. Densities of sei whales were high from June to August, but then declined continuously until December. Sightings data from Durban, Donkergat and Costinha are illustrated in Figure 4.5.

From the Antarctic, sightings densities have not been published monthly and Table 4.1 is an extract from the Japanese sightings data of the Far Seas Fisheries Laboratory. It is from the 1969/70 season and covers the longitudes of 0-180°E. In this area the sei whales appear to be distributed evenly from 20-40°S in October, and from 30-50°S in November with few south of 50°S. As the season progresses relatively more whales are found south of 40°S and the density between 50° and 60°S increases consistently until March.

Table 4.1: Sightings density of sei whales per 10000 n. miles steamed for whaling Areas III+IV combined in 1969/70, by month and latitude.

	Oct	Nov	Dec	Jan	Feb	Mar
20-30°S	286	-	-	-	-	-
30-40°S	318	644	191	175	0	-
40-50°S	-	753	1052	579	209	75
50-60°S	-	0	7	40	79	114
60-70°S	-	-	-	-	105	-

<u>Marking</u>. The mark recoveries from Tables 3.9-3.11 give information about the movement of marked whales in the Southern Hemisphere and Figure 4.6 gives a representation of the latitudinal movement of the sei whale using selections from the tables. Significant recoveries have been included but not all recoveries have been plotted to avoid over-complicating the figure.

Except for catches from land stations the catches were taken mainly from south of 40°S and it can be seen that, before September, sei whales can be found at about 30°S and, over the austral summer, move towards the higher latitudes. This general tendency seems to continue until the end of January, after which movement is more random. The Soviet return (Table 3.11) at about 35°S was recovered from Cape Town and shows that there may be only limited latitudinal movement over the September to November period. From Table 3.10 it can be seen that several whales were marked north of 30°S, particularly in the Indian Ocean, and that over several seasons there was only a restricted longitudinal movement.

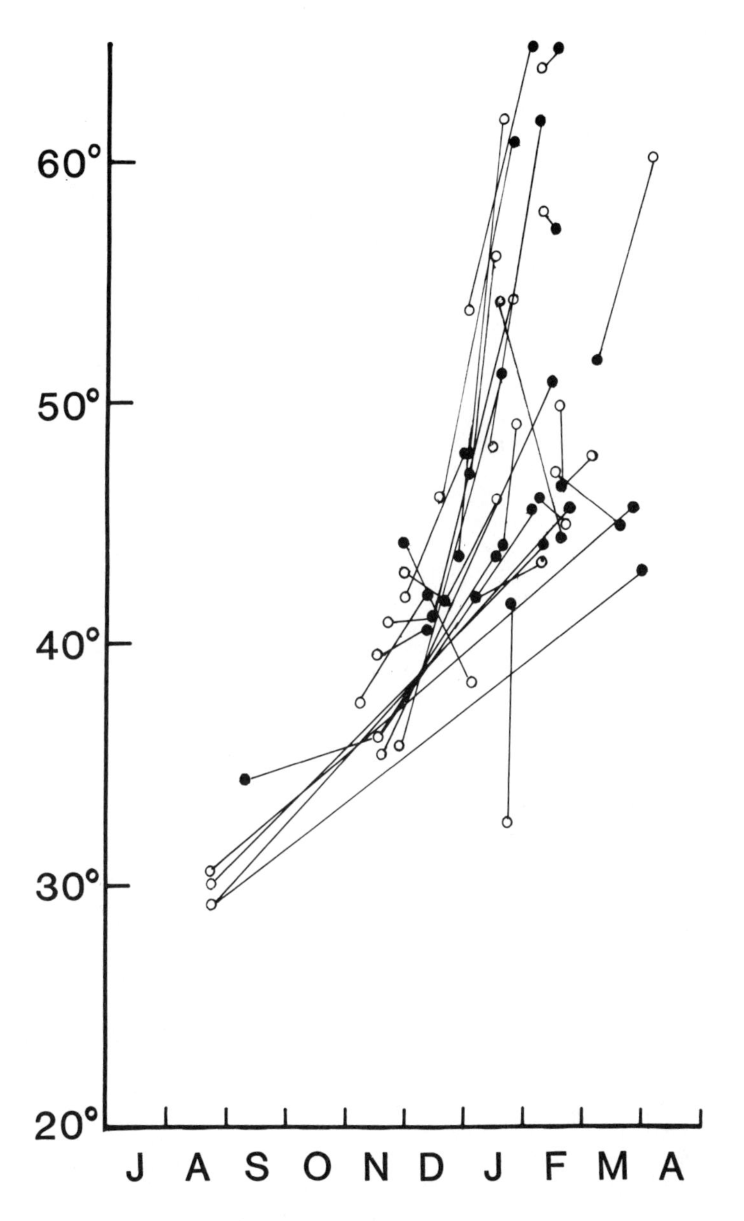

Figure 4.6: Seasonal and latitudinal movement of selected marked sei whales from the Southern Hemisphere. Circle: marking, dot: recovery

Direction of movements. The direction of movement of schools of sei whales was recorded from the spotter aeroplane at Durban, which is particularly useful since such information is more reliable than comparable data from vessels, as the whales do not seem to respond to aircraft, and frequently do respond to ships. Data from the period 1961-63 are summarised by Bannister and Gambell (1965) but the total number of sightings was only 41. The tendency is of a north and eastwards movement in May to August and to the south and east in September; the coastline of Natal runs about 30°E of N. The data series was extended by Gambell (1968) to cover the years 1961-66, and his results show that in May and June the overall movement is predominantly north-east, in July the movement is to the southeast, there is no movement in August, and it is to the east in September. From Donkergat similar data were presented by Best (1967a) for the years 1963 to 1964. From April to July the overall direction is predominantly to the north, in August similar numbers were seen swimming north and south and in September, October and March the main direction was south.

Diatom film. Bennett (1920) noted that fat whales in the Antarctic had a greater infection of diatoms on the skin than did thin whales, which he suggested was because the thin whales were recent arrivals from the warmer waters to the north, and consequently the extent of the diatom infection might help in the description of migrations. The main diatom involved is *Cocconeis ceticola* Nelson and can give the white parts of the whale the appearance of a deep yellow-orange. A general review of the incidence of diatom infestation in relation to the migration of blue and fin whales is given by Lockyer (1981a). Hart (1935) and Nemoto (1959) estimated that the diatom film took a month or less to develop on the skin of sei whales after they had entered the cold waters. Hart described the infection on sei whales in February to April over the years 1928-1931 and considered that the low incidence implied arrivals late in the season. Data from 1960 to 1965 were given by month by Gambell (1968); in November and December about 25% had a diatom film, suggesting that a quarter of the whales had been in the Antarctic for about a month or more. From January to March the occurrence is 3-6% and this indicates a large influx of sei whales from warmer waters in these months.

INTERPRETATION OF MIGRATION

The information above allows conclusions to be drawn on the migration of the sei whales. Each type of data alone is probably insufficient since they are all subject to problems of interpretation. Whaling does not take place in areas and at time of low abundance and consequently catches only confirm that whales can be found at a location and at a time; what happens elsewhere, or outside the whaling season, is strictly undefined. The catches and CPUE are more useful if trends can be found within the season. However, both these indices of abundance are usually distorted by the fact that sei whales were seldom the preferred species of the whalers. Sightings data, and particularly those from the extensive surveys by Japanese scouting vessels, are more easily interpreted, but again the vessels tend not to have surveyed areas in which it was thought few whales lived at a particular time. The marking and recovery information is particularly useful since it shows that individual whales conform to the behaviour of the population as inferred from the gross indices of catches and sightings; however in themselves the important recoveries are too few to draw firm conclusions about the movement of the entire population. The direction of movement of whales and the incidence of diatom infestation found in the Antarctic provide additional supporting material.

The above information clearly shows that the sei whales winter in the warmer waters and can be found near the equator in mid-winter. The wave of migration north and south past the east and west coasts of South Africa can be seen from the catch and sightings data. In the South Atlantic the sei whales pass the Cape in May and June and are found off the northeast of Brazil in July to September. By October sei whale numbers have declined in the waters north of 30°S and from November to February the sei whales appear to be moving steadily to the higher latitudes. Complementary movements are found in the Northern Hemisphere. No movements across the equator have been established. This pattern is similar to that seen in most of the other rorquals as described in general by Kellogg (1928), Gambell (1976) and Lockyer and Brown (1981), and specifically for humback whales by Townsend (1935) and Dawbin (1966), Southern Hemisphere blue and fin whales by Mackintosh (1966) and North Pacific whales by Nishiwaki (1966). Minke whales also undergo such a migration; however it should be noted that the seasonal movements of the Bryde's whales are much more restricted.

The reasons for the migration are associated with feeding and reproduction. As will be shown in the next chapter incidence of feeding is greater in the polar waters compared with the warmer waters and quantities of food in the stomach are much greater in the former. The greater feeding, particularly in the Antarctic, is reflected in the seasonal changes in blubber thickness and oil yield. Lockyer (1981a) reviewed blubber thickness in blue and fin whales and showed low levels in October to December with rapid increases thereafter. Catches from South Africa in May to July were of fat whales but there was a decline in blubber thickness from August. For sei whales Gambell (1968) showed blubber thickness reducing from May to September at Durban and increasing from January to March at South Georgia. It is clear that even in March the sei whales still would have had much feeding to do to obtain the levels of blubber thickness in May. However Lockyer (1981a) considered that different sites of measurement on the whale meant that such a direct comparison was not justified. From Norway, Ingebrigtsen (1929) noted that sei whales caught in April did not yield more than 10 barrels of oil but those caught in July yielded about 20 barrels. Matthews (1938) also compared blubber thickness from the two catching areas and found thickness to be greater in the Antarctic. His samples were from late summer in the Antarctic and from July to September at Durban. Consequently, it can be seen that one reason for the migration is to feed on the greater standing stocks of the polar and sub-polar waters. In addition to this strong seasonal cycle of feeding, reproductive activity is also seasonal for all mysticetes with calving taking place on the return to warmer waters and lactation extending into the start of the feeding season. Gestation is usually about twelve months. Whether the warmer waters are necessary for calf survival is not known, but as the adult baleen whales are not thought to suffer from heat loss (Lockyer et al 1984) the reason for their long migration to the sub-tropics and tropics is most likely to be associated with the physiology of the calves.

The above description is complicated by the fact that within the seasonal migrations a substantial segregation by age and sexual groupings is found. In the Antarctic pelagic catches are of mainly mature animals and Doi, Ohsumi and Nemoto (1967) and Horwood (1986) demonstrated that smaller sei whales tend to stay north of the Antarctic Convergence. Horwood (1986) found that larger females arrived first in the Antarctic waters. The proportion of the different sexual classes of sei whale in the catch from South Africa and the Antarctic have been described by Matthews (1938), Bannister and Gambell (1965), Gambell (1968), Nasu and Masaki (1970), Lockyer (1977a) and Best and Lockyer (1977 ms). In the Antarctic waters Lockyer (1977a) showed that older adults tended to arrive before younger adults, pregnant females were more frequent in the early part of the season compared to the other groupings, and resting (non-pregnant or non-lactating) females comprised an approximately constant proportion of the catch. From Durban immature sei whales preceded the mature whales in early winter but in later winter, on the return to the Antarctic, mature males dominated the catch (Gambell 1968, Bannister and Gambell 1965). From Donkergat, Best and Lockyer (1977 ms) showed immature males more prevalent than mature males in May to August and mature males more prevalent in September and October. For females they showed a more striking segregation; from May to November the density of immature females was about constant, but pregnant females arrived in late July and densities declined in September, with few lactating females before September and peak densities in October. This trend however not only reflects a migration by sexual class, but from June to September conceptions are still happening which will increase the numbers of recently pregnant. Births would have been largely completed by the end of August and so the final increase in lactating females cannot but be real.

The frequency of the different sexual classes in the North Pacific catches was described by Masaki (1976a). From May to July there was a high percentage of sexually mature sei whales north of 40°N but there is a suggestion that in the central North Pacific the percentage drops in August. Although there may be differences with longitude Masaki found that the pregnant females comprised the highest percentage of the catch in the early months, as was found in the Antarctic. The larger females often dominate the catch at Iceland, (Martin 1983), but the recent Icelandic fishery is on the northern edge of the range for sei whales and the sex ratio in the catch alters in years of high availability of sei whales. In the North Pacific, Nemoto (1959) described the segregation of sei whales by sex but neither of these last two studies can be strictly related to the migration. Segregation of the age and sexual classes has also been described in fin and humpback whales (Laws 1961, Dawbin 1966). The distribution of catches from land stations (Chapter 2) shows sei whales to be found on the outer edges of the continental shelf, as they migrate past the coastal whaling stations, or when they are on the more tropical grounds. The relative concentrations of sei whales along the shelf edge compared with distributions offshore during the migration has not been well described but in the North Pacific sei whales are caught and seen across the ocean. Nevertheless it is possible that the shelf edge is used as a migratory route for sei whales and this is suggested by Slijper (1962) for fin whales.

LOCAL MOVEMENTS

On the feeding grounds the distribution of the baleen whales is largely associated with oceanic frontal systems, of which Nasu (1966) described three major types. These are firstly ocean fronts, such as the meeting of the southern Kuroshio with the northern Oyashio off Japan or the Antarctic Convergence, secondly there are the eddy systems of either a dynamic type, such as the eddies shearing off the Kuroshio or Gulf Stream or topographic eddies such as that about South Georgia and thirdly there are the upwelling systems associated with ocean gyres or topographic features. Most of these fronts are of a fairly permanent character but even these are subject to fluctuation and the dynamic eddies are more erratic in their distribution than the others.

The distribution of sei whales off Japan in relation to these fronts has been described by Uda (1954), Uda and Dairokuno (1957) and Nasu (1966). The sei whales are found along the major mixing zone and in eddies that have broken away from the front, with the sei whales tending to be found in the warmer waters. Depending on the relative strengths of the two ocean currents the front may be off Hokkaido or Sanriku and this affects the whaling from those regions, and as the front moves throughout a season the sei whales will follow the frontal system.

Early records noted the speed with which sei whales could leave a whaling ground in mid-season. Ingebrigtsen (1929) considered the sei whale sensitive to weather and noted that a northerly cold wind in April or May would drive them from the main whaling ground in the North Atlantic, and Haldane (1907) reported that the manager of the Buneveneader whaling station as saying that sei whales were being regularly caught about St Kilda in 1906, but then, after a storm, there were none on the grounds. Haldane reported a similar experience from Shetland, and similar comments were made by Japanese whalers (Nemoto 1959). Mitchell and Kozicki (1974) considered this reported behaviour as quite realistic and possibly associated with feeding tactics. Nemoto (1959) and Kamamura (1974) remarked on the ability of plankton swarms to withstand rough weather but thought that if sei whales did move, following storms, it was because of associated behaviour or redistribution of prey species. In contrast Uda and Nasu (1956) calculated the whales caught before and after the passage of cyclonic storms and for sei and fin whales they found depressed catches before the storm and higher catches one day after.

Marking and recovery data allow investigation of whale movements over short periods. Figure 4.2 shows that once the sei whales in the North Pacific have reached the feeding grounds in May there is very little additional latitudinal movement and Masaki (1976a) also presented the longitudinal movement of sei whales from the North Pacific mark recaptures (his figure 64). Within the same season longitudinal movements are restricted to less than 20°, but of those marked and recovered on the northern grounds, the longitudinal movements are more typically 10°. Most of the information presented in Tables 3.9-3.11 was reviewed by Brown (1977b) who found that of whales marked in the same season there was little longitudinal movement. The tabular data show that longitudinal movements average less than 5° per day, and usually much less which is more restricted than those exhibited by fin whales. One notable exception was the whale marked on 28.12.69 at 40°S 39'W and recovered ca 6.1.70 at ca 42°S 12'E, and this movement of about 2200 miles within 10 days is much faster than any records from blue and fin whales. It certainly shows that a migration from 20°S to 50°S can be made within two weeks and Lockyer (1981a) gives a range of speeds for sei whales of 2-16 knots; nevertheless migratory speeds are usually much less than 9 knots.

SUMMARY

The sei whales undergo annual migrations in much the same way as fin, blue, humpback and minke whales. Winter is spent in the tropics and sub- tropics and mating and birth takes place at this time. In summer the sei whales move towards the polar regions to feed. The distribution in winter is diffuse but is latitudinally more concentrated in summer. As the summer progresses the sei whales move more pole-wards, but they seldom penetrate to the icy waters. During the migration there is a pronounced segregation of groups of different sexual status. Generally pregnant whales arrive earlier on the feeding grounds and mature whales stay longer. The information on migration comes from indices of population abundance that are interpreted with the aid of recoveries from marked individuals. The specific reason for such a migration is unknown for although they obviously feed intensely during the summer there is no known reason why the adults should leave the feeding areas; one is then led to suspect the reason for the migration as being associated with the physiology of the calves.

On the feeding grounds sei whales are associated with physical discontinuities which allow the development and concentration of food species. Some of these are transitory and their disappearance causes a movement of local aggregations of sei whales. Longitudinal movements on the feeding grounds are somewhat limited although one marked whale moved 2200 miles in 10 days.

Chapter Five

FEEDING ECOLOGY

INTRODUCTION

The feeding ecology of the baleen whales is an integral part of their population biology. It is their demand for food that sends them on their impressive migrations to the feeding grounds, although it is by no means obvious why they return. Annual variations in their food supply may affect reproduction and mortality, and the distribution of the prey species influences the distribution of the whales. The extent of sharing of prey species with other whales will affect the intensity of any interspecific competition. Reviewed below are aspects of feeding behaviour, prey type, time and quantity of feeding and the similarities and differences between the feeding of sei and other baleen whales.

BEHAVIOUR

Based on their feeding behaviour Nemoto (1970) grouped the baleen whales into three categories: (i) swallowers, (ii) skimmers and (iii) skimmers and swallowers. Those whales assigned to each group are shown in Table 5.1. The skimmers are the right whales; these have the finest fringes on the baleen and by far the greatest filtering area per body length (Nemoto 1959, Kawamura 1980). They feed by swimming near the surface and straining plankton through the fine sieve formed by the mat of bristles. The swallowers exhibit a variety of feeding behaviours, tending not to strain plankton over large periods but to search for concentrations of plankton and then taking large mouthfuls. The ventral buccal area expands like a bullfrog as the ventral grooves allow a large amount of water to be

Table 5.1: Feeding categories

swallowers	skimmers	swallowers and skimmers
blue	right	sei
pygmy blue	bowhead	gray
fin	pygmy right	(minke)
Bryde's		
humpback		
minke		

taken into the mouth. The mouth is shut, forcing the water out through the baleen, and the food is retained on the coarser filters. This group has developed a range of methods for obtaining a concentrated food supply before ingestion, (Ingebrigtsen 1929, Gaskin 1976, Jurasz and Jurasz 1979), including bubble-nets to trap plankton and lunging into fish shoals.

The sei whale uses both these feeding strategies but they are less well developed in both extremes. The baleen fringe is fine but not to the degree found in right whales and similarly the relative size of the filtering area is intermediate between that of the right whales and the other balaenopterids. Ingebrigtsen (1929) described the sei whale skimming for food as swimming at a great speed through swarms of copepods, with a half-open mouth, and its head above the water to just behind the nostrils. Water continually goes into the mouth and passes out through the baleen leaving the plankton trapped on the inner bristles. After a while the mouth is closed and the whale dives and swallows the collected food. Nemoto (1957) and Watkins and Schevill (1979) recorded a similar feeding behaviour. The description by Watkins and Schevill is as follows, "The feeding motion of the sei whale within a patch of plankton was that of irregularly opening its mouth quite widely, and then over as much as 20 sec to one minute closing it slowly (but often not completely) before the next mouth opening. The throat of the sei whale was slightly distended each time the mouth was almost closed. The throat area contracted to its usual shape prior to opening the mouth again. The whale swam within a metre or two of the surface and fed almost continuously throughout the observations. Breathing was irregular and often coincided with more complete mouth closings." Although the minke whale is described as a swallower I have observed minke whales behaving in a similar way in low densities of euphausiids. The ventral grooves and morphology of the tongue of the sei whale (Lambertsen 1983) also allow it to take advantage of the swallowing strategy, and this is also suggested by the presence in the stomach of fish and other agile, shoaling prey, but as far as I am aware no description of a sei whale feeding in this manner exists.

When feeding, the rorquals tend to swim on their side, left flipper exposed, or else roll onto their side at surfacing, (Gaskin 1982), and it has been argued that this gives them greater horizontal manoeuvrability (Rice and Wolman 1971). Many authors have noted that the sei whale does not do this (e.g. Andrews 1916a, Millais 1906) and presumably it is less specialised for a chasing and swallowing type of feeding than the other rorquals. Gaskin (1982) observed one sei whale swimming in a latero-ventral position, but if feeding this is uncommon. When chasing shoals of small fish the sei whale can swim quickly, and agilely twist and turn its body if necessary, (Andrews 1916a). A more smooth rolling behaviour was observed by Gill and Hughes (1971) of a sei whale feeding on saury. From their text it is not clear whether this "rolling" is to be interpreted as a rocking from side to side or the more pronounced rolling onto one side to feed as seen in the other rorquals. It would therefore appear that the sei whale is an active swallower but lacks the extremes of movements seen in the other rorquals. There is no description of the sei whale feeding on squid, which it does extensively in the North Pacific, and this must involve an active swallowing technique. Liouville (1913) also described sei whales feeding, in the fashion of a skimmer, but I am not convinced of the validity of his identifications.

Direct observations have indicated that the sei whale feeds most commonly at evening or early morning, associated with the diurnal concentrations of surface zooplankton, (Ingebrigtsen 1929), but they have also been found feeding on species which aggregate in daylight (Nemoto 1957). The time of feeding and the types of prey have been illuminated by examination of stomach contents. Sei whales tend to feed in the surface 100-150m and as Brown (1968b)

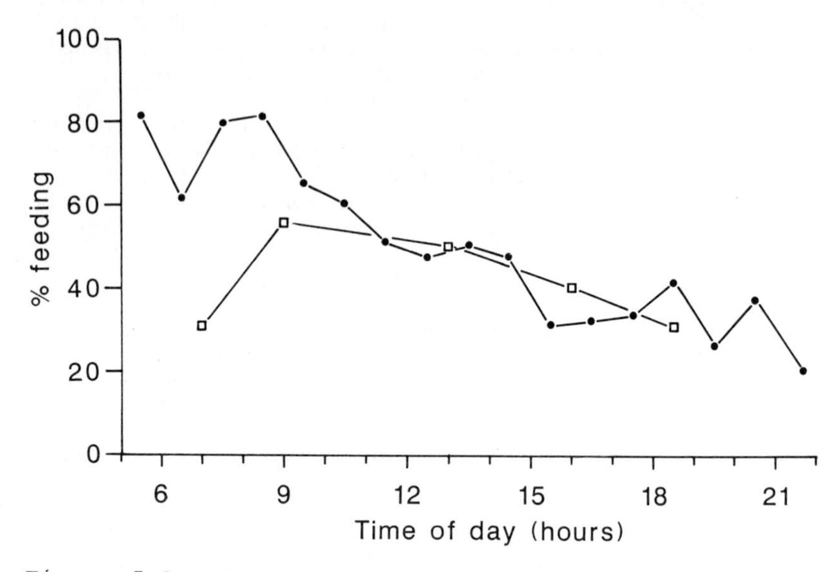

Figure 5.1: Percentage of whales with a full, first stomach against time of day. Circles: Southern Hemisphere pelagic samples. Squares: South African samples.

found myctophid fish in stomachs at South Georgia, and because of the diurnal migratory behaviour of these fish, he considered that the sei whale fed mainly during the evening. Klumov (1963) also noted the frequent occurrence in stomachs of sei whales of bathypelagic fish that came to the surface at night and concluded that sei whales fed at least during the night. A different perspective has been given by Kawamura (1974) in his extensive description of sei whales feeding in the more open waters of the Antarctic. He found a general tendency for feeding to be greatest in the early morning, but for the species composition in the first stomach to be closely related to the vertical migration patterns of the locally abundant prey. Figure 5.1 shows the proportions of the first stomach full by time of day from the Antarctic (after Kawamura 1974) and from South Africa (after Best 1967a). Kawamura concluded that in a given locality the sei whale will feed at any time that there is a sufficiency of suitable prey. This is also true throughout its annual feeding pattern; feeding mainly takes place in the summer, in the polar regions, but when food is available on the winter grounds it is taken, (Best 1967a, Budker 1958).

After, and sometimes during, feeding the sei whales as well as other species are often described as easy to approach and oblivious to vessels. They are easy to catch and mark. However whales difficult to catch and of a skittish disposition are often described by whalers as "hungry whales". Ingebrigtsen (1929) found that in waters where copepods are scarce sei whales may be chased all day without being caught.

As will be shown later the sei whale is essentially a feeder on copepods, but within its broad feeding strategies it appears as very selective. The prey need to occur in aggregations and in the surface waters, and this is well illustrated in their common prey of *Calanus tonsus* and *C. finmarchicus*, and in the rarer forms such as the fish *Myctophum asperum* and the squid *Ommatostrephes sloanei*. Kawamura (1974) found that not all swarming plankton were taken by the sei whale in the Southern Hemisphere, including surface swarms of young euphausiids and amphipods. Somewhat in contrast Brown (1968b) considered that samples from South Georgia showed that feeding was largely nonselective and Bannister and Baker (1967) considered that the sei whale took whatever swarming plankton was available. Both Nemoto (1959) and Best (1967a) thought that the sei whale could feed on loosely scattered prey. However the extensive studies of Klumov (1963) and Kawamura (1974) indicate that the usual behaviour on the major feeding grounds is to select a particular swarming prey; in addition, daily feeding is usually monospecific which would indicate a hunting and selecting of plankton swarms. From his studies in the northern North Pacific Klumov (1963) remarked that there was "not a single case where there was a mixed food", and Kawamura (1974) found that typically over 80% of sei whales had only one species of food in their first stomach. Consequently we see the behaviour of the sei whale as actively seeking out the swarms of preferred prey, and for some unknown reason neglecting others. The observations by Watkins and Schevill (1979) showed the sei whales feeding on a plankton swarm, but ignoring a nearby school of fish. It is possible that the sei whale has to be less discriminating in the poorer waters about South Africa and the southern coasts of Japan.

FOOD TYPES AND INCIDENCE OF FEEDING

The prey that is eaten by sei whales depends upon several factors, the most important of which is the location, with different prey species inhabiting different oceans and different regions along the migrations of the sei whales. As described before the prey tend to be of a swarming character and occur in the surface waters. The food types are described in detail below by geographical areas; Budylenko (1978b) tabulated 82 species of prey found in the stomachs of sei whales taken in the Southern Hemisphere, and this is not duplicated here, but Table 5.2 gives a list of those species found from samples in the Northern Hemisphere. Some of the obviously aberrant forms, such as marine birds (Olsen 1913) have been excluded. Many of the species listed by Budylenko and in Table 5.2 are taken only rarely and incidentally.

The stomach of baleen whales is composed of four chambers, but the first is a large dilated sac at the lower end of the oesophagus. The second and third are the real gastric chambers. The anatomy and morphology of the stomachs of the sei whale are described by Hosokawa and Kamiya (1971). Digestion appears fairly rapid and consequently it is the contents of the first stomach that are described in feeding studies. The locally monospecific nature of feeding means that records of relative incidence of occurrence do largely reflect the importance in the diet of a prey species and as a consequence much of the original data has been presented in the form of incidence of occurrence.

Antarctic and Sub-Antarctic.

The Antarctic provides the luxuriant summer feeding grounds of the Southern Hemisphere baleen whales and the importance of the krill, *Euphausia superba*, has often been described. However the summer distribution of the sei whale, and particularly that of the smaller individuals, is more northwards than those of the blue, fin, humpback and minke whales, with the main catches, from 60°W eastwards to 140°E, taken to the north of 55°S latitude and north of the Antarctic Convergence. In the remaining sector both catches and the Convergence are further south, (Horwood 1986). Consequently only a limited proportion of the population visits the areas rich in *E. superba*, and these tend to be the larger whales,

Table 5.2: Food organisms found in the stomachs of sei whales in the northern hemisphere. (Repeated references to species have not been given).

Euphausidae		
Euphausia pacifica	Klumov 1963	Nemoto 1957
Euphausia similis	"	Nemoto 1959
Thysanoessa inermis	"	Collett 1886
Thysanoessa longipes	"	Nemoto 1957
Thysanoessa raschii	"	
Thysanoessa gregaria	"	
Euphausia recurva	Kawamura 1973	
Euphausia diomedeae	"	
Euphausia tenera	"	
Thysanoessa spinefera	"	
Thysanoessa difficilis	"	
Meganyctiphanes norvegica	Jonsgard and Darling 1977	
Cephalopoda		
Ommatostrepheses sloanei	Klumov 1963	Nemoto 1959
Loligo opalescens	"	
Watasenia scintillans	"	
Gonatus fabricii	"	
Octopus sp	"	
Decapoda		
Sergestes similis	Kawamura 1973	
Pleuroncodes planipes	Matthews 1938	
Copepoda		
Calanus plumcrus	Klumov 1963	Nemoto 1957
Calanus cristatus	"	"
Calanus glacialis	"	
Calanus pacificus	"	Kawamura 1973
Eucalanus elongatus	"	
Metridia ochotensis	"	
Pleuroncodes planipes	"	
Calanus finmarchicus	"	Collett 1886
Temora longicornis	"	
Metridia lucens	Nemoto 1957	
Pisces		
Sardinops sagax	Klumov 1963	Nemoto 1959
Engraulis japonica	"	"
Engraulis mordax	"	Rice 1977
Mallotus villosus	"	
Yarrella microcephala	"	Nemoto 1959
Argyropelecus sp	"	
Polyipnus sp	"	

Table 5.2 (cont'd)

Myctophium asperum	Klumov 1963	
Cololabis saira	"	Nemoto 1957
Boreogadus saida	"	
Theragra chalcogramma	"	
Trachurus japonicus	"	
Trachurus declivis	"	
Ammodytes sp	"	
Pneumatophorus sp	"	
Sebastodes sp	"	
Pleurogrammus monopterygius	"	Nemoto 1957
Ranzania typus	"	
Scomber japonicus	Nemoto 1959	
Maurolicus muelleri	Kawamura 1973	
Pseudopentaceros richardsonii	Kawamura 1973	
Trachurus symmetricus	Rice 1977	
Sebastes jordani	"	

perhaps because of thermal restrictions or because of a behavioural advantage associated with their greater size.

The difference in character in the feeding of the various species can be seen in Table 5.3, from the data of Nemoto (1970). These data are from whales caught between 1961 and 1965 by the Japanese fleets and are of sei whales caught mainly north of the Convergence. The blue whales are mainly pygmy blue which inhabit more northerly regions than the blue, (Ichihara 1966a). From the large numbers of samples it can be seen that the sei whale has more catholic tastes than the other whales, with less euphausiids and more copepods, and from these data a higher incidence of empty stomachs.

Kawamura (1974, 1970) described in great detail the data on feeding from catches taken over the whaling seasons 1967/68-1971/72. Again these catches were mainly taken from between the Antarctic and Subtropical Convergences and from 60°W eastwards to 150°W. Those sei whales caught further south tended to have been taken about the Kerguelen Island chain, or the southern penetration of the Antarctic Convergence. From his data the most common prey was the copepod *Calanus tonsus*, with some lesser importance given to *C. simillimus*, *Parathemisto gaudichaudii*, *E. superba* and *E. valentini*. When found most of these species occurred monospecifically. The distribution of the prey changes with latitude with *C. tonsus*, *C. simillimus* and

Table 5.3: Percentage occurrence of prey found in stomachs from catches south of 40°S.

	sei	blue	fin	minke
Euphausiids	22.7	43.0	46.0	89.8
Euphausiids & others	0.0	0.3	0.1	0.0
Copepods	9.4	0.2	0.0	0.0
Amphipods	5.8	0.5	0.0	0.0
Decapods	0.3	0.0	0.0	0.0
Fish	0.1	0.0	0.2	0.0
Empty	61.7	56.0	53.7	10.2
Nos. examined	26182	1203	35139	98

Table 5.4: Percentage of prey found in the stomachs of large whales caught in the North Pacific.

	sei	blue	fin	Bryde's	humpback	right
Euphausiids	12.6	97.6	64.1	88.9	77.3	0.0
Eu. and others	0.0	1.1	3.7	0.0	4.5	0.0
Copepods	82.7	1.3	25.5	0.0	0.6	100.0
Cop. and others	0.1	0.0	0.0	0.0	0.0	0.0
Fish	3.1	0.0	5.0	11.1	17.2	0.0
Empty	44.5	51.9	34.0	75.2	32.8	0.0

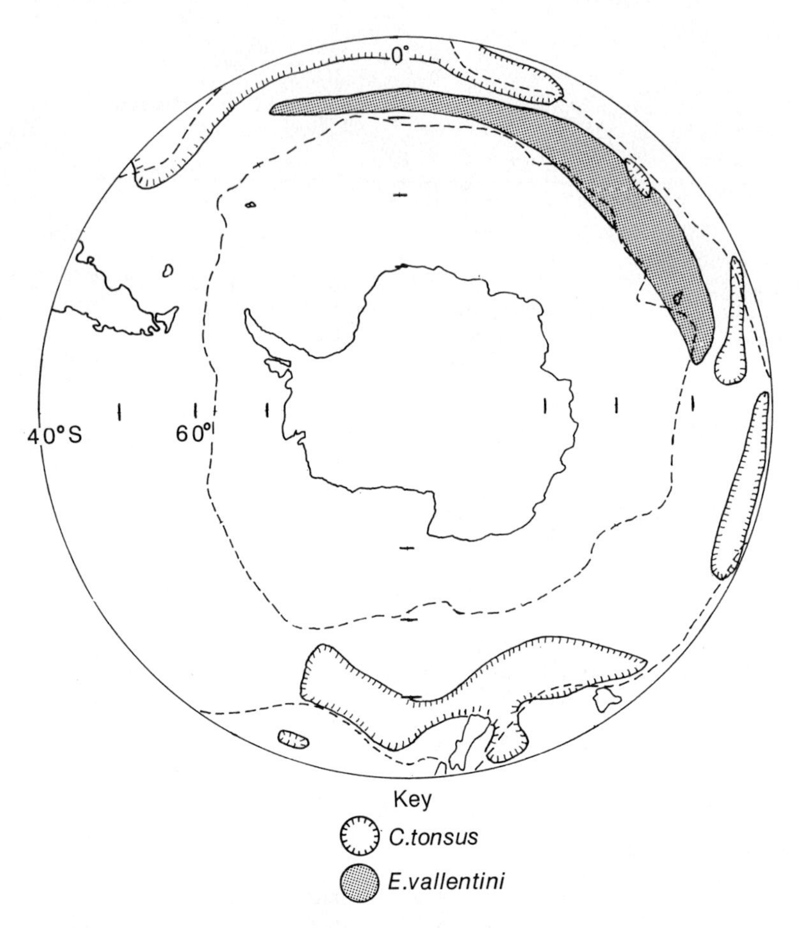

Figure 5.2: Distribution of two main prey species in the Southern Hemisphere, south of 40°S.

C. laticeps occurring between the Subtropical and Antarctic Convergences, but with *E. vallentini* and *P. gaudichaudii* occurring in the more open regions of the Southern Ocean. Near to the Antarctic Convergence *P. gaudichaudii* dominated. South of the Antarctic Convergence only *E. superba* was found in the stomachs. Figure 5.2 shows the distribution of the prey species (after Kawamura 1974), and the whaling grounds in this Antarctic sector reflect the distribution of *C. tonsus* and *C. simillimus*. Kawamura

(1974) reported that the copepodite stage five of *C. tonsus* was the most preferred and this is found in waters of 9.5-13°C; it is about 3.0-3.6 mm long. After January *C. tonsus* begins its vertical migration out of the surface waters and the sei whale turns to *C. simillimus*, which is found in waters further south of temperatures of 5-9°C. As the season progresses the sei whales tend to move further south to take *P. gaudichaudii* and *E. valentini*.

Pelagic whaling has been restricted to south of 40°S but the distribution of sei whales in the early summer extends further north to coincide with the distributions of *C. tonsus* and Budylenko (1978b) reported sei whales feeding in these regions in November. Around the islands of South Georgia and in the southwest Atlantic he found that the sei whales were feeding on euphausiids but with a greater diversity than is found further south. Matthews (1938) and Brown (1968b) found that sei whales at South Georgia fed on *E. superba* but also took copepods and fish. In the subtropical South Atlantic, Golubovsky et al (1972) recorded sei whales as being "well fed" on the fish *Tetraganurus cuvieri*. The findings of Kawamura (1974) and Budylenko (1978b) are in agreement, and around the important whaling grounds of the Kerguelen Islands they both reported *C. tonsus* and *C. simillimus* taken in the north and *E. vallentini* and *P. gaudichauii* taken in the south. As in the North Pacific, samples taken from near to coast or shelf regions show a slightly greater diversity.

South Atlantic

In the South Atlantic sei whales are found along the entire coast of South America, the coast of West Africa, and off the Gabon, Congo and Angola. South of 30°S the sei whales have been observed to feed on large aggregations of *C. tonsus* as previously described. Catches from Brazilian land stations from July to October show the sei whales are not feeding, but Best (1967a) has demonstrated that some feeding does occur in the winter months, and Budker (1958) described that, from one sei whale caught off Gabon, virtually on the equator, a very small quantity of fish, possibly clupeoids, was found in the stomach. The whale was very thin. Similar occurrences have been found off the Indian Ocean coast of South Africa.

Best (1967a) sampled stomachs from 900 sei whales caught at Donkergat, just N.W. of Cape Town, and the prey species he identified are included by Budylenko (1978b). He found that even in this region 47-55% were feeding and, as in other regions, many had fed monospecifically. Euphausiids, mainly *E. lucens* and *E. recurva*, constituted the main food by volume and as a monospecific occurrence. Copepods were of next importance with equal weight given to *C. tonsus*, *Calanoides carinatus* and *Clausocalanus arcuicornis*. Although largely monospecific in food intake more of a mixture of copepods and euphausiids were found than in the more southern waters and some 45 different prey types were identified. Best considered that the area was not an important feeding ground and the occurrence of the sei whales did not coincide with the peak abundances of plankton. Their positive association with gradients of sea temperature off South Africa however does indicate a general interest in feeding.

North Atlantic.

There is not an extensive literature on feeding of sei whales in the North Atlantic. Collett (1886) examined sei whales caught off the coast of Finnmark, North Norway, and found they predominantly fed on the copepod *C. finmarchicus*, but he also found that the euphausiid *Thysanoessa inermis* was also taken at some locations. At the entrance to Romsdal Fjord, Hjort and Ruud (1929) found the baleen whales feeding on *Meganyctiphanes norvegica*, *T. inermis* and *C. finmarchicus*, but the sei whales fed almost exclusively on the last. Hjort and Ruud also sampled zooplankton and discovered that the sei whale arrived on the feeding grounds about one week after the peak numbers of copepods; a similar timing was found by Nemoto (1959) of sei whales in the North Pacific. Off More, sei whales were usually caught in waters of over 200 m depth, but Ruud (1932) reported that they rarely extended into the Atlantic waters as *C. finmarchicus* was there in only small numbers compared to the concentrations found on the shelf. Ingebrigtsen (1929) also found the sei whales feeding on copepods with some euphausiids but reported that the presence of sei whales was influenced by the weather; cold north winds drove them from the whaling grounds. Storms also cause movements of sei whales in the North Pacific

(Chapter 4), and the dispersions are probably associated with dispersion of the shoals of plankton.

More recent information on feeding is provided by Jonsgard and Darling (1977). 52 stomachs were sampled from catches from off West Norway in 1952 and 1953, over the months April to September. Of these, 35% had empty stomachs, 48% fed exclusively on copepods, 10% on euphausiids and 8% on a mixture. The empty stomachs occurred later in the season and no new prey species were identified. Mitchell (1975) sampled stomachs of whales landed in Nova Scotia, where the whales fed mainly on copepods from June to October with only euphausiids taken in May and November. In contrast to the situation in the North Pacific the local fin whales took a more diverse food than did the sei whales but the fin whales took no copepods. Kapel (1979) reported that two sei whales taken off West Greenland had fed on krill.

In the western North Atlantic, Watkins and Schevill (1979) observed sei whales feeding on zooplankton and ignoring nearby schools of fish.

Indian Ocean

Sei whales have been caught from the land stations at Durban and feeding was investigated by Bannister and Baker (1967). Food species found were the euphausiids *E. recurva*, *E. spinifera* and *T. gregaria* and *megalopa* larvae. Samples were too few to distinguish any differences from fin whale feeding. The quantities found were all small, as also described by Gambell (1968). In 1962 and 1965 it was found that 35-37% had been feeding whereas a significantly fewer 10% were feeding in 1963.

North Pacific

Nemoto and Kawamura (1977) extended the earlier and outstanding studies of Nemoto (1957, 1959, 1962) and summarised the occurrence of different prey in the stomachs of the large whales of the North Pacific. The information in Table 5.4 shows a more striking difference between the sei whale and the other species than is found in the Southern Hemisphere, but as in the other areas copepods are the predominant food with some euphausiids and fish.

These data were compiled from information gathered over the whaling seasons of 1952-1971. Over this time the whaling operations moved southwards as blue and fin whale stocks were depleted, but under the regulations of the International Whaling Commission whaling was restricted to the region north of 40° or 45°N, depending on longitude. So the above table shows the pattern of feeding in the northern North Pacific. Data from the earlier years of 1952-1958 from just south of the Aleutian Chain showed an even greater reliance on copepods, (Nemoto 1957,1959). *Calanus cristatus* and the less densely swarming *C. plumchrus* are the dominant species with the euphausiids *Thysanoessa inermis* and *T. longipens* taking a secondary role. Thompson (1940) also found that *C. cristatus* was taken by the one sei whale sampled at Port Hebron, Alaska.

From near the Kuriles, Klumov (1963) reported that from a sample of 48 sei whales, 29% were eating copepods, 43% squid and 19% fish, but in the years of sampling Klumov claimed that zooplankton was scarce, and all the whales suffered from a reduction in food. He described the distribution of sei whales in the northern North Pacific as connected to the aggregations of, firstly, the calanoid copepods and, secondly, the pelagic squid *Ommatostrephes sloanei-pacificus*. These squids occur in the surface waters in vast shoals and in one stomach 3100 squid were found weighing 620kg. The samples from the Kuriles also tended to be monospecific. Klumov cited the occurrence of 38 prey species which include many fish; some of these were taken regularly in large numbers but some only rarely. Again it is possible that the larger range of food items is associated with catches nearer the shallow sea areas.

Feeding by sei whales in the more southern regions has been reviewed by Kawamura (1973). In 1972 international agreements and Japanese domestic legislation allowed the North Pacific whaling grounds to be extended to the south, but few catches were taken south of 40°N since this latitude roughly defines the boundary of the Subarctic. However in the west the major ocean currents of the Oyashio and Kurushio mix, to produce an extended productive frontal zone in which many balaenopterids are found, and particularly the sei whales. From catches south of 41°S the stomachs of 884 sei whales were examined with an average of 33% feeding; to the north the figure was 50%. Kawamura (1973) found that although the incidence of feeding was less in the more southern waters those feeding usually had fuller stomachs than the whales in the Subarctic. To the south copepods comprised only 22% of the diet compared with 80% in the north, and fish 70% compared with 5% in the north, (Kawamura, corrected in litt to his table 5). More than 20 food species were identified but the main ones were *C. pacificus*, *Scomber japonicus*, *Sardinops melanosticta* and *Engraulis japonica*. The sei (seje) whale, somewhat misnamed in the North Atlantic, can correctly be called the "sardine whale" in Japanese - iwashi kujira.

Further south around the Bonin Islands both sei and Bryde's whales took fish and euphausiids; in this region copepods are scarce, (Mizue 1951, Nemoto 1959, Kawamura 1980). Catches from Sanriku, north-east Japan, have shown that the sei whale takes both large numbers and a great variety of fish, but in the original data (Mizue 1951) catches of sei and Bryde's whales were not separated. I have assumed that the sei whales referred to by Nishimoto et al (1952) were Bryde's whales.

Off the coast of Japan, Omura (1950a) found that over the months May to October sei whales, (easily distinguished from Bryde's whales in his text), had an incidence of empty stomachs of 31-58%. Off central Japan over the months June to August there was the highest incidence of rich feeding. Off Hokkaido the peak feeding was a little later, from August to October.

From off the coasts of Mexico and California the sei whale has been shown to have catholic tastes. Matthews (1932, 1938) reported that sei whales took the Pacific red crab larvae, *Pleurocodes planipes* from off Mexico, and Rice (1963, 1977) showed that off California the sei whales, which arrived in later summer, fed mainly on *Engraulis mordax* with *E. pacifica* also important; several other species of fish were also taken, but blue and fin whales in the same area were mainly feeding on euphausiids.

FOOD INTAKE

Lockyer (1981a) reviewed data on seasonal changes in blubber thickness and has confirmed that feeding takes place mainly in the summer with little if any in the winter, and as we have seen, most of the annual food is taken on the summer feeding grounds; nevertheless the stomach samples show that at any one time a high percentage are not eating. In the

Antarctic Nemoto (1970) found that only 38% of sei whales were feeding and the more extensive survey of Kawamura (1974) found that between 31 and 61% were feeding, depending upon the area. In the North Pacific Nemoto and Kawamura (1977) found that in general 55% were feeding, and Kawamura (1973) found that north of 40°N 50% were feeding, compared with 33% to the south. One explanation for such a low incidence of feeding would be a very rapid digestion rate, but this does not appear to be the case.

The sei whale does not show such a pronounced bimodal diurnal feeding as is seen in the other large whales, and Kawamura (1974) has shown that feeding is related to the vertical availability of the prey; *Calanus simillimus* is taken crepuscularly, whereas *C. tonsus* can be taken continuously. Over a day the incidence of feeding, measured by the presence or absence of food in the first stomach, typically falls from 80% at dawn to 30% at dusk. This may be interpreted as the whales finding abundant food early in the day and digestion reducing this figure to daily averages of about 50% of whales feeding. There is a continuous feeding on zooplankton but its contribution is small compared with the dawn intake. It should be remembered though that in this category of feeding whales not all stomachs are full. Klumov (1963) says that of all the sei whales examined in the North Pacific only two had full stomachs.

In the polar regions some information on full stomachs is available. Klumov (1963) reported that one whale had 620 kg of squid and another had 500 kg of saury; up to 370 kg of calanoid copepods were noted. Brown (1968b) measured up to 305 kg of euphausiids. Kawamura (1974) gave a list of typical stomach contents from feeding sei whales in the Antarctic. These are: copepods <100kg, amphipods 150-250kg, euphausiids 150-200kg. Samples from 14 sei whales ranged between 116-197 kg.

In the warmer waters off S.W. South Africa Best (1967a) found that 47-55% were feeding but only a few had obtained much food. An average over all whales gave 3.7-4.5 litre which we may suppose is no more than 5 kg. Gambell (1968) and Bannister and Baker (1967) also found 4-37% feeding but the maximum amount in any stomach at Durban was 16 kg. None or negligible feeding was found off Brazil and Gabon. Off Mexico over July to October Rice (1963) found 40% of the sei whales to have full stomachs and 15% had stomachs half full.

Various theoretical studies have been undertaken to calculate the daily intake of food of whales. Klumov (1963) considered that they needed 30-40 g per kg of whale per day; for a mature 18 t sei whale this gives 540-720 kg per day. Sergeant (1969) assumed that feeding data from small whales could be extrapolated to large whales on the basis of a constant heart weight to body weight ratio and Kawamura (1973) calculated that this predicted a ration of 4.4% per day with a ratio of heart weight to body weight of 0.00487. This is close to the figure given by Klumov. Kawamura considered that typical densities of zooplankton would not allow 800kg per day to be taken by skimming. Gaskin (1982) thought these rates too high and suggested that 2.5 to 3% was more appropriate for rorquals. Lockyer (1981a) presented an extensive study on the energetics of large whales including sei whales; she found that pregnant females were the best fed, as they arrived at the feeding grounds first, and lactating females were lean. Lockyer concluded that for rorquals a feeding rate of 1.2% of body weight per day was likely. This figure is based on assuming 85% feeding per day at a rate of 30-40 g per kg per day for 120 days of the summer feeding period and at a tenth of this for the rest of the year.

If we assume that the sei whales feed on copepods then the average stomach content in the Antarctic is about 100kg. At the begining of the day the content will be greater and 80% will have fed on one day. If it is assumed that initially 200kg were present, or were taken initially and over the course of a day, then for an 18 t sei whale the daily intake is: 0.8x200x100/18000 = 0.9%. To obtain the 3-4% speculated above the whale would need to take 3-4 times the amount and even the rate of 1.2% may be high. A greater intake of food is not supported by Kawamura's data and a better resolution of the field data and theoretical studies is needed.

From Durban it was shown that in 1962 and 1965 incidence of feeding was much greater than in 1963; 35% compared with 10%. Data from Mackintosh (1942) show that feeding levels were high in 1930. Klumov (1963) also reported that 1963 was a poor year for feeding in the northern North Pacific. The significance of possibly good and bad years of feeding, in summer or winter, to the population dynamics of the sei whales has yet to be examined, but Lockyer (1986) has shown strong associations between plankton abundance, body condition and fecundity in North Atlantic fin whales.

INTERSPECIFIC COMPETITION IN FEEDING

As we shall see later there is a persuasive amount of indirect information that suggests that interspecific competition exists amongst the baleen whales. The interpretation of the data is not without argument, but the significance of such competition to the population dynamics of the baleen whales is great and without question. This section presents information on the direct feeding of sei whales to investigate whether competition for food is likely.

On the feeding grounds there is a general segregation of the species. In the North Pacific the bowhead whale extends well into the Arctic seas and fin, humpback and minke whales are found above the Aleutian Chain whereas sei and blue whales tend to be more to the south. Few sei whales were taken north of the Aleutian islands by the Japanese fleets over the period 1952-1972 and sightings records confirm this distribution. Catches of black right whales have been described by Townsend (1935), and more recently by Omura (1958) and Berzin and Doroshenko (1982); they show that the right whales are found in more marginal seas than the sei whale, but with an overlap particularly in the northeastern North Pacific where all the large baleen whales are found. Ohsumi (1977) described the catches of sei, fin and Bryde's whales in relation to sea surface temperature, and as shown in Figure 5.3, most sei whales were caught in waters of less than or equal to 18°C whereas almost all Bryde's whales were caught in waters at or above 19°C. Peak fin whale catches can be seen at much lower temperatures.

Less detailed information is available for the North Atlantic but the distribution shows similarities with that of the North Pacific. Few sei whales have been found near the ice and they have a more pelagic disposition than do the other species. However in many localities catches of blue, fin and sei whales have been made in the same months, and sei and right whales were caught together off Ireland (Burfield 1912,1913,1915).

The sei whale has long been regarded in the North Atlantic as a species occurring erratically on the off-shore whaling grounds with "sei whale years" a feature of their catch history (Collett 1886, Jonsgard and Darling 1977). Changes in relative abundance of fin and blue whales, described as "blue or fin whale years" has also been described from the Southern Hemisphere and North Pacific (Mackintosh

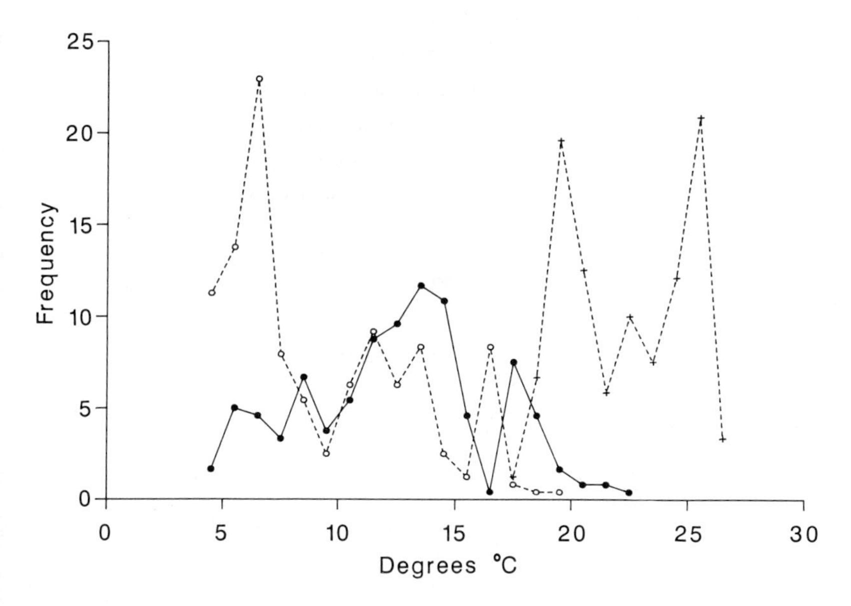

Figure 5.3: Distribution of catches by species, in relation to temperature, from the North Pacific. Open Circles: fin whales, closed circles: sei whales, crosses: Bryde's whales.

and Wheeler 1929, Nemoto 1959). Nemoto noted that other rorquals were leaving the whaling grounds as sei whales arrived. These changes in distribution are probably related to the environmental determination of the distribution of prey species.

In the Southern Hemisphere a clear spatial segregation exists which has been characterised by Laws (1977b) and is shown in Figure 5.4. Blue, minke and sometimes humpback whales feed near the edge of the pack ice, with latitudinal concentrations of fin whales, then sei whales and last right whales to the north. Catches of fin and sei whales are well segregated although there is some overlap (Nasu and Masaki 1970). Budylenko (1978b) noted that sei whales feed in the regions occupied by the right whales before the sei whales continue their annual southern migration, and

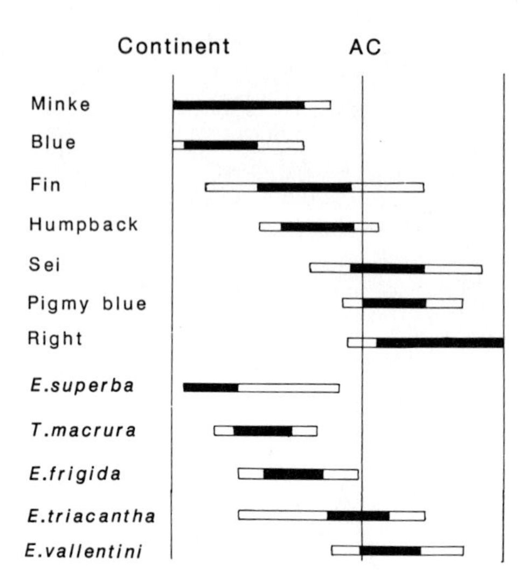

Figure 5.4: Distribution of Southern Hemisphere whales and Euphausiids with latitude, from the continental shelf northwards, AC: Antarctic Convergence. The dark areas are regions of highest density. (After Laws).

Horwood (1986) has shown that a large proportion of sei whales stay north of the Antarctic Convergence, with only the larger sei whales penetrating further south. Consequently it appears the sei whale has its own spatial niche but with substantial overlap with other species. Nemoto (1962) and Bannister and Gambell (1965) suggested that the sei whale had extended its southwards distribution in more recent years and that this may be associated with the decline of blue, fin and humpback stocks. However the evidence for a more general southern distribution is not strong, and occasional invasions of sei whales into the seas north of the Aleutian Islands are found along with a late seasonal southern movement in the Antarctic associated with less cold waters.

Laws(1962, 1977a,b) and Bannister and Gambell (1965) considered that a temporal segregation also serves to reduce any competition. In the Southern Hemisphere a progression of species with time is seen with blue and humpback the first to arrive in the Antarctic, then fin, and sei whales arriving late in the season. In addition to geographical and temporal segregation the species tend to feed at different depths; minke, sei and right whales tend to feed in the surface waters with blue and fin capable of feeding at greater depths. However right whales have been observed feeding at depths and fin whales are frequently observed feeding at the surface.

The sei whales do have a distinct niche in time and space but this is best seen in the comparison of stomach contents described earlier. In most places the diet of the sei whale is copepods, with some preference for euphausiids and amphipods. A similar feeding is seen only in the right whale. Harmer (1931) found it surprising that all baleen whales at South Georgia fed largely on one species of krill and his speculation that this would not be found throughout the Antarctic is born out by the above descriptions. The food preferences of the baleen whales has been summarised by Nemoto (1959) and Mitchell (1974b), and Mitchell's scheme is illustrated in Figure 5.5.

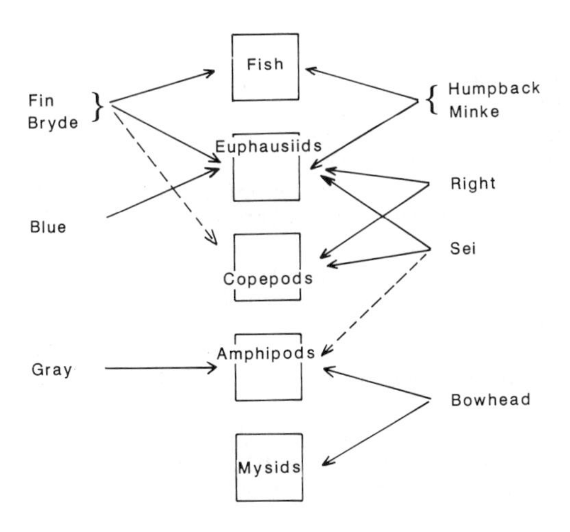

Figure 5.5: Mitchell's scheme of feeding interrelationships.

The above arguments have led Mitchell (1974b), Kawamura (1980) and Gaskin (1982) all to consider that if competition exists amongst the mysticetes then it is more between the sei and right whales than between any other pair. In summary the sei whales have a specific feeding niche the central part of which is exclusive to them but there is overlap at the edges. A reduction of competition by a reduction in the numbers of other species might well make these areas of overlap more attractive to and exploitable by the sei whales.

SUMMARY

The sei whale employs a swallowing or skimming behaviour to filter its prey, the preferred type of which are copepods. The filtration apparatus is intermediate between that of the right whales and that of the fin and blue whales. A list is provided of the prey species that have been found in the stomachs of sei whales, but most of them occurred only occasionally. Those taken most often and in the greatest quantities are of surface swarming plankton that live in waters of about 5-15°C. The sei whale usually takes only one prey species on any day, and that the most suitable locally abundant, but there is limited evidence of some active selection. Feeding is concentrated at dawn with some limited intake during the day and this is possibly associated with the diurnal migrations of the zooplankton. At most 85% feed on one day and most do not have full stomachs. The field data indicate that intake is less than 1% of body weight per day, but this is low compared with estimates from theoretical considerations.

The feeding ecology shows that the sei whale feeds on prey usually not consumed by blue, fin, humpback and minke whales. It is thought that the concentrations of copepods are typically too sparse to sustain these species. The greatest overlap is with the right whales. It can be concluded that the sei whales have a specific feeding niche, the central part of which is exclusive to them, but there is overlap at the edges which allows competition and expansion.

Chapter Six

AGE DETERMINATION

INTRODUCTION

The correct determination of age is vital to the calculation and interpretation of demographic parameters and to the use of these parameters in understanding fluctuations in population size. Various techniques have been used to age whales and these are presented below. Much of the early methodology was developed with material from blue and fin whales, and particular attention has been given to ageing by counting laminations seen in the ear plug since this is the main technique now used to age baleen whales. There are difficulties in attributing absolute age to any one index and in this section reference is made to relating indices to the number of corpora. As will be shown in later sections this is useful since, rarely in mammals, the corpora albicantia do not completely regress and remain as a record of ovulatory behaviour. Consequently one may assume that those whales with more corpora are older than those with less. Given the understanding derived from these techniques the later sections present information specifically about sei whales.

METHODOLOGIES

Baleen plates

As early as 1820 Scoresby had suggested that transverse lines found on the baleen plates might be related to age and quantitative studies were undertaken by Ruud (1940, 1945) and Tomilin (1945). They were able to identify particular steps in the pattern of undulations on the surface of the baleen which might be related to age. Subsequent studies

(Ruud 1959, Ruud et al, 1950 and Nishiwaki 1950b, 1951, 1952) confirmed that these age classes correlated with size of and corpora number in blue and fin whales. However, and notwithstanding the correct calibration of age at that time, it was found that the baleen plate was worn from the distal end such that little information could be extracted after about six age classes. Ichihara (1966b) concluded that it was impossible to determine the age of fin whales over four years since the neonatal mark on the baleen plate begins to wear off at about three years of age; these ages are approximately correct as determined by the studies on ear plugs.

Tympanic bullae

Boney tissue with laminations, particularly mandibular bones and the tympanic bullae (Laws 1960, Klevezal' 1980), have been used to age marine mammals. Christensen (1981) used the bullae to age minke whales, since other techniques had not proved successful, and obtained a good correlation between the numbers of layers counted in the outer, or periosteal, zone of the bullae and ear plug laminae and corpora counts. The bullae of sei and fin whales were investigated by Klevezal' and Mitchell (1971). On the basis of very few samples, (nine), they found that counts from ear plugs were larger than those from the bullae but concluded that periosteal layers were laid down annually.

Lens colour

Nishiwaki (1950a) noticed that the colour of whales' eyes appeared to change with size and consequently examined the absorption of light by the lens in relation to body length and corpora count. For both blue and fin whales there was a linear relationship between the absorption of light and length, but with a wide scatter. An even better relationship was found with corpora number. Nishiwaki reasonably concluded that the change in coloration of the lens was a far more accurate and reliable method of estimating age than previous methods; however at that time the absolute units of age were speculative. It is probable that this method was a precursor of the next technique.

Aspartic acid racemization

Ageing of marine mammals based on aspartic acid

racemization in teeth and lens nuclei has been described by Bada, Brown and Masters (1980). Most amino acids can exist in two different isomeric forms called D and L enantiomers. In living protein only the L form exists and is positively maintained as in a chemical equilibirum D and L forms would exist equally. After death the process which maintains the disequilibrium stops and a process called racemization starts to convert the L forms to D forms until there is an equal ratio. In teeth the protein is not maintained and it has been found that the L to D ratio is a good predictor of age. Additionally, it was found that racemization occurred in eye lens nuclei (Masters et al 1977). This technique was applied to mysticete whales for the first time by Nerini (1983) who demonstrated a high correlation between the number of ear plug laminations and the magnitude of the aspartic acid racemization.

Ear plug

It was discovered by Purves (1955) that the keratinised and fatty ear plug of the fin whale, when sectioned longitudinally, showed a pattern of alternating light and dark bands. On theoretical grounds it was assumed that the light laminae were laid down during periods of active growth and that the dark laminae were formed during periods of reduced growth. The migration of the baleen whales were thought to give rise to two growth layers per year, where a growth layer group is considered as a pair of one light and one dark bands. A sei whale ear plug is shown in Figures 6.1 and 6.2.

The ear plug was first described by Lillie (1910) and subsequently in more detail by Purves (1955) and Ichihara (1959). The plug is composed of an outer covering derived from the lining of the external auditory meatus, and is made up of folds of squamous epithelium, and the core. The core is composed of a series of columns of keratinised epithelial cells in light and dark bands and are derived from the glove finger. Following on from the previous work, Roe (1967a, b) described the development of the light and dark laminae of the core. The difference in the laminae is caused by different types of cell formation. The light laminae have a high fat content, and the cells, formed from the glove finger epithelium, are round, containing intracellular fat. As the cells get pushed, from below, into the core they become

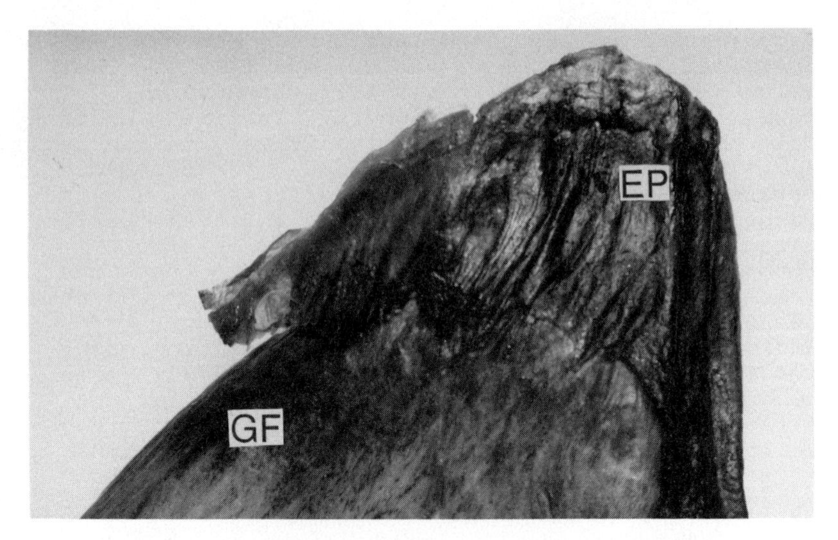

Figure 6.1: Plate of an unsectioned sei whale ear plug showing the ear plug (EP) sitting on the glove finger (GF).

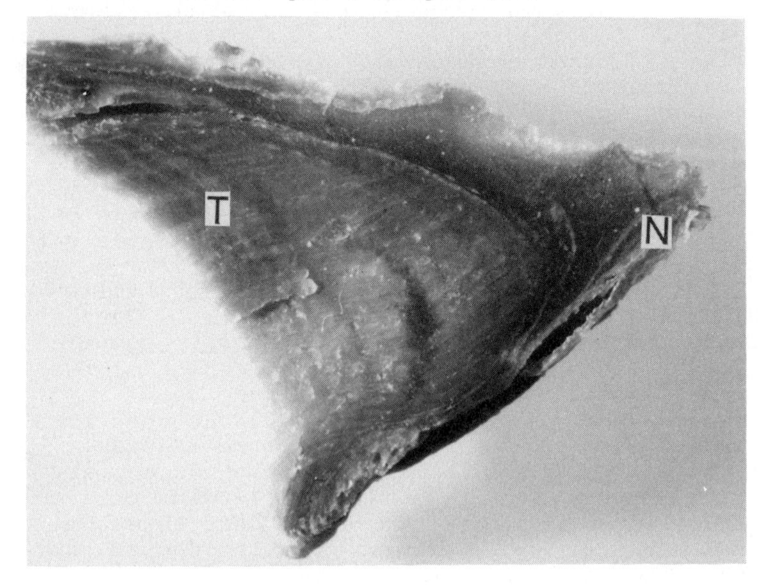

Figure 6.2: Plate of a sectioned sei whale ear plug showing the alternating light and dark laminae and the positions of the neonatal line (N) and the transition layer (T).

flattened and keratinised and the fat is forced out of the cells to form a band. In the dark band a similar process takes place but the cells are bigger and contain little fat. The bands are visible to the naked eye and staining for fats emphasises the difference just described.

The calibration of the growth layers into years proved difficult and initial interpretations based on readings of baleen plate ridges and perceived rate of accumulation of corpora albicantia led to a belief in two layers per year (Nishiwaki 1957, Nishiwaki, Ichihara, Ohsumi 1958, Chittleborough 1959, Purves and Mountford 1959, Laws 1961). This conclusion was supported by the recovery of a humpback whale that was marked, probably as a yearling, and recovered six years later with an ear plug showing 12 growth layers (Chittleborough 1960). Supportive evidence from marked humpback whales was provided by Dawbin (1959). However, with the possible exception of the humpbacks which still present a problem, it has now been resolved that each pair of laminae (one growth layer) represents about one year. For fin whales Roe (1967a) looked at the type of the most proximal layer by month using ear plugs from catches at Durban, and from the Antarctic land stations and pelagic catches. Figure 6.3 shows the results replotted from his Table 1. It is clear that the fatty light laminae are formed during the intensive feeding period of October to March and that the dark laminae are formed in the winter June to September. Similar findings were described by Ichihara (1966b) for Antarctic fin whales. Recoveries of marked fin whales have been made up to 30 years after marking and dividing the number of laminae by the time the whale had been marked will give a maximum number of laminations per year. Such studies are reported by Ohsumi (1962, 1964a), Ichihara (1966b) and Roe (1967a) and many returns show that there cannot be two pairs of laminations per year and long-term recoveries indicate much nearer to one pair per year. These two sources of information have led to the conclusion that in general one pair of laminations (one growth layer group) per year is laid down. Young whales may have additional fainter laminae and not all whales may migrate every year, but nevertheless, for fin whales at least, the conclusion is relatively robust. Lockyer (1984a) noted that the record from the marked humpback whale is controversial, not least because of the problems in interpreting the laminae of very young whales, and at an IWC meeting on age

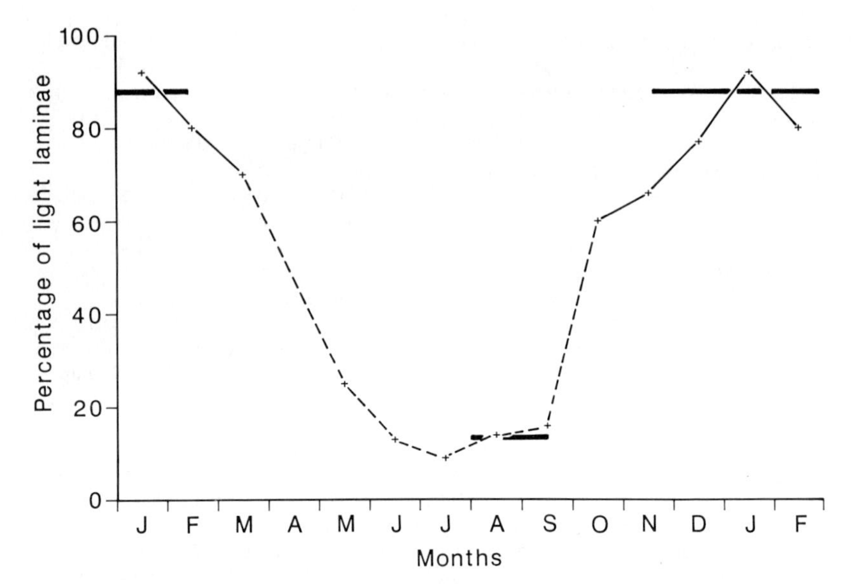

Figure 6.3: The percentage of ear plugs in which dark laminae are forming by month from fin whales caught from May to September at Durban and from October to March in the Antarctic (+). The horizontal bars show similar data from Southern Hemisphere sei whales.

determination in Oslo in 1967 (Anon 1969) it was agreed that for fin whales one growth layer group was formed per year.

Lockyer (1984a) considered that absolute evidence for any single deposition rate does not exist for any baleen whale but marks involving quinacrine, which is absorbed by tissue (Kozicki and Mitchell 1974), could provide a greater detail of information.

AGEING OF SEI WHALES

Following his work on fin whales Roe (1968) investigated the laminations in the ear plugs of the sei whales collected at Durban and in the Antarctic. Standard staining failed to highlight the laminations and bleaching, which was used to reduce the heavy dark pigmentation did not improve the contrast. Readability was found to be very difficult. In addition the germinal layer had rotted away in most samples. Over half the sample had areas that were keratinised but in which no layering was detected and Roe suggested that the layering may have been squashed out during growth of the plug. From a large number of samples Roe investigated variability in counting from whales of different sex and size. Whereas for fin whales there was no difference with either sex and with length there was a dramatic decline in readability of plugs from male sei whales from 44 to 57 ft. Histological differences were found between the plugs of the sei and fin whales. Pike (1968) also found, for some sei whale plugs, substantial variability amongst readers. The report from the meeting on age determination (Anon 1969) concluded that the value of ear plugs to ageing sei whales was limited.

In order to overcome the problems of identifying laminae, X-ray techniques were applied (Masaki 1968, Lockyer 1974). The preparation is laborious and as little additional information was achieved over readings by eye the techniques have not been pursued.

However Lockyer (1974) developed a bleaching technique that has allowed successful counting by eye. The main problem in recognising the layers is the dark pigmentation. After cutting longitudinally with a razor the cut surface was smoothed down on a wet stone and the washed plug put in a 1.02 g per ml solution of hydrogen peroxide. The bleaching process is somewhat destructive. From Durban, South Georgia and pelagic operations in the South Atlantic, Lockyer examined 1127 ear plugs. Before bleaching 50% were satisfactorily readable and after bleaching 93% were. A satisfactory reading was determined as three readers being within ± 1 for plugs with less than 20 laminae or ± 2 for plugs with a greater number. Consequently the plugs can be consistently read.

A sample of 198 ear plugs was examined to identify the seasonal development of the light and dark laminae by examining plugs collected in November to February from South Georgia and in July to September from Durban. The germinal epithelium and basal laminae were often damaged or lost so that only 50% could be analysed. Of these, the younger (less than 15 laminae) suffered less damage. Lockyer found that at South Georgia, the terminal lamina was mainly of a light type (86%) whereas at Durban it was of the dark type (81%), Figure 6.3. This is very similar to that found in fin whales and is suggestive of a basic one growth layer group per year. Mark recoveries at that time, or the information given earlier (Chapter 3) are insufficient to shed any light on the question of numbers of laminae per year.

As an independent evaluation of the numbers of growth layers per year, Lockyer (1974) related the number of layers to the number of corpora for animals with over 7 layers. For numbers of growth layers between 8 and 30, a regression gave a slope of 0.68 corpora per layer. From considerations of the production of corpora throughout the year, and assuming a two year reproductive cycle, Gambell (1968, 1973) arrived at an estimated rate of 0.61-0.69 corpora per year. For fin whales, Lockyer (1972a) gave a comparable figure of 0.69. These estimates do not explicitly use age data. Consequently the evidence supports a general assumption of one growth layer per year.

Lockyer (1974) noted that whales of over 30 years (growth layer groups) had more corpora than was predicted from the above regression. She considered that in the older physically mature adults the ear plug core grows very slowly and some growth layers could be squashed out. She argued that in the older females corpora number may be a better index of relative age than ear plug laminations. Mizroch and Breiwick (1984) investigated relationships of age and length in sei whales and found that for both the young males, of less than 12 growth layers, and old whales, over 30 growth layers, variability in length at age and age at length increased. This is possibly due to the problems of identifying and interpreting laminae in very young or very old individuals.

Lockyer and Martin (1983) reported on the ageing of sei whales caught at Iceland over the period 1967-1981. In the later years most of the catch were sampled and over the entire period between 70-75% of the samples were readable. Initial concerns that readability was more difficult in smaller whales (Lockyer 1978a) were dispelled but it was shown that readability was significantly better in males. A comparison with corpora counts supported the interpretation of one growth layer per year, as suggested by Lockyer et al (1977). Best and Lockyer (1977 ms) reported on ageing of sei whales from Donkergat and from the catch 43-48% were sampled and of those 63-73% were read successfully. Lockyer (1984a) tabulated values of percent readability by species and for sei whales these were: North Pacific 63% of sampled ear plugs were readable,

Antarctic 93%, West Indian Ocean 75% and from Iceland 66-80%. From California, Rice (1977) recorded that he obtained readable plugs from 80% of the males and 72% of the females examined and that errors in counting may amount to ± 20%.

GENERAL PROBLEMS WITH AGEING

Although the above material indicates that for sei and fin whales ear plug readings will give reasonable estimates of age one must take great care that biases do not enter from a variety of sources. Of particular concern over the early 1980s has been the evaluation of trends in age at maturity with year class, or year of birth. The year of birth of a whale is obtained by subtracting the age of the whale given by ear plug data from the year of capture and the age is needed to a particular degree of accuracy. The IWC considered this problem in relation to Southern Hemisphere minke whales (Anon 1984).

Comparisons of estimates of number of laminae from several readers gave a wide range of correlations, some extremely poor. However minke whale plugs are very small and can be difficult to read; in addition, some readers were not experienced with minke plugs. A higher correlation was found amongst schools of readers. Lockyer (1974) reported a good degree of consistency amongst readers of sei whale plugs and the minke whale exercise does not throw doubt on this aspect of the determination. The meeting considered the readability of plugs with "age" or size. It was noted that readibility might suffer, or underestimation of laminae occur with older sei whales. This would have some effect on studies of age at maturity but would be particularly significant in some techniques to estimate mortality rates.

Finally, there is the problem of relating the information in the sample to that in the population. If readability is a function of size this poses special problems but in addition the collection of the plug from the whale may be related to age; collection is difficult in young minke whales (Kato 1984). Usually only a proportion of the catch is aged and an age-length key is constructed from which the age distribution in the catch is obtained. If the original sample was collected randomly or proportional to numbers at length and the readability of the plugs was random, then the key is also representative of the population. However, changes occur over time in the age-length composition of both population and catch. If mortality is dependent only on age the relative distribution of length at age is invariant. If there is some degree of selection by length then length at age will vary with time. The segregation of whales in space and time also means that the age-length composition will vary if whaling alters its season or location. If particular year classes are abundant this will affect the key. Consequently an age-length key is needed for each season and whaling fleet. This is difficult to achieve and often one age-length key has been provided for a long time period. The use of such keys to obtain distributions of age and length should be viewed with caution.

SUMMARY

General techniques of ageing whales are described, and for most baleen whales the most useable method is that of counting the numbers of growth layer groups in a longitudinally cut half of the ear plug. Each year a dark lamina is formed in the winter and a fatty light lamina in the summer. Of those ear plugs successfully extracted from the whale between 60 and 80% can be aged after suitable treatment. Variability, and possibly bias, may be greater in the reading of very young or very old ear plugs. If these data are used to estimate age-compositions of the catch, care should be taken to ensure that bias is not introduced at the time of collection of the ear plug, or at reading. For instance a greater number of plugs may be less readable from larger whales (which tend to be older), and the age distribution of readable plugs is then not the same as that of the sample.

Chapter Seven

GROWTH AND SIZE

INTRODUCTION

The sei whale is by no means the biggest of the whales; the blue whale adults will average 26 m (86 ft) and 120 t, but the sei whales will reach an average length of about 15 m, weighing 19 t. The largest recorded lengths of sei whales in the Antarctic were 18.6 m (61 ft) for males and 19.5 m (64 ft) for females. From off Kamchatka lengths of 18 m and 18.9 m were recorded for a male and female respectively. From the North Atlantic Thompson (1928) reported catches of males of 18.9 m and 19.5 m from Scottish whaling stations, and he mentioned that in Haldane's notes, there was a record of a male of 19.8 m (65 ft). However, the current standardised method of measurement is from the tip of the snout to the centre of the notch of the flukes, measured horizontally and recorded to the nearest foot. Thompson observed in his section on right whales that at least some of the length measurements were done along the curvature of the body, giving an inflated length. No such comment is made about sei whales but it is probable that he was referring to the measurements of all whales. Martin (1983) did not consider these early, North Atlantic measurements reliable. Although Clark (1983) found that in the BIWS data differences in measurements by different countries may be as much as 5%, the maximum lengths from the Antarctic can be accepted as relatively accurate (Horwood 1986).

The growth of whales is rapid over the first few years but quickly reaches an asymtote. Growth, as in all mammals, occurs by intercalary growth of the vertebrae at the cartilaginous zones joining the epiphyses to the diaphyses. At physical maturity these parts of the vertebrae fuse until all signs of the cartilage or a previous join are absent, and growth in length effectively ceases. Laws (1961) says that growth in other dimensions may continue and that the skull may grow a little further. Ohsumi, Nishiwaki and Hibiya (1958) demonstrated that, for fin whales, ossification began at both ends of the vertebral column ending in the thoracic region and this is considered true for sei whales also (Matthews 1938). Laws (1961) also noted that in a single vertebra, fusion was slowest in the centre so that the deeper any section was cut through the vertebra, the better would be the result. Consequently physical maturity should be determined by examinations of deep sections through thoracic vertebrae.

Data on the increase in length with age have been acquired through catches, but because of selection for the largest animals and because of regulations governing the minimum size at capture, the younger ages tend to be represented only by the larger individuals. Because of the great difficulties in weighing, few records exist of individual weights. Fetal growth is examined through samples from pregnant whales. Finally, it is possible that growth rates have varied with time and this is examined.

FETAL GROWTH

Growth rates of whale fetuses are obtained through a consideration of size of the fetus and date of capture of the pregnant mother; however the description of change of size with time has presented problems. Huggett and Widas (1951) proposed that fetal weight could be described by the equation $W = (a\ (t-t_o))^3$, where W is the weight in kg, t is the time in days from conception, a is the rate of growth in units of the cube root of W per day and t_o is a constant number of days. They considered that t_o was about 20% of the gestation period, which it will be shown later is just less than a year. In general values of a for balaenopterids are in the region of 0.35-0.60, and Frazer and Huggett (1974) and Lockyer (1981a) show that these are the highest rates found in mammals.

A difficulty exists with the interpretation of length data from the period of late gestation. Late pregnant females tend to leave the feeding grounds earlier and are thus not represented in samples from those areas. In addition, if samples can be taken from near term whales, underestimation of lengths occurs because of parturition and protection of calves. Nevertheless, the data provide smooth trends in growth and imply a somewhat restricted and annual time of conception and parturition. Lengths of sei whale fetuses can be over 500 cm and, notwithstanding the bias mentioned above, the mean length at parturition will be smaller. Lengths of sei whale calves, thought to be newly born, were given by Mizue and Jimbo (1950) and Tomilin (1967) as 440 and 427 cm respectively. Masaki (1976a) considered the mean length at parturition to be 440 cm in the North Pacific and Gambell (1968) accepted a length of 450 cm for South Atlantic sei whales. Tomilin (1967) cited a range of 427 to 532 cm but he considered that the length of new born calves in the Southern Hemisphere may be bigger than in the Northern Hemisphere; there is insufficient evidence to support this claim.

Southern Hemisphere

The growth of fetal sei whales from the Southern Hemisphere has been described by Matthews (1938), Gambell (1968) and Lockyer (1977b). Matthews gave tabulated data on about 200 fetuses collected from South Africa, South Georgia and by South Atlantic expeditions. From South Georgia recorded lengths were approximately constant from late February to April at about 300 cm with a maximum of 488 cm. From August to November mean lengths increased from "0" to 80 cm. One fetus of 335 cm captured in October was considered to be a result of a very late pairing. Gambell (1968) considered additional material from the same areas and the pattern was the same as that found by Matthews. There was no difference in mean size by month with sex of fetus, and Gambell described the growth in length by two straight lines, one from July to October and the second from October to June. The second line gave a length at birth of 450 cm at the end of May. The lines representing the mean size against data are shown in Figure 7.1. Gambell investigated the length frequency distribution by month and found little change in the skewness of the distribution, different from that seen in fin whales by Laws (1959). Additionally, Gambell found that the growth in the later months was described better with a linear relationship than an exponential one although Rice (1977) claimed an accelerating growth rate on the basis of the modes rather than the means. From

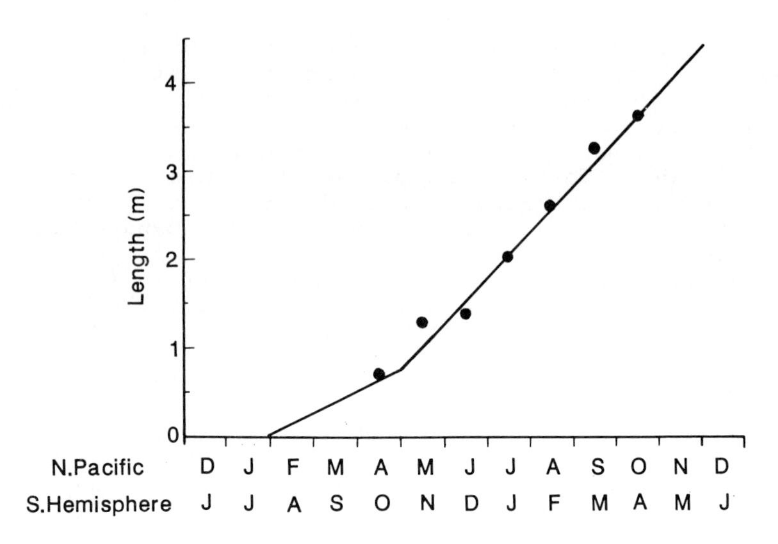

Figure 7.1: Mean lengths of sei whale fetuses (cms) against month. Dots: North Pacific, Lines: Southern Hemisphere.

Costinha, Brazil, a fetus of 4.6 m has been reported in August as well as two of less than 0.15 m (Paiva and Grangeiro 1965).

Lockyer (1977b) considered growth in weight. From 134 samples from South Africa and the South Atlantic she found a relationship,

$$\ln(W) = -11.176 + 2.9198 \ln(L),$$

where W is the weight in kg and L is the length in cm. Based on Gambell's (1968) graphs of mean lengths by month Lockyer constructed a mean weight by month curve. Using the assumption of a 342 day gestation period and parturition at 450 cm (780 kg) she obtained values for parameters of the fetal weight equation of $a = 0.034$ kg^{-3} d^{-3} and $t_o = 74$ days.

North Pacific

Growth rates of North Pacific sei whales have been described by Laws (1959) using information from Mizue and Jimbo (1950), and by Masaki (1976a) and Rice (1977). Despite confusion over identification between sei and Bryde's whales before 1950, Laws (1959) considered that, from the information given by Omura and Fujino (1954), the fetuses described by Mizue and Jimbo were from sei whales. Laws found that the monthly mean lengths could be very well described by an exponential model. However, Masaki has been able to use data from 4573 pregnant sei whales and his data provide the most information. These data were collected over the period 1952 to 1972 from Japanese pelagic and coastal operations over the months April to October. Masaki shows the frequency distributions by month of fetal lengths and in many cases a lack of symmetry is seen. Those from September and October are distorted by possible early departure of late-pregnant females from the whaling grounds. Figure 7.1 shows the plot of mean lengths by date along with the linear growth curves from Gambell (1968), from the Southern Hemisphere. Masaki considered that, with an adjustment of 6.2 months, the comparison between the lengths of older fetuses was sufficiently good that Gambell's data could be used to infer the size of fetuses in earlier months, and hence the time of conception.

Rice's data are of only 58 whales from California but are of interest in that in each month from June to September his mean lengths are approximately 30 cm less than those of Masaki. This appears to be a real feature rather than one of statistical variability. This suggests a different time of conception for these whales, which might be due to the behaviour of a different stock, or of an earlier migration past California.

North Atlantic

The size of the sei whale fetuses from the North Atlantic have been described by Mitchell and Kozicki (1974 ms) from Canadian catches and reviewed by Jonsgard and Darling (1977) for Norwegian catches and by Lockyer and Martin (1983) for Icelandic catches.

Collett (1886) described that sei whale fetuses from Finnmark ranged from 91-122 cm in the beginning of July, 183-213 cm in mid-July and 244-366 cm in August. Goldberg (1886 - cited in Jonsgard and Darling 1977) reported lengths of fetuses taken off Norway as increasing from 100-130 cm in early June to 180-300 cm in the second half of July. Risting (1928) gave information on fetuses from several species and from 95 sei whales from the North Atlantic. Most samples were from More but some were from Shetland and Finnmark. They show a regular increase with month and he considered that length at parturition was 480 cm. Ingebrigtsen (1929) described sei whales arriving at More in April, accompanied by calves, which he considered to have been born some one or two months earlier, these seem out of phase with the others, and he noted that in April females are caught with 10 to 40 cm fetuses.

From Icelandic catches Lockyer and Martin (1983) described the fetal lengths sampled from mid-June to the end of September. Late-term females leave the Icelandic feeding grounds early but not before September. At Iceland and at many of the other land stations, the belly of the whale is slit open at sea to avoid the meat spoiling. Consequently, fetal records are not common but from over the period 1948 to 1981 enough material was found to describe a linear growth in length with time as, $L = 151.1 + 1.80\ t$, where L is the length in cm and t the days from June 17.

A similar scarcity of data occurs from eastern Canada but Mitchell and Kozicki (1974 ms) gave the lengths of 13 fetuses from June to October. Those in June are about 140 cm and those in September and October about 320 cm; these are similar to the values calculated from the above Icelandic equation.

NEONATAL GROWTH

Length

The previous section shows that average length at birth is about 4.5 m. Table 7.1 gives the frequency distributions of length in the catch of Southern Hemisphere sei whales from 1961/62 to 1974/75, and the average size in the catch is about 15 m with a maximum of 19.5 m. Substantial numbers of sei whales have been aged and one might then suppose it would be a straightforward process to describe growth in length with time. Unfortunately this is not so, due to three aspects. Firstly, collection and preparation of ear plugs from the youngest individuals is difficult and (Ohsumi and Masaki 1978; Mizroch and Breiwick 1984), the youngest individuals may be under-represented in the sample. Problems of ageing the oldest individuals do not much matter since growth after about 25 years is very little. Secondly, IWC whaling regulations prohibit the taking of sei whales under 40 ft (12.2 m) by pelagic operations and under 35 ft (10.7 m) by coastal operations, and so few samples from whales under 40 ft were obtained (Table 7.1) and thirdly, and of most importance the whaling gunners will select the largest individuals when it is possible. These considerations mean that only the largest

Table 7.1: Length (ft) frequency distribution of pelagic catches of Southern Hemisphere sei whales 1961/62–1964/65 and 1971/72–1974/75. 61 = season 1961/62 etc. Upper: females lower: males.

Season	61	62	63	64	71	72	73	74
≤ 39	2	–	1	9	7	8	8	5
40	1	21	19	67	76	47	30	25
41	7	15	23	78	90	61	43	53
42	4	18	22	107	92	58	52	41
43	24	29	31	142	150	86	67	55
44	26	45	30	170	146	91	67	64
45	31	59	58	221	161	78	76	81
46	64	62	78	271	153	108	110	95
47	93	129	149	357	199	148	160	123
48	158	191	255	478	201	172	183	149
49	253	268	392	703	243	171	208	178
50	370	370	543	1035	289	201	242	230
51	462	429	597	1190	248	182	248	202
52	505	523	797	1509	278	191	270	201
53	383	339	486	1040	106	99	124	80
54	283	290	315	725	62	48	63	34
55	170	157	173	397	19	16	17	13
56	83	100	62	150	4	12	7	6
57	26	36	17	48	5	3	1	2
58	15	4	5	9	1	–	–	–
≥ 59	6	–	1	4	–	–	–	–
≤ 39	–	–	–	4	7	6	2	5
40	7	14	25	69	95	41	29	39
41	8	19	31	146	131	75	69	73
42	10	25	33	119	130	73	50	48
43	31	48	68	217	195	118	111	75
44	41	57	66	304	194	131	103	90
45	61	122	134	476	232	149	171	123
46	139	208	333	773	340	216	297	215
47	201	290	517	1335	408	252	410	345
48	278	391	773	1888	404	326	459	409
49	299	390	854	1998	379	303	319	373
50	298	354	696	1927	234	207	232	264
51	196	238	345	1055	117	114	114	117
52	105	142	244	600	49	53	34	38
53	38	59	58	149	4	16	8	7
54	27	29	17	47	5	1	3	1
55	8	9	5	17	–	1	1	–
≥ 56	3	2	1	3	1	–	–	–

individuals of the young age classes are represented in the catch and give a distorted picture of growth in length.

Notwithstanding the above, growth curves are often derived from catch data and are most often described by a von Bertalanffy (1938) relationship of the form,

$$l(t) = L_{\infty} (1.0 - \exp(- k(t - t_o))),$$

where $l(t)$ is the length at time t, L_{∞} is the maximum average length reached by the population, k is the annual growth rate, t_o a constant giving a hypothetical time of zero size and t the age in years. Although there are theoretical reasons to accept this form of relationship for growth in the population it is only justified in representing size in the catch because of its flexible and asymtotic character. Best (1982) for instance, fitted an asymtotic curve by eye for minke whales. For fin whales Laws (1961) showed that growth in length was seasonal with a relatively rapid increase during the feeding season. The annual growth curves do not attempt to describe this feature which in any case has not been investigated for sei whales.

Parameters for the above curve are estimated by a range of methods. If only three sets of age and length are used then the three parameters are uniquely defined. If there are more data then statistical fits are obtained and these are usually done by plots of $l(t+1)$ against $l(t)$, a Ford-Walford relationship (Ricker 1975), or $l(t+1) - l(t)$ against $l(t)$ (Gulland, 1983); with modern computers full non-linear minimisations are easily accomplished. An assessment method which involved the estimation of growth parameters was applied to sei whales by Jones (1978), but its use is limited and not discussed further. As growth and size possibly vary with location the more detailed descriptions are given by ocean regions.

Southern Hemisphere. The difference between the apparent growth curve derived from the catch data and one adjusted so as to be more representative of the population is illustrated in Figure 7.2 from Ohsumi and Masaki (1978). Age-length keys were constructed from the raw data of some 21,018 sei whales which were caught and aged over the period 1959/60 to 1975/76, from the Southern Hemisphere. This key had irregularities in the distributions of length at age and age at length, and so the raw key was smoothed, to what they termed a refined key,

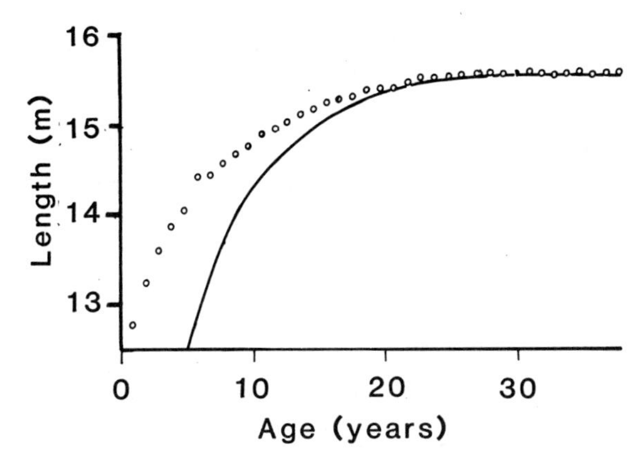

Figure 7.2: Comparison of the sample mean lengths at age from Antarctic catches (o), and from an assumed growth curve (line).

before the key was used as a description of length at age. In addition to the problem of selection by whalers Table 7.2 (from Mizroch and Breiwick 1984) shows that there is a strong trend in the proportion of sei whales aged with length and this will tend to further increase the apparent growth rate over and above that due to selection. Figure 7.2 shows the example given by Ohsumi and Masaki for female sei whales from Antarctic Area IV, wherein the mean length at age from the catch data is much higher than that predicted using a fixed length at birth of 4.5 m and an estimated length at weaning (about six months) of 9.0 m. Table 7.3 gives the parameters of the von Bertalanffy curve for the Antarctic catch data from Ohsumi and Masaki and it can be judged that L_∞ may be representative of the population but that the values of k and t_0 will only assist in indicating the length in that proportion of the catch that was aged.

In order to overcome the problems discussed above, Ohsumi and Masaki devised their "ideal" age-length key. A necessary first step was to construct a growth curve for the population. This was done assuming 4.5 m at birth, 9 m at six months and calculating the length at physical maturity to be given by the average of those whales 25 years and older. In addition Masaki (1978) also determined the relationship between age and length at sexual maturity (although it should be noted he found substantial changes in age at sexual maturity with

Table 7.2: Numbers sampled and percentage of those aged of the catch of Antarctic sei whales, by length (feet) group from 1966 to 1978, from Japanese material.

Length	Males	% Aged	Females	% Aged
36	1	0.0	1	0.0
37	2	0.0	12	16.7
38	23	0.0	13	7.7
39	23	0.0	23	8.7
40	758	10.7	597	6.2
41	961	10.2	791	10.2
42	907	12.6	779	11.4
43	1356	15.6	1022	11.7
44	1593	21.0	1079	13.0
45	2088	29.5	1155	16.6
46	3208	34.6	1342	19.6
47	4294	40.8	1605	24.6
48	4924	45.7	1968	27.7
49	4406	47.2	2339	37.7
50	3141	50.0	2751	41.3
51	1409	49.7	2614	46.6
52	655	46.1	2822	48.9
53	115	56.5	1292	52.4
54	26	69.2	609	55.0
55	6	50.0	291	51.9
56	0		111	49.5
57	0		26	57.7
58	0		8	62.5

Table 7.3: Parameters of the von Bertalanffy growth equation calculated from smoothed age and length keys for the Antarctic, 1959/60-1975/76, by Area.
Area II: 0-60°W, III: 0-70°E, IV:70-130°E,
V: 130°E-170°W, VI: 120-170°W.

Area	males			females		
	L_∞	k	t_o	L_∞	k	t_o
II	49.12	0.131	-14.09	51.82	0.136	-9.53
III	48.10	-	-	50.86	-	-
IV	48.28	0.266	-3.43	50.99	-	-8.11
V+VI	48.69	-	-	50.95	0.170	-

Table 7.4: Parameters of the von Bertalannfy growth equation calculated from four pairs of ages and lengths from the Antarctic, by Area.

Area	males			females		
	L_∞	k	t_o	L_∞	k	t_o
I	47.36	0.160	-5.48	50.45	0.147	-5.53
II	49.03	0.169	-4.68	51.70	0.135	-6.78
III	47.86	0.170	-5.20	50.48	0.171	-4.76
IV	48.41	0.167	-5.20	50.76	0.170	-4.52
V	49.08	0.163	-4.98	51.22	0.162	-6.54
VI	48.50	0.152	-5.46	51.22	0.172	-4.40

time). These four sets were used to get estimates of the growth parameters by whaling Area and they are given in Table 7.4. These growth relationships then had combined with them a variance that increased from 1.0 (ft squared) at age 1 to 6.25 at age 25, and an age structure determined from a mortality rate of $M = 0.08\ y^{-1}$. The resulting "ideal" age-length relationships are illustrated by Ohsumi and Masaki. It must be emphasised that these distributions are derived using several specific assumptions and although they could generally reflect the age and length structure in a natural undisturbed population they must not be used as conventional age-length keys to obtain age distributions from length distributions.

Growth rates of sei whales were described by Christensen (1968) but ageing was done using baleen plates, and as described in Chapter 6 this is only useful for young whales. Growth rates based on samples aged with ear plugs, from both the Antarctic and South Africa, were calculated and reviewed by Lockyer (1974, 1977b, 1981a). Lockyer (1974) fitted von Bertalanffy curves to length at age data for males and females sampled at Durban over the years 1960 to 1965. Only a small proportion of the catch was sampled, but most plugs from these samples were read, and the data are probably representative of the catch. The minimum size limit of 35 ft (10.7 m) meant that more small whales were sampled, but Lockyer (1977b) noted that maximum lengths were greater at South Georgia than at Durban and immature animals are usually not caught in the Antarctic because of segregation. She considered the relative scarcity of data of one year old animals at Durban to be due to their protection as undersized animals. The relationships she found have the following parameter values.

males L_∞ : 14.8 k : 0.145 t_o : -9.36 (n = 152)
females L_∞ : 15.3 k : 0.134 t_o : -10.00 (n = 92)

Lockyer (1977b) calculated the sizes of fin whales at particular ages relative to fetal length. For sei whales the ratios of lengths at weaning (which she took to be 8.0 m in sei whales), sexual and physical maturity were very similar to that of fin whales. If this relationship between fin and sei whales also holds for one year olds, then it predicted that a one year old male would be 10.4 m and a female 10.7 m. As expected the fitted equation predicts a somewhat higher value but nevertheless it can be recognised that growth over the first five or ten years is very rapid.

Growth rates were calculated from sei whales caught at Donkergat in 1962 and 1963 by Best and Lockyer (1977 ms). Between 63 and 76% of the catch was sampled for ear plugs and of these 63 to 78% proved to be readable. However, no trends of readability with length were found, the sampled lengths were similar to those in the catch, and the data were considered to be representative of the catch. They believed that their relationship of mean length with age, although overestimating size for the first two years, was a close approximation to the population growth curve. Growth curves for males and females were drawn by eye and show that average maximum adult size was about 14.8 m for males and 15.8 m for females. The average maximum size from South Africa does not differ much from those reported by Ohsumi and Masaki (1978) but their values of the growth rate, k, (Table 7.4) are a little higher than those calculated for Durban.

Mizroch and Breiwich (1984) studied the variability in age and length of sei whales that were caught by the Antarctic Japanese fleet over the years 1966 to 1976. They found that the pattern of the coefficient of variability of mean age at length was similar to that of fin whales, and decreased steadily from 0.09 at 12 m to a minimum of 0.01 at 15 m before a steep rise. A similar U-shaped distribution is seen in the coefficient of variation of length at age. They ascribe the initial declines to aspects of poor readability but the later increases to a real natural variability.

Matthews (1938) recorded data on lengths from sei whales from South Georgia and South Africa but, because of problems in ageing whales at that time, only measurements of length distributions are of interest. However, he did record the relative lengths of different body proportions and reviewed their growth rates relative to body length. His data showed that the relationship of various parts of the body (y) could be related to the total length (x) with the equation, $y = ax^b$. Values of b over unity indicate a relative growth rate increasing with time and this is found in only three of the 19 measurements, indicating a slight relative elongation of the anterior part of the whale with age; however the magnitudes are sufficiently small that the basic shape does not change much over time. Aspects of growth of the skull of sei whales are discussed also by Lockyer (1974).

North Pacific. Masaki (1976a) provided growth curves for the North Pacific from samples collected from 1952 to 1972. The catches before 1967 were taken from the western part of the North Pacific, north of 50°N, and from the Gulf of Alaska. From 1967 catches were also taken from 40-50°N. At the time of writing Masaki considered that methods of age determination were unreliable, with less ear plugs readable from immature sei whales, and he considered that for the younger ages the number of laminae might be underestimated. This last feature would tend to overestimate growth rates. His data are illustrated in his figure 24 and show the apparent high growth rates due to selection and minimum size limits. Maximum average body lengths were about 15.0 m for females and 14.0 m for males.

Masaki (1976a) also investigated the relationship between physical maturity and length, where physical maturity was determined by the fusion of the epiphyses to the centra in the thoracic vertebrae. Samples were collected between 1969 and 1973 and were of 97 males and 92 females. The relationship between the proportion of the samples fused at length is shown in Figure 7.3. A wide scatter is seen but 50% of the sample had reached physical maturity by 14.3 m for males and 15.2 m for females. These values are higher than the ones calculated from the asymptote of the size at age curves, even considering the variability found in Figure 7.3. If anything, it would have been anticipated that inspections for epiphyseal fusion would have overestimated the proportion fused at any length. There may be two explanations for this. The first is that

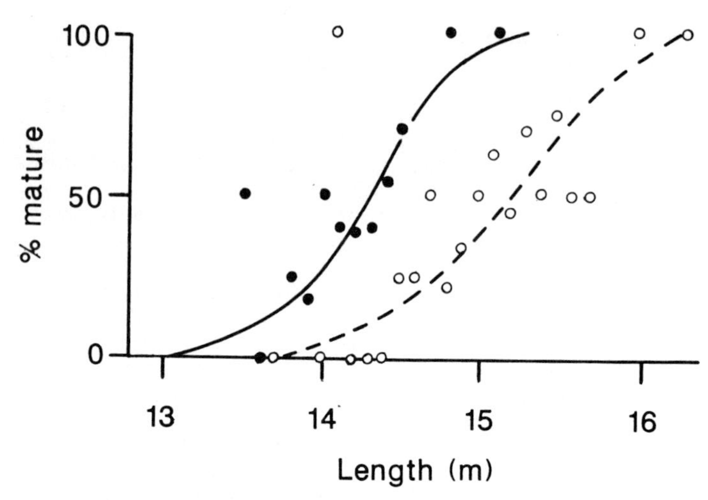

Figure 7.3: Proportion of North Pacific sei whales physically mature by length (m), 1969-1973. Dots: males, Circles: females.

the above values of L_∞ (15.0 and 14.0 m) were from data which Masaki (1976a p29) described as, "The mean body length shown represents the figure of the mode of body length composition to each age class". Presumably these values represent the average maximum modal length and if the length frequency distribution at age is positively skewed the mode will be less than the mean; however, Ohsumi and Masaki (1978) argued that it was reasonable to consider that body lengths at each age class are normally distributed. It would be useful to re-examine these distributions in light of the second possible explanation which is that the largest animals were not well represented in the age-determined samples. This could have arisen by the largest sei whales having been selected by the whalers before the start of sampling (Chapter 10 shows that large numbers were taken between 1910 and 1951 by land stations in the North Pacific) or else collection and readability of ear plugs was much less satisfactory for large whales at a given age.

Information was also given by Masaki (1976a) on the size of a very young whale. Under a scientific permit a cow and calf were caught at 44°N 175°E on 2 June. The cow's mammary glands contained some milk, but a limited amount of milk and some copepods in the calf's stomach indicated that it had started to feed. Masaki considered that the period of lactation was seven months and weaning would ease when the calf was 9.0 m.

North Atlantic. The lengths at age from catches from Nova Scotia are given by Mitchell and Kozicki (1974 ms). For reasons that they could not specifically identify, few females over 50 years were caught whereas there were numerous males. Average maximum lengths are about 13.7 m for males and 14.3 for females. Several small sei whales were taken including one of 10 m with eight growth layers. No growth rates were calculated.

From Iceland, growth has been described by Lockyer (1978a) and Lockyer and Martin (1983). As described in relation to mortality rates, the later study showed their samples to be representative of the catch, and collection and readability was not dependent upon length. Their data, from the seasons 1967 to 1982, show average maximum lengths of about 13.6 m for males and 14.5 m for females.

All areas. The growth rates based on catch data do not usually represent growth in the population, but the average lengths from the oldest ages should be comparable from all locations. From the Antarctic these values are about 14.4-15.0 m for males and 15.4-15.8 m for females. Comparable sizes were found at Durban and Donkergat. However, from the North Pacific the lengths are about 14.0 m for males and 15.0 m for females, but probably based on modes rather than means. The median length at physical maturity in the North Pacific is also less than the average length of the older whales in the Southern Hemisphere. In the North Atlantic the average maximum lengths are 13.6-13.7 m for males and 14.3-14.5 m for females. As noted by Matthews (1938) and Tomilin (1967) this indicates that the sei whale in the Southern Hemisphere grows to a larger size than in the Northern Hemisphere.

WEIGHT

Weighing balaenopterid whales, of which blue whales grow to 100 t, is not easy and is accomplished by weighing bits at a time. In this process there tends to be a loss of body fluid. Weights of sei whales from the North Pacific were measured by Omura

(1950b). The whales taken from Kamaishi, in Sanriku, are taken to be sei whales whereas those from the Bonin Islands might have been Bryde's whales. Sixteen sei whales were measured and give a relationship of, $W = 0.0242\ L^{2.43}$, where W is the weight in metric tonnes and L is the length in metres. For a 15.2 m whale the proportion of the weight coming from the different body parts is meat-34%, blubber-33%, bones-9%, internal organs-7% and others-17%. However, because of the pattern of annual feeding, the blubber thickness and oil yield vary with season (Lockyer 1981a). For humpback whales Ash (1953) showed that in summer yields increased by about 3% per week. Over the season the increase may be as much as 50% (Lockyer 1972b). Recent studies of sei whales from Iceland (Lockyer et al 1985) have described the seasonal increments in weight in relation to different tissues and sexual status.

Lockyer (1976) used Omura's data but corrected them for fluid loss. Addition weights were provided by Tomilin (1967), Bjarnason and Lingaas (1954) and from records of the Whales Research Institute, Tokyo and the Union Whaling Co. of Durban. The following relationship was obtained, $W = 0.0258\ L^{2.43}$. This can be used to give weight at age from the growth rate equations and a 15.5 m sei whale would be predicted to have a weight of 20 t. However Lockyer (1977b) has said that weight may increase after maximum body length has been achieved and that the maximum average body weight for Southern Hemisphere sei whales is likely to be nearer 20.5 t for males and 22 t for females. The above relationships provide a good approximation to weight at age but it is based on only 16 whales and seasonal variations are not considered. In addition the equation has been solved for a logarithmic transformation and consequently the re-transformed estimates will be slightly negatively biassed.

In a further recent study Lockyer and Waters (1986) reported the results of weighing eight fin and three sei whales at Iceland. A sei whale fetus was also weighed. They found that sei whales from the North Atlantic were substantially heavier than North Pacific sei whales at the same length. This is associated with a greater girth. In fact they found that a single relationship could be used to predict the weights of fin and sei whales, from all locations, if girth was included. They proposed the formula,

$$W = 0.0469\ G^{1.23}\ L^{1.45},$$

where W is the total weight in tonnes, L the length in metres and G is the girth measured at 55-57% along the body, from the snout.

TEMPORAL AND SPATIAL VARIATIONS

In the previous sections growth with age was described and differences between ocean areas were found. However within ocean areas sizes of whales in the catch have varied spatially and temporally. As exploitation occurs on a stock the length at age distribution may remain unaltered to give the same mean length at age but the numbers at age get less and result in a lower mean age in the population and hence a lower mean length. Cooke (1984) modelled the effect of relative depletion of a stock on the mean length in the catch for a hypothetical fin whale population. The model generated mean lengths in the catch for stocks reduced to between 50 and 0% of initial numbers over a twenty year period. He demonstrated that, over the twenty years, only the stock being fished to extinction would show reductions of over 10% in the mean length in the catch. This is due to the very rapid growth to near asymtotic lengths. His conclusion was that trends in mean length with time would be so small that, unless the stock was being fished to near extinction, one could not use such data to estimate exploitation rates. Nevertheless, substantial changes in mean length with time have been detected but these trends have been affected by changes in fishing locations.

The difference in size of sei whales north and south of the Antarctic Convergence was described by Doi, Ohsumi and Nemoto (1967) from Japanese pelagic catches, and from all pelagic catches by Horwood (1986). An example of the length frequency distribution north and south of the Convergence is given in Table 7.5 for catches in 1968/69. Figure 7.4 shows the distribution of average lengths of females in the Antarctic catches, by latitude, from 1968/69 onwards and the increase of size with latitude can be seen. For sei whales caught in and after 1968/69, following a period of intensive exploitation, the difference of the mean lengths north and south of the Convergence ranged from 0.15-1.16 m for males and 0.27-1.19 m for females. Consequently latitudinal position of the Antarctic catches will affect the mean lengths. Horwood also found that between January to March, at a given latitude and longitudinal sector, there was no difference in mean

Table 7.5: Length (ft) frequency distribution of females caught from the 1968/69 season and after in January between 60-90°E. N: north of the Antarctic Convergence, S: south of the Convergence. © Crown 1986

	<40	40	41	42	43	44	45	46	47	48	49	50	51	52	53	54	55	>55 ft
N:	2	30	41	47	43	65	52	80	85	84	85	91	84	76	25	12	6	2
S:	0	0	0	1	2	2	3	4	2	7	5	11	13	14	7	0	3	0

Table 7.6: Percentage of the Antarctic pelagic catch of sei whales by Area and Series. (100% is made up with catches from Series C). © Crown 1978

Area	I			II			III		
Series	D	A	B	D	A	B	D	A	B
1961/62	0	13	86	22	46	32	39	51	10
1962/63	0	23	77	7	40	53	83	17	0
1963/64	0	4	96	11	89	1	65	34	1
1964/65	13	87	0	45	55	0	74	25	1
1965/66	0	100	0	93	6	1	93	4	3
1966/67	66	33	0	93	7	0	99	0	1
1967/68	0	100	0	0	100	0	5	95	0
1968/69	0	100	0	100	0	0	97	2	1
1969/70	-	-	-	100	0	0	100	0	0
1970/71	-	-	-	91	9	0	100	0	0
1971/72	0	27	73	82	18	0	98	2	0
1972/73	93	7	0	97	3	0	100	0	0
1973/74	7	29	64	95	5	0	100	0	0
1974/75	96	1	3	100	0	0	100	0	0

	IV			V			VI		
	D	A	B	D	A	B	D	A	B
1961/62	12	85	3	0	0	100	0	0	94
1962/63	28	20	53	0	0	100	0	0	100
1963/64	64	12	24	22	26	52	-	-	-
1964/65	87	9	4	23	41	36	-	-	-
1965/66	33	0	67	13	24	63	18	82	0
1966/67	18	76	6	1	0	99	0	0	100
1967/68	43	54	3	63	3	34	7	61	31
1968/69	73	26	1	70	14	16	1	49	50
1969/70	94	6	0	3	18	80	0	9	91
1970/71	97	3	0	55	0	45	0	9	91
1971/72	99	1	0	100	0	0	-	-	-
1972/73	95	5	0	96	1	3	0	43	57
1973/74	100	0	0	100	0	0	2	86	12
1974/75	96	4	0	99	0	1	27	72	1

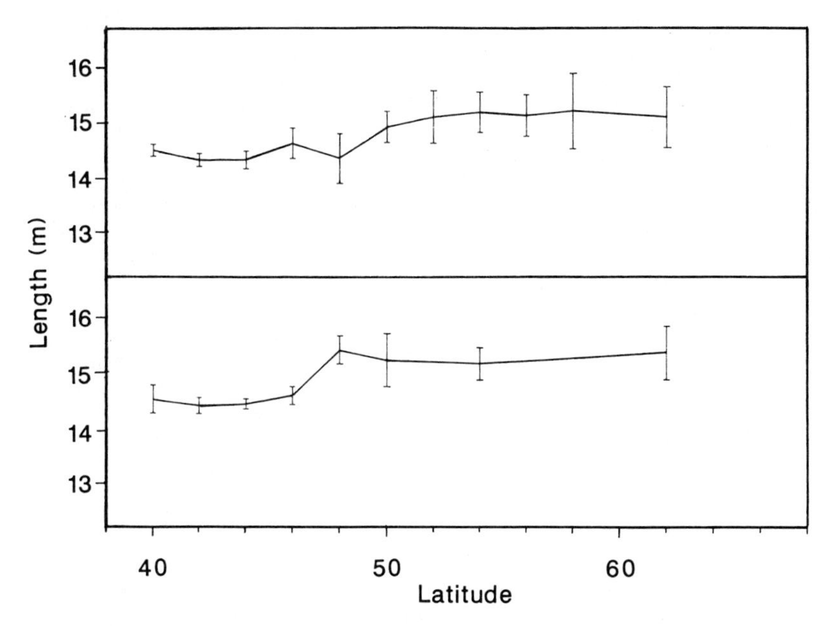

Figure 7.4: Mean length (m, ± 2 s.e.) by latitude of female sei whales caught from 1968/69 onwards. 40 represents those caught between 40 and 42°S. Upper: January, 60-90°E, Lower: February, 30-60°E.
© Crown 1986

lengths of males, but females taken in January were usually longer than those caught later in the season. This probably reflects the earlier arrival of mature females.

The average size of sei whales caught by the Antarctic pelagic fleets over the period 1956/57 to 1974/75 and the numbers in the catch are given in Figure 7.5 after Horwood (1978). If coastal catches are included, the 1964/65 catch exceeded 22000. A steady decline in the mean length is found from 1956/57 to 1964/65. A large reduction of about 0.5 m is seen in 1965/66 and 1966/67 after which a further slow decline is found. Subsequently, it will be demonstrated that the Southern Hemisphere sei whale was relatively heavily depleted and that the two periods of gentle negative trend are quite consistent with Cooke's (1984) findings. The large reduction is also only about 6% and could also be consistent with Cooke's results. Over this period though there was a change in the distribution of whaling effort. The basic BIWS data are tabulated by Omura (1973) but Doi, Ohsumi and Nemoto (1967) show the latitudinal distribution of the catch from 1959/60 to 1965/66 and this is extended in Table 7.6. From 1959/60 to 1962/63 most of the catch was taken between 50-70°S, but by 1965/66 almost all the catch was taken between 40-50°S; both north and south of the Antarctic Convergence. Consequently at least some of the large changes were probably due to distributional features. In order to overcome this confounding of effects, distributional and fishing, one can inspect trends in mean length for the different whaling areas in the IWC zones; series D(40-50°S), A(50-60°S), B(60-70°S), C(70-80°S). These are given in Table 7.7, and Figure 7.6 shows that from 1960/61 to 1974/75 there was a reduction in mean length of about 0.6 m in males and 0.9 m in females. This is probably due to exploitation. In addition Woolner and Horwood (1980) analysed the trends in mean length by Series and found that significant differences in mean lengths with time existed between different whaling Areas.

Variations in mean length have also been described by Martin (1983) from Icelandic catches. Within a season he found that mean lengths of females declined from mid August to the end of September and this was due to the seasonal distribution of immature females. A decline in mean lengths of males also occurred at the end of the season. Within the whaling ground no segregation by size was found.

Notwithstanding the above changes in mean size due to distributional characteristics of the sei whales and the whaling fleet, and of declines due to exploitation, there is indirect evidence to believe that there has been temporal changes in growth rate. For fin, sei and minke whales of the Southern Hemisphere studies of the position of the transition layer of the ear plug (Lockyer 1972a) in relation to year class (date of birth) have indicated a decline in the age of maturity since about 1910. The formation of the transition layer is believed to be directly related to the onset of sexual maturity in both males and females. Although, as we shall see later, this interpretation is not without argument, there are strong reasons to believe such a decline has occurred in sei whales of the Southern Hemisphere (Lockyer 1974, 1979) and of the North Atlantic (Lockyer and Martin 1983), and in fin whales of both localities (Lockyer 1972a, 1979, 1981b). However, Masaki (1978) has shown that in the Southern Hemisphere the length at sexual matu-

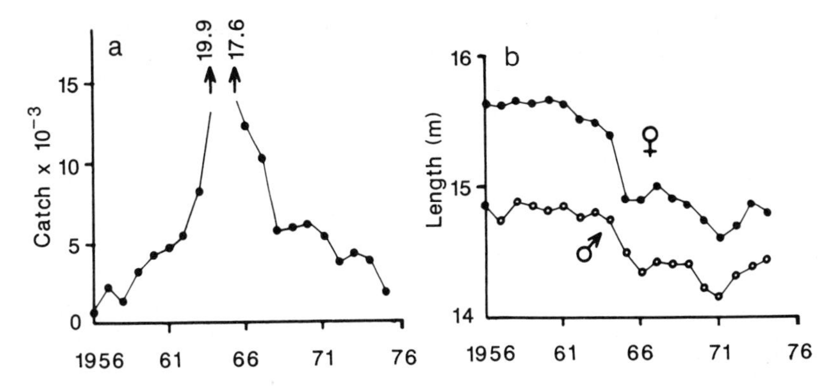

Figure 7.5: Numbers of sei whales caught by pelagic fleets and the decline in the mean length in the catch from 1956/57 in the Antarctic. © Crown 1978

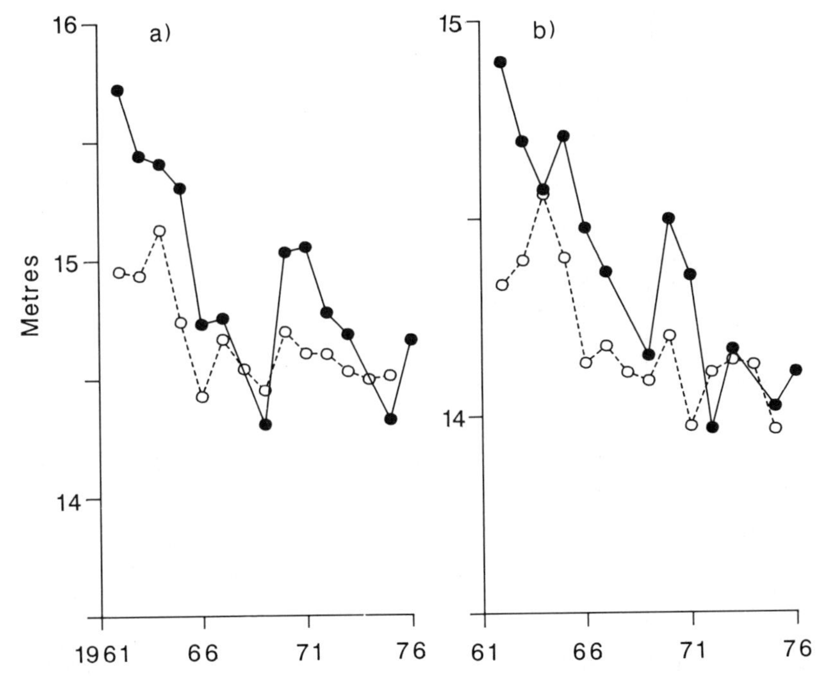

Figure 7.6: Declines in the mean lengths in the Antarctic catch with season and Area from Series D (40-50°S). a: males, b: females. Dots: Area II, Triangles: Area III, Circles: Area IV.

Table 7.7: Average length at capture by Area and Series. D: 40-50°S, A: 50-60°S, B: 60-70°S. Males first.

	D	D	A	A	B	B	Total	Total
				AREA I				
1961/62	-	-	48.6	51.1	49.3	51.8	49.2	51.7
1962/63	-	-	48.8	51.3	48.1	51.1	48.3	51.1
1963/64	-	-	-	52.3	-	-	-	52.3
1964/65	-	-	48.4	49.3	-	-	48.4	49.8
1965/66	-	-	48.9	50.9	-	-	48.9	50.9
1966/67	-	-	-	-	-	-	-	-
1967/68	-	-	48.9	50.8	-	-	48.9	50.8
1968/69	-	-	-	-	50.3	50.6	50.3	50.6
1969/70	47.6	49.4	-	-	-	-	47.6	49.4
1970/71	-	-	-	-	-	-	-	-
1971/72	-	-	-	-	-	-	47.5	51.5
1972/73	48.0	49.8	49.2	-	-	-	48.1	49.9
1973/74	47.3	48.9	47.5	49.3	48.0	50.5	47.8	50.0
1974/75	47.6	49.5	-	-	48.6	50.9	47.7	49.6
				AREA II				
1961/62	48.9	51.6	48.7	51.3	48.9	51.9	48.8	51.6
1962/63	48.2	50.7	48.3	51.1	49.8	52.2	49.0	51.7
1963/64	47.8	50.5	48.7	51.3	49.1	51.6	48.6	51.2
1964/65	48.3	50.2	48.3	50.8	-	-	48.3	50.5
1965/66	47.5	48.3	49.0	51.4	48.8	51.3	47.6	48.6
1966/67	47.1	48.4	48.4	50.9	-	-	47.2	48.6
1967/68	-	-	48.9	50.8	-	-	48.9	50.8
1968/69	46.4	46.9	-	-	-	-	46.4	46.9
1969/70	47.6	49.3	-	-	-	-	47.6	49.3
1970/71	47.1	49.4	47.3	49.9	-	-	47.1	49.5
1971/72	45.8	48.4	47.9	50.1	-	-	46.0	49.1
1972/73	46.5	48.2	-	-	-	-	46.8	48.1
1973/74	-	-	-	-	-	-	46.2	49.4
1974/75	-	-	-	-	-	-	46.8	47.4
				AREA III				
1961/62	47.0	49.1	48.1	51.1	-	50.2	47.6	50.3
1962/63	47.2	49.0	47.6	49.8	-	-	47.3	49.1
1963/64	47.8	49.6	48.4	50.5	48.5	-	48.0	49.9
1964/65	47.2	48.4	48.2	49.4	-	-	47.5	48.6
1965/66	46.4	47.3	48.6	50.9	49.5	51.4	46.5	47.7
1966/67	46.5	48.1	49.9	-	48.6	51.0	46.5	48.2
1967/68	46.3	47.7	47.4	49.3	-	-	46.3	47.8
1968/69	46.2	47.4	48.3	49.9	-	51.1	46.3	47.6
1969/70	46.6	48.2	-	-	-	-	46.6	48.2
1970/71	45.8	47.9	-	-	-	-	45.8	47.9
1971/72	46.3	47.9	46.5	-	-	-	46.3	48.1
1972/73	46.3	47.7	-	-	-	-	46.3	47.4
1973/74	46.4	47.6	-	-	-	-	46.3	48.3
1974/75	45.6	47.6	-	-	-	-	46.0	47.5

© Crown 1978

Table 7.7 (cont'd)

	D	D	A	A	B	B	Total	Total
				AREA IV				
1961/62	47.5	49.0	47.8	50.7	-	50.0	47.8	50.5
1962/63	47.4	49.7	48.2	49.8	49.2	51.5	48.5	50.7
1963/64	47.8	50.1	47.4	50.1	48.0	49.1	47.8	49.9
1964/65	48.4	50.3	48.1	50.5	46.9	50.8	48.3	50.3
1965/66	46.4	48.8	-	-	48.2	50.5	47.6	50.0
1966/67	47.7	48.2	47.7	50.1	48.5	50.9	47.8	49.8
1967/68	47.6	49.6	47.4	49.8	47.6	50.1	47.5	49.7
1968/69	47.1	48.8	46.9	49.7	-	-	47.1	49.0
1969/70	47.1	48.0	48.1	49.2	-	-	47.2	48.1
1970/71	46.6	48.0	46.9	49.2	-	-	46.6	48.0
1971/72	46.4	47.7	47.4	48.2	-	-	46.4	47.8
1972/73	47.2	48.2	46.6	48.0	-	-	47.1	48.2
1973/74	46.6	47.6	-	-	-	-	46.6	47.6
1974/75	46.0	47.0	48.3	49.3	-	-	46.1	47.1
				AREA V				
1961/62	-	-	-	-	48.6	50.6	48.6	50.6
1962/63	-	-	-	-	50.3	52.6	50.3	52.6
1963/64	47.8	48.5	48.7	51.1	49.2	51.8	48.8	50.9
1964/65	49.3	50.7	48.7	50.9	48.5	51.1	48.8	50.9
1965/66	46.5	48.5	48.7	51.3	48.4	50.9	48.1	50.8
1966/67	-	46.4	-	-	47.9	50.8	47.8	50.6
1967/68	46.8	47.5	47.7	50.0	48.7	51.2	47.4	48.9
1968/69	47.4	48.7	48.3	50.8	48.7	51.2	47.7	49.4
1969/70	-	-	48.9	49.4	49.2	50.9	49.0	50.6
1970/71	46.6	47.8	-	-	47.1	50.0	46.8	48.9
1971/72	46.5	47.7	-	-	-	-	46.5	47.7
1972/73	46.1	47.4	-	-	47.7	49.1	46.2	47.5
1973/74	46.5	47.9	-	-	-	-	46.5	47.9
1974/75	47.9	48.3	-	-	-	-	47.9	48.4
				AREA VI				
1961/62	-	-	-	-	48.6	51.2	48.7	51.3
1962/63	-	-	-	-	48.6	50.9	48.6	50.9
1963/64	-	-	-	-	-	-	-	-
1964/65	-	-	-	-	-	-	-	-
1965/66	-	-	50.4	52.5	49.0	51.8	49.3	51.9
1966/67	-	-	-	-	48.6	51.1	48.6	51.1
1967/68	46.6	49.4	47.5	49.2	48.8	51.4	47.8	50.0
1968/69	-	-	48.2	50.9	48.7	51.5	48.4	51.2
1969/70	-	-	-	-	49.0	51.4	48.3	51.2
1970/71	-	-	-	49.3	48.1	49.8	48.2	49.8
1971/72	-	-	-	-	-	-	-	-
1972/73	-	-	48.6	50.0	47.8	49.3	48.2	49.5
1973/74	49.3	-	48.0	50.1	48.0	50.5	48.0	50.1
1974/75	48.0	49.4	47.7	49.2	-	-	47.8	49.3

rity, has not altered with time. This was based on direct examination of testes and ovaries of sei whales from samples collected from 1956/57 to 1975/76, and from earlier data recorded by Matthews (1938), Mackintosh (1965) and Gambell (1968). It is then concluded that growth rates have increased over the period. Using the von Bertalanffy equation Lockyer (1978b) calculated that some of the estimated declines in age at maturity would have required an increase in the growth parameter (k) of between 50-100%. One might add that if this has occurred then the observed distributions of length at age will be in a far from stable distribution.

SUMMARY

Fetal growth of sei whales has been described and whereas Laws (1959) found an exponential increase with time, the subsequent greater amount of data available from both the North Pacific and Southern Hemisphere indicate that growth in length is better described by two linear relationships. Growth is slow at first, for the first three months, at which time the fetus is about 50 cm. From then to birth, 11 to 12 months later, length has increased to about 450 cm. Growth in both hemispheres is nearly identical. A length-weight relationship has been obtained.

Estimates of post-natal growth are affected by biassed sampling of the younger ages because of selection and minimum size limits. Various relationships have been obtained to describe growth in length with age or, rather, length in the catch with age. In some localities readibility and collection of ear plugs has been found to be a function of length so not all growth curves are even representative of the catch. Maximum average lengths are about 14 m for males and 15 m for females but the values are greater in the Southern than Northern Hemisphere. The largest, reliably recorded length is of a female of 19.5 m. Attempts have been made to describe the length distribution in the population, rather than the catch, but these are subject to various assumptions, and no good descriptions of growth over the first few years are available. The relationship between length and weight of adults has been described and adult sei whales in the Antarctic are likely to attain weights of 20-22 t with rare individuals reaching 35-40 t.

Growth

Differences in mean lengths in catches vary greatly with location, year of capture and, to a lesser degree, month within a season. In the Antarctic, sei whales are larger south of the Antarctic Convergence than to the north and the intensive exploitation during the 1960s reduced the mean length in the catch. Declines in the mean age of sexual maturity indicate that since the turn of the century growth rates of sei whales increased.

Chapter Eight

NATURAL MORTALITY

INTRODUCTION

In whales we are used to thinking of two sources of mortality, natural mortality and fishing mortality due to exploitation. Within a species natural mortality may vary with sex within a species and is likely to vary with age. This is especially true of the younger ages which are likely to be subject to greater hazards. The pattern of mortality in mammals has been reviewed by Caughley (1966) who found a typical "U" shaped relation with a greater mortality rate for both younger and older mammals; however this was not true in all cases. Natural causes of death in large cetaceans are largely unexplained but various instances of death and illness have been recorded in a variety of whales. Still-born calves have been found in the breeding lagoons (Swartz and Jones 1983). Killer whales have been observed to kill right, humpback, gray, minke and blue whales (Shaler 1873, Baldridge 1972, Handcock 1965, Tarpy 1979) and also to have attacked sei whales (Rice 1968, Gaskin 1982). Aggressive interactions between large whales and swordfishes have been frequently reported (Brown 1960, Major 1979) and a broken off sword was found in the musculature of a sei whale caught in the North Pacific (Machida 1970). Broken bones and deformed jaws are found in apparently healthy whales and even severe reduction of the baleen through disease has not seemed to affect sei whales (Morch 1911, Rice 1961). The large whales suffer from a range of endo- and ectoparasites, briefly reviewed by Harrison (1979), but from a study of the pathology of several thousand animals Cockrill (1960) concluded that whales were probably the healthiest of living creatures. Nevertheless endoparasites do cause severe lesions, and may be the most important source of natural mortality, but the degree of response to a particular parasite varies between balaenopterids (Ulys and Best 1966, Dailey 1985). In recent years small numbers of whales have become entangled and drowned in floating or fixed fishing gears, mainly in more coastal areas. Nevertheless, between 5 and 10% of adult sei whales die annually through causes other than direct exploitation. Fishing mortality is affected by two main components. Firstly, there is the magnitude which is dependent upon the relative sizes of catch and stock, and the selection pattern which determines how fishing affects particular ages. The age specific selection pattern has been affected by the preference of the whalers, they often avoided shooting at small whales because of economic constraints, and also by the availability of particular age components of the stock.

In the following, techniques are presented that allow estimates to be found of average or age specific mortality rates. Usually this is the total mortality but if the data are from the beginning of a fishery then the total mortality can be regarded as the natural mortality.

Consider some basic definitions of mortality and define:

$N_{i,t}$ as the numbers of whales age i in year t,

$C_{i,t}$ as the catch in numbers of age i in years t,

M_i as the age specific natural mortality rate,

f_tq_i as the age specific fishing rate where

f_t is the fishing effort in year t and

q_i is the age specific selectivity or catchability.

$Z_{i,t}$ is the total mortality rate, $M_i + f_tq_i$

S is the annual survival rate which is exp(-Z).

It is then usual to describe the within year change of numbers with time as, $dN_i/dt = -(M_i + f_tq_i)N_i$

and hence

$$N_{i+1,\ t+1} = N_{i,\ t} \exp -(M_i + f_tq_i)$$

and

$$C_{i,t} = (f_tq_i/Z_{i,t})\ (1 - \exp(-\ Z_{i,t}))N_{i,t}.$$

The above relationship is used throughout the following chapter with variations such as mortality contant with age.

For almost all uses of the estimates of natural mortality rate it is necessary that the rate is applicable to the stock as a whole and not just the component of the catch. Because of the existence of minimum allowable sizes at capture and of the preference for larger whales, the catch curves (the frequency distribution of numbers at age in the catch) show an increase in numbers with age for the first few years, but at some age it is assumed that the stock becomes fully recruited to the fishery, and is then sampled proportionate to the numbers at age; Figure 8.4 shows the catch curve from Southern Hemisphere sei whales where this feature can be seen. Estimates of mortality are made from after the age of full recruitment.

Nevertheless, the distribution of the population itself may well be such that the catch will not be representative of the population. From the Antarctic we have seen that smaller sei whales stay north of the Antarctic Convergence, and that on the migrations there is also segregation by age and sexual classes and in the Antarctic whales of a similar age tend to be caught together (Lockyer 1977a). From the northeast Atlantic minke whale fishery the larger animals are not thought to be available, perhaps due to a greater degree of cautious behaviour (Christensen and Rorvik 1981). De la Mare (1985a) reviewed many problems associated with obtaining estimates of natural mortality in whales and demonstrated that it was only too easy to obtain biassed results. Particular attention should be paid to exactly what parameter is needed, (e.g. age specific mortality or an average mortality) before any estimation. Additional problems arise if there is variability in ageing (Barlow 1984), and de la Mare (1985a) and Eberhardt (1985) stress pit-falls that can exist if age-dependency of mortality is neglected.

COMMON ESTIMATION TECHNIQUES

Chapman - Robson

The following assumptions are made, (i) that survival rate is contant over all recruited ages, (ii) that the catch at age is random and is representative of the population and (iii) that the age distribution is stationary. From these assumptions the catch at age will follow a geometric distribution, and Chapman and Robson (1960) and Robson and Chapman (1961) calculated a maximum likelihood estimator for the annual survival, $\hat{S}$. This is a minimum variance and unbiased estimator and gives,

$$\hat{S} = T/(n + T - 1),$$

where, when x is the age of first full recruitment,

$$T = C_{x+1} + 2C_{x+2} + 3C_{x+3} + \ldots,$$

and

$$n = C_x + C_{x+1} + C_{x+2} + \ldots .$$

This has a variance,

$$\mathrm{var}(\hat{S}) = \hat{S}(\hat{S} - (T - 1)/(n + T - 2))$$

The estimate, $\hat{S}$, can be transformed into an estimate of Z, which is not unbiased, of the form,

$$\hat{Z} = - \ln(\hat{S}).$$

Further modified relationships are provided by de la Mare (1985a) and Cooke and Beddington (1981) also provided an estimate based on maximum liklihood.

<u>Catch and effort</u>
Beverton and Holt (1957) considered the ratio of catches from two adjacent time periods, or more usually of catches from two adjacent years, of the same cohort, and developed a regression method of estimating the natural mortality, M, and the catchability, q, with known effort, f. For small mortality rates and unit time periods this model reduces to that of Paloheimo (1961) where,

$$E(y_i) = M + k\ (f_i + f_{i+1})/2$$

and

$$y_i = \ln(C_i/f_i) - \ln(C_{i+1}/f_{i+1})$$

Variances are given by Seber (1973), and Allen and Chapman (1977) give various methods for calculating total mortality if data of several seasons and age classes are available.

Heincke

The estimator of survival developed by Heincke (1913) was intended to overcome the problem that usually few older animals existed and that the ratios of their numbers at age were very variable. Heincke's estimator is,

$$\hat{S} = \sum_{i=x+1} C_i / \sum_{i=x} C_i$$

where x is at least the age of first full recruitment and S is then the average survival rate over all ages from age x. It provides an unbiased average per capita survival rate over the sampled age classes even if there are changes in survival rate with age. A feature strongly in the favour of Heincke estimates is that the older age classes are grouped and the technique is therefore robust to higher variability in the ageing of old individuals. The variance of the estimate is greater than that for the Chapman-Robson estimator and for this reason the latter is often preferred; the variance is given by,

$$\text{var}(\hat{S}) = \hat{S}(1 - \hat{S}) / \sum C_i .$$

Logarithmic Regressions

With the exponential model described above a linear regression of logarithm of numbers in the catch at age, against age will yield an estimate of the total mortality rate. This method is very easy and has been frequently applied. Unfortunately, the estimate is biassed downwards for small sample sizes and strictly other techniques should be used (see Chapman and Robson 1960). Often the regression relationship does look good but de la Mare (1985a) has shown that the logarithmic nature of the model can often visually hide significant variability and consequently tests of goodness of fit should be applied.

Interspecific Relationships

Over a very wide range of sizes there are strong relationships between size and biological indices such as metabolic rate and longevity (e.g. Lockyer 1981a). In fish, Beverton and Holt (1959) showed a relationship between maximum size and longevity; longevity is obviously related to the natural mortality rate. Consequently Ohsumi (1979b) examined the relationship between the average maximum body length

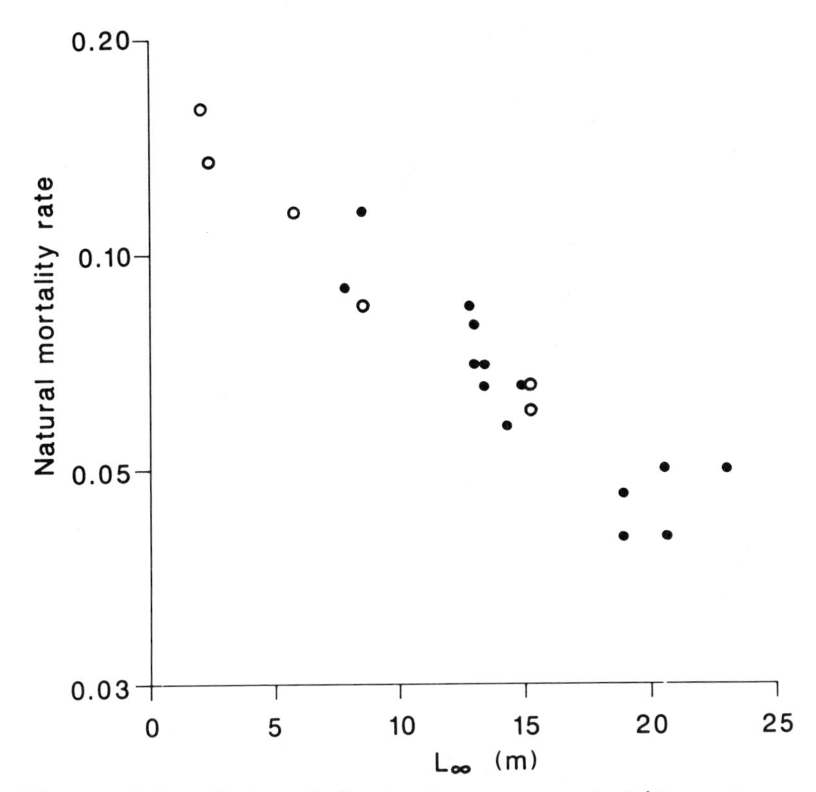

Figure 8.1: Natural instantaneous mortality rate, plotted on a scale of natural logarithms, against average maximum lengths of males in metres, for different species and stocks of Odontocetes (dots) and Mysticetes (circles).

(L_∞) of whale stocks and their natural mortality rate. He found that for males and females separately there was a good linear correlation of the natural logarithm of natural mortality and average maximum length. The males provided a better fit and so Ohsumi advocated the use of the relationship,

$$\ln(M) = -1.842 - 0.06\ L_\infty,$$

where L_∞ is from the males and is in metres. This relationship was based on data from 16 mysticete and 15 odontocete stocks and Chapman (1983) revised the relationship to include a re-evaluation of gray whale mortality and to include only baleen whales. The resulting regression equation was,

$$M = 0.113 - 0.00317\ L_{\infty}$$

Figure 8.1 shows Oshumi's original relationship redrawn from data of his Table 1. This relationship is potentially extremely useful, for ageing has not yet proved possible for some species, such as the bowhead, and for others exploitation has been such that selectivity presents a peculiar pattern of catch at age, as in the northeast Atlantic minke stock. In the Southern Hemisphere the stocks of minke whales were thought to have been expanding before exploitation, due to interspecific effects, and so direct estimates of natural mortality could not be obtained. In fact, problems of calculation and of interpretation exist for most or all of the mortality values contributing to the above regressions; these were reviewed by Mizroch (1985). She concluded that the original data did not provide an adequate basis for construction of such a regression. In principle the technique may still prove valuable when previous estimates of mortality have been re-evaluated.

The longevity of whales, defined as the maximum age found for a species, was predicted by Sacher (1980) on the basis of body and brain weight from regressions derived from other mammals. He claimed good agreement for the few odontocetes for which maximum ages have been given, but his estimates for the mysticetes are much too low.

ESTIMATES FOR SEI WHALES

North Atlantic

Frequency of numbers at age in the Icelandic catch have been provided by Lockyer (1978a) and Lockyer and Martin (1983). Lockyer (1978a) collected material from most of the 1967 to 1975 whaling seasons, and over these years 532 sei whales were landed. From the males, ear plugs were taken from 37% of the catch and 80% were readable. From the females 40% were sampled and 72% were readable. A linear regression of the logarithmically transformed data, grouped over all years, gave Z = 0.0773 for males, and Z = 0.0727 for females.

The values have wide confidence intervals and the material was re-analysed by Lockyer and Martin who noted that in these samples readability

Table 8.1: Estimates of proportional survival (S) and total instantaneous mortality rate (Z) for males and females caught at Iceland, 1967-1981. 95% confidence limits are given in parentheses.

	S	Z
males (tr= 5)	0.910 (0.015)	0.095 (0.049)
males (tr=10)	0.912 (0.019)	0.092 (0.062)
females (tr=10)	0.902 (0.014)	0.103 (0.043)

increased with length and such a phenomenon would distort the age-frequency distribution. Lockyer considered that as the sei whale fishery, at that time, was small and opportunistic the mortality rates would be near to the natural mortality rate.

Lockyer and Martin (1983) presented additional data from whales sampled at Iceland from 1977 to 1981; only 3 sei whales were caught in 1976. In this later period, where possible, both plugs were collected and this helped to improve the overall proportion of the catch that was aged. From 1977, 426 sei whales were landed and for males 83% were sampled and of these 75% were readable. For females 86% were sampled and of these 67% were readable. An important finding was that there was no trend of readability with size and it was thought that the earlier concerns had been due to small sample sizes. Provided that both collection and readability is random with respect to size (or specifically age) the sampled data can be considered representative of the catch. Figure 8.2 shows the frequency distribution from data from Table 3 of Lockyer and Martin. Three techniques were used to estimate the total mortality and they were presented as S (the survival rate) and Z, both using the Chapman and Robson (1960) method, and Z using the relationship of Cooke and Beddington (1981). For the males two ages at first full recruitment (t_r) of 5 and 10 years were used and for females a value of 10 years was used. The results are given in Table 8.1, but as the results for Z were very similar with the two techniques only those from Chapman and Robson are given. The values range from 0.09 to 0.10 and Lockyer and Martin considered that the increase over the previous values to be mainly due to the improved estimation procedure. Nevertheless, they caution that these rates are unlikely to be representative of the population which may have its centre of

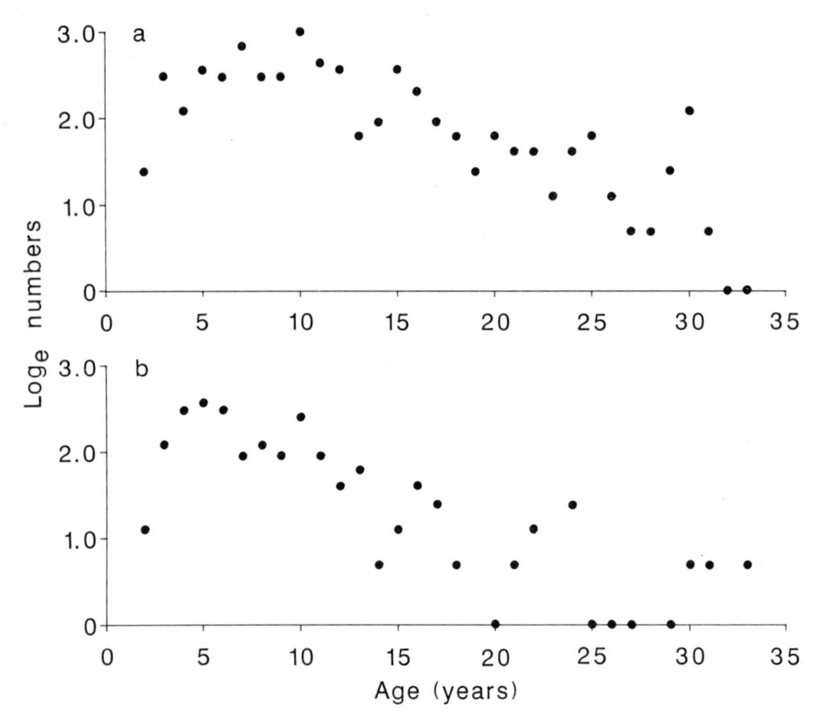

Figure 8.2: Logarithmic frequency of an aged sample against age (years) from Icelandic catches. a: males. b: females.

distribution elsewhere. They make no comment as to the relationship between total mortality and natural mortality.

North Pacific

Total mortality rates of North Pacific sei whales appear to have been first estimated by Doi, Nemoto and Ohsumi (1967a). They obtained estimates from the Kuril Islands, Hokkaido and Sanriku for males and females, from age frequency distributions, and from following cohort density using a measure of catch per unit of effort. Their findings are given in Table 8.2. Doi, Nemoto and Ohsumi (1967b) summarised results from pelagic data as giving a natural mortality rate of 0.06-0.08 for males and 0.10-0.12 for females. Allen (1969b) used a value of $M = 0.05$ y^{-1} in a sei whale stock assessment, which was derived from inspection of a catch curve, but Doi and Ohsumi (1970) accepted the earlier values in

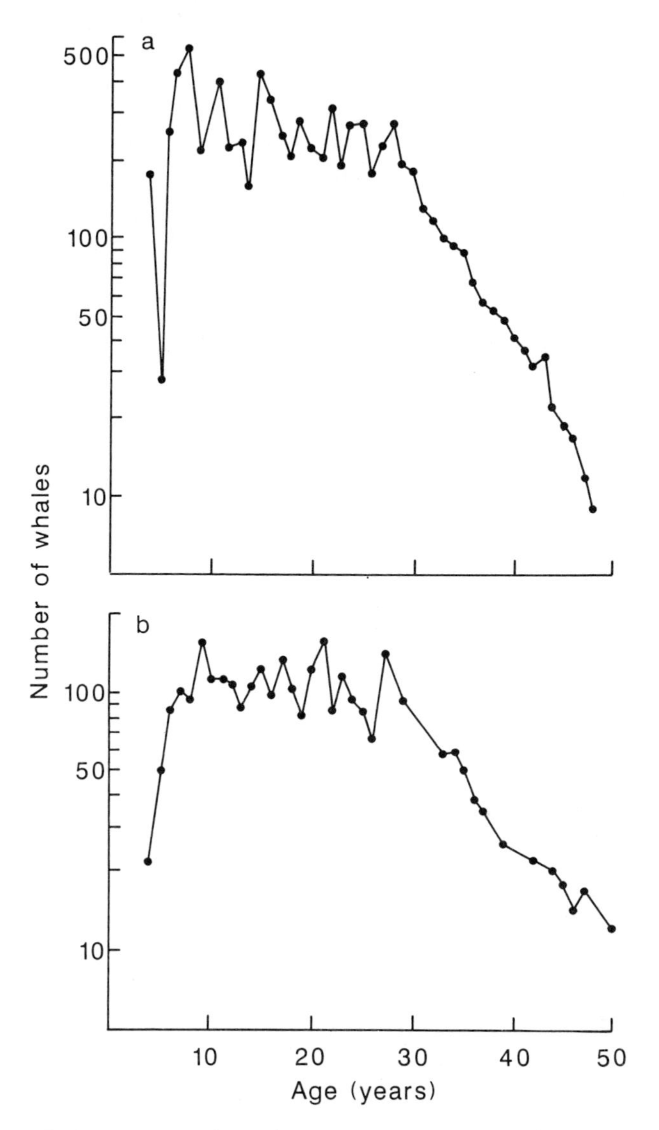

Figure 8.3: Logarithmic frequency of catches of sei whales from the North Pacific against age in years. a: males caught between 140-160°W in 1968. b: males caught between 160-180°W in 1971.

Table 8.2: Estimates of total mortality (Z) summarised from Doi et al (1967a).

a) Age frequency data

	Kuril Is. (1952-1960)	Hokkaido[1] (1955-1964)	Sanriku[1]
males	0.090	0.070	0.067
females	0.121	0.101	0.111

1: values calculated from the "older ages"; probably over 20 years

b) CPUE by cohort
Kuril Is (number in parentheses calculated excluding 1962 data)

	> 20 years	> 30 years
males	0.132 (0.208)	0.152 (0.193)
females	0.222 (0.240)	0.149 (0.164)

Hokkaido + Sanriku

males (8-28 y)	0.019
females (6-26 y)	0.065

Table 8.3: Instantaneous mortality rates estimated by Masaki (1976a) for older animals from the North Pacific. Areas II: 120-140°W, III: 140-160°W, IV: 160-180°W, V: 160-180°E. m: males, f: females.

Area	II		III		IV		V	
Year	m	f	m	f	m	f	m	f
1965	0.028	0.042	0.019	-	0.023	-	0.020	-
1966	-	0.061	-	-	0.046	0.040	0.057	-
1967	-	-	-	-	0.058	-	0.058	0.096
1968	-	-	0.067	0.065	0.079	0.090	0.043	0.048
1969	-	-	0.078	-	0.059	0.094	0.072	0.064
1970	-	-	0.062	-	0.061	0.023	0.070	0.024
1971	-	0.053	-	-	-	-	-	-

their review of biological parameters. For these analyses it is not evident how age compositions were obtained and details of the estimation techniques used are not clarified, and the quality of these estimates is uncertain.

Catch at age data, and subsequent estimates of mortality rates, for the period 1965 to 1972 were reviewed by Masaki (1976a) for locations in the North Pacific, and examples of the age distributions can be seen in Figure 8.3. As can be seen from Chapter 11 whaling from the coasts of America and pelagic whaling was light until about 1965, and Masaki considered that the frequency distribution from the older animals represented the pattern of natural mortality. He gave estimates each year, by sex for four Areas, II (120-140°W), III (140-160°W), IV(160-180°W) and V (160-180°). For most Areas he gave two estimates, one for younger ages and one for older ages, but his figures indicate that full recruitment may not have occurred until about age 30 and so Table 8.3 gives estimates from only the older ages. It can be assumed that in most Areas and years several hundred animals contributed to each estimate which I assume was calculated using a simple regression. For the entire North Pacific Masaki (1976a) gave estimates of annual natural mortality of 0.054 for males and 0.060 for females.

Southern Hemisphere

Estimates of total mortality from catches of sei whales in the Southern Hemisphere were first provided by Doi, Ohsumi and Nemoto (1967). Catches at age were calculated for the six Antarctic whaling areas with the use of an age-length key but the authors noted problems in getting adequate ear plug data for the youngest and oldest ages. Ages 20-45 were used in the subsequent analysis. Table 8.4 gives their estimates of total mortality from regressions. They claim that catches before 1961 were small and that these estimates can be interpreted as natural mortality rates. Rates for females are greater than for males and it is argued that this is due to a segregation of females and that the rates from the males should be used to give a rate for natural mortality. This is not very convincing as females may well have a greater mortality rate but nevertheless we have already seen that there is a segregation during migration and this may affect the age composition of the catches.

Additionally, Doi, Ohsumi and Nemoto (1967)

attempted to estimate total mortality from the density of whales from a cohort (those born in the same year) in two adjacent years. Unfortunately, the densities are higher in the second year, presumably due to effort being incorrectly assessed because of factors such as change in the preferred species by the whaling fleet. From a consideration of the sei whales actually seen compared with the catch, a fishing mortality of 4-28% was found but the authors rightly consider that this was not a reliable technique. On the basis of regressions of numbers at age they considered that the rate of natural mortality was about 0.059 to 0.079 and a value of 0.065 y^{-1} could be accepted. Given this natural mortality rate they calculated rates of exploitation based on the numbers of marked whales and recoveries to give estimates of 10-22%.

A more sophisticated exercise using marking data was carried out by Garrod and Brown (1980). They used information from 395 marked whales and 51 recoveries, aggregated from the entire Southern Hemisphere, and because of the relatively few recoveries grouped the data by pairs of years. Robson's method (Jones 1976) was used to calculate survival rates (exp(-Z)). From 1961/62 to 1964/65 the annual survival was about S = 0.97 (Z = 0.031) and from 1965/66 to 1976/77 S = 0.75 (Z = 0.28). These rates were used to re-evaluate the numbers of marked whales available for capture and from these data Petersen estimates of stock size were obtained. The relationship of catch to the stock estimate gives another, somewhat independent, estimate of the exploitation rate. If a natural mortality of 0.06 was assumed (or even double that) the analysis indicated a large additional loss rate. Garrod and Brown reported that there were many possible explanations for this additional loss rate associated with mortality at marking, under-reporting of recaptures and mark shedding. Difficulties in utilising mark and recapture data are reviewed by de la Mare (1985b) and are discussed in detail later. One must conclude that these estimates of survival cannot be interpreted in a useful manner.

Lockyer (1977c) calculated estimates of mortality rates from samples from catches of sei whales at South Georgia over the period 1960 to 1965 and from Durban over the period 1962 to 1965. Catch frequencies were obtained using an age-length key from each location. In the season 1960/61 ear plugs were collected from 40% of the South Georgia catch but for the rest of the period readable ear plugs

Table 8.4: Instantaneous mortality rates from Southern Hemisphere sei whales by sex, season and Area after Doi et al (1967). 57: season 1957/58 etc. Areas II: 0-60°W, III: 0-70°E, IV: 70-130°E, V: 130°E-170°W, VI: 120-170°W. m: males, f: females.

Area	II		III		IV		V		VI		Total	
	m	f	m	f	m	f	m	f	m	f	m	f
57	-	-	-	-	-	.086	.065	.085	.096	.084	.063	.085
58	-	-	-	-	-	.083	.093	.084	.089	.080	.064	.084
59	-	-	-	.088	.069	.088	.070	.088	.087	.083	.069	.088
60	-	-	.056	.081	.063	.088	-	-	.088	.082	.069	.086
61	-	-	.063	.095	.062	.087	-	-	-	-	.067	.088
62	.080	.082	.064	.094	.064	.092	-	-	-	-	.070	.088
63	.079	.085	.066	.090	-	-	.085	.098	-	-	-	-
64	.061	.105	.059	.112	-	-	-	-	-	-	-	-

Table 8.5: Estimates of instantaneous mortality rates, Z, from catches from Durban and South Georgia, 1960-1965, after Lockyer (1977c). Values in parentheses are from catches raised using an age-length key. m: males, f: females.

Season Locality	Sex	Ages Used (years)	Z	95% limits
Durban				
1965	m	8-24	0.065	0.048
1962-1965	m	8-23	0.066	0.064
	m	12-24	(0.047)	0.006
South Georgia				
1963/64	m	20-36	0.099	0.054
		20-30	(0.094)	0.020
1964/65	m	20-36	0.115	0.053
		19-30	0.073	0.005
1960/61-64/65	m	20-36	0.111	0.037
1963/64	f	18-34	0.106	0.051
		21-30	(0.101)	0.023
1964/64	f	18-34	0.068	0.057
		21-30	(0.064)	0.010
1960/61-64/65	f	18-34	0.085	0.039

averaged 82% of the catch. From Durban in 1965, 73% of the catch was sampled but in the earlier years collections averaged 38%. Simple regressions were used to estimate mortality rates and the values are summarised in Table 8.5. Lockyer concluded that from South Georgia the most reliable estimates of total mortality rate were 0.073-0.094 for males and 0.064-0.101 for females and that although there had been whaling in this region the estimates were probably near to that of the natural mortality rate. From Durban the rates were 0.047-0.066 which she considered to be of natural mortality. Analyses from catches at Donkergat taken over the years 1962 and 1963 were done by Best and Lockyer (1977 ms) and from regressions they found a range of total mortality rate of 0.063 to 0.077 which was similar for both sexes.

Ohsumi (1978) calculated mortality rates from material collected from the pelagic operations over the seasons 1959/60 to 1975/76. Estimates were calculated using regression techniques from three sets of data. The first uses only the age sampled data, the second converts the aged data to the full catch data using a range of age-length keys and thirdly the catch curve is calculated using an "ideal" age-length key which was designed to overcome the problem of selection for larger whales at young ages (Ohsumi and Masaki 1978). His values are given in Table 8.6. Estimates of total mortalities based on a virtual population or cohort analysis were also obtained but with the low mortalities experienced in the whale fisheries such techniques do not give satisfactory answers and they have not been included. In providing a synthesis Ohsumi noted that there were differences by sex and Area and that for Areas II and III previous exploitation may have given rise to higher values. Overall estimates of total mortality were 0.0785 for males and 0.0901 for females. Given the short history of exploitation in Areas I and IV-VI he concluded that the natural mortality coefficients are about 0.065 for males and 0.080 for females.

As has been mentioned earlier the regression techniques give a biassed estimate if relatively small numbers are used. Cooke and Beddington (1981) recalculated estimates from Lockyer (1977c), who used samples of size 33 to 189 to estimated rates for fin and sei whales. Their revised estimates were all between 30 to 80% higher and indicate that any of the above estimates, based on regression and with less than several hundred catches, are likely

Table 8.6 Instantaneous mortality rates from Antarctic catches by sex and whaling Area, 1959/60-1975/76, after Ohsumi (1978). A: aged data only used. B: data raised by age-length key. C: "ideal" age-length key used. m: males, f: females.

Area	A		B		C	
	m	f	m	f	m	f
I	-	-	0.099	0.084	-	-
II	0.075	0.121	0.070	0.108	0.082	0.112
III	0.091	0.105	0.100	0.088	0.102	0.087
IV	0.070	0.106	0.068	0.087	0.067	0.084
V	0.076	0.090	0.071	0.088	-	-
VI	0.073	0.096	0.065	0.083	-	-
V+VI	-	-	-	-	0.073	0.085

to be much underestimated.

De la Mare (1985a) investigated catch at age data for the period 1966 to 1978 combined, which were obtained from Mizroch and Breiwick (1984), (Figure 8.4), and he demonstrated a significant difference between the sample catch curves of male and female sei whales. He also calculated Heincke estimates from each age, and for males there was a steady increase to about 55 years and for females to 30 years. De la Mare concluded that the Heincke estimates, calculated for a range of species, supported the hypothesis of natural mortality increasing with age. However for sei whales, the increase could be explained by selection and problems in reading older ear plugs, and Mizroch and Breiwick (1984) show an increase in the percentage of the catch aged with increasing length, and such trends could have been generated by a progressive difficulty in reading older ear plugs.

It is possible that in the Southern Hemisphere sei whale numbers had been increasing prior to the period of sampling in the 1960s. This is suggested by increases in sampled reproductive parameters and observed indices of abundance which are reviewed later. Given that the population was increasing prior to sampling Horwood (1980a) considered what effect this would have on estimates of mortality rate, and it was found that if a natural mortality rate 0.06 was assumed, then by 1960 the age composition, over ages 20 to 50, would yield by regression an estimate of 0.065, which is positively biassed. Ohsumi (1978) and Best and Lockyer (1977 ms) argue that their estimates, being based on older animals,

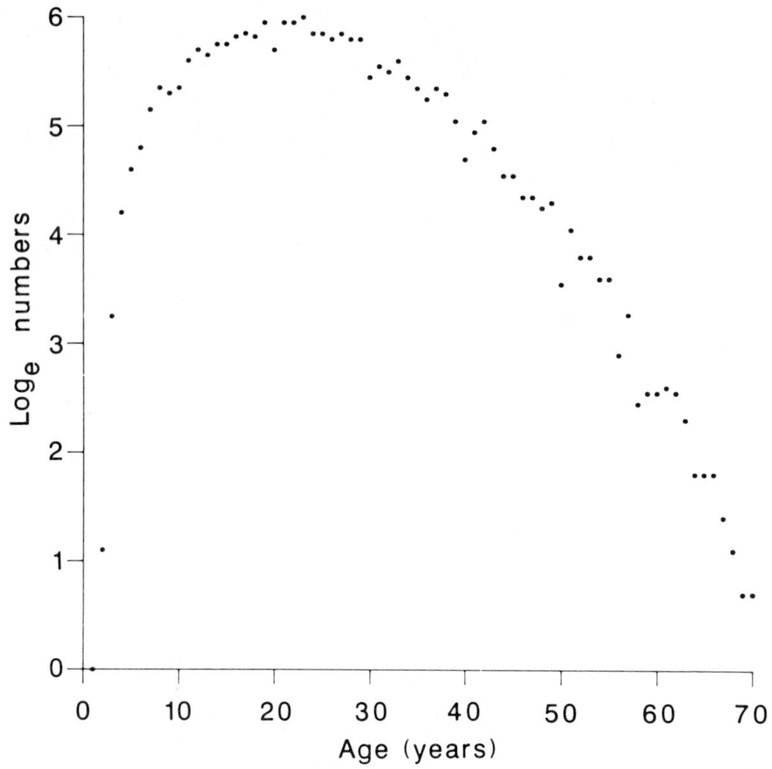

Figure 8.4: Logarithmic frequency of catch of Southern Hemisphere male sei whales against age, from Japanese catches 1966-1978.

would avoid the problems caused by the rapid exploitation in the late 1950s. However they are confounded by any population increase and in addition Horwood (1980b) showed that in Areas II and III, which include South Georgia and South Africa, 20% of the catch ever to have been taken in those Areas had already been taken by 1956/57. Consequently the catch distributions would have been affected by exploitation.

Finally, Horwood (1980a) provided estimates of mortality from the period 1957 to 1965. The technique used assumed that fishing mortality was directly proportional to catch per effort and that the recruited male and female stocks were the same size. Given this assumption, natural mortality can be calculated using the average total mortality rates from Doi, Ohsumi and Nemoto (1967), and this resulted in estimates of 0.0645 to 0.0823 per year.

SUMMARY

Most mortality rates have been calculated using data on catch at age and regression techniques that are likely to underestimate the rates. In addition it has not been possible to show that catches were representative of the population even if sampled ages were representative of the catch. It has not been established whether natural mortality is or is not dependent upon age. The confidence limits of the estimates are usually large and a wide range of estimates, from different locations, have been presented. Estimates of natural mortality have been presented from the North Pacific and Southern Hemisphere and indicate values about 0.06-0.07 y^{-1}, but these are far from reliable because of the problems outlined above. From the Icelandic sei whales, not intensively exploited, total mortalities of 0.091-0.10 y^{-1} were found. There is some suggestion that females may have a slightly greater natural mortality rate but this has not been established. The data are totally inadequate to consider whether there have been any changes in natural mortality with time or to provide estimates of juvenile mortality rates.

Chapter Nine

REPRODUCTION

INTRODUCTION

The sei whale is a fast moving, open ocean species and this, combined with the difficulty of identifying individuals, has resulted in little knowledge of its social behaviour associated with reproduction, or indeed any aspects of its natural history. However other facets of its reproduction have been extensively studied, both because of the importance of reproductive studies to understanding how numbers vary and how the whales might be managed, and because many of the studies have been possible from histological and morphological examination of material from catches. From these sources have been described the development of the testes and ovaries, the process of maturation with length and age, and the pregnancy rates with age and time. It is shown that trends with time have been found in sampled pregnancy rates and ages at maturity, and that these occurred at the same time as reductions in the stocks of sei and other baleen whales. These aspects are reviewed in detail, but more general descriptions of the reproductive biology of Cetaceans are given by Slijper (1966), Harrison (1979), Gambell (1973), Harrison and Weir (1977) and Lockyer (1984b).

MALES

Testes

The testes occur as a pair of similar sized, smooth, elongated organs found near the posterior abdominal wall, and each testis can weigh up to 10 kg; nevertheless they have a similar morphology and histology to that of most mammals. It will be demonstrated that there is a rapid increase in the size of the testes at maturity and so the weight or volume of the testes is often used to determine maturity in large scale field studies. The identification of a mature male could be made more precisely by the recognition of the presence and quantity of spermatozoa, but this can be difficult. Seasonal cyclic activity of spermatogenesis might exist and if so would be reduced during the main summer whaling season, but a significant practical problem is that the necessarily long post-mortem times result in degeneration of tissues and consequent difficulties in recognising spermatozoa and spermatogenesis. Nevertheless where fresh gonad samples can be collected, the identification of spermatozoa and spermatogenesis is practicable. Associated with maturity is an increase in the diameter of the seminiferous tubules and this is an easily recognised and useful diagnostic character. However, in sei whales the maturation of the testis does not occur homogeneously throughout the organ (Masaki 1976a), as has also been reported in some odontocetes (Best 1969, Kasuya and Marsh 1984). Whales are usually categorised as immature, pubertal or mature on the basis of testicular material. In immature tissue there are small, closed tubules with numerous interstitial cells; in mature tissue the tubules are large with an open lumen, possibly recognised spermatozoa and developed spermatogenesis. Intermediate or pubertal tissue has tubules of both or intermediate sizes with spermatogonia and Sertoli cells. From 29 specimens, sampled from 11 sites within the testis, Masaki (1976a) illustrated the development of maturation within the testis; maturation begins in the centre and proceeds to the posterior and dorsal sections, near the cauda epididymus, and last to mature is the ventral surface and the anterio-ventral section.

Development of maturity

The presence or quantity of spermatozoa in sei whale testes has been recorded by Matthews (1938) and Mitchell and Kozicki (1974 ms) from Southern Hemisphere and Canadian catches respectively, but the data are not presented in relation to the size of whales; however several studies have described the mean diameter of tubules with size and sexual condition. Unlike the case in humpback whales Gambell (1968) found that mean diameter of tubules of sei whales landed at Durban increased with body

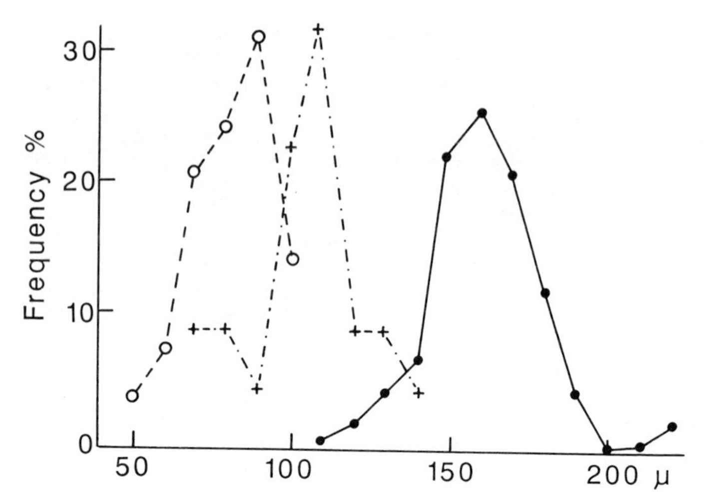

Figure 9.1: Frequency distribution (%) of mean diameter of seminiferous tubules from immature (o), pubertal (+) and mature (●) North Pacific sei whales. Diameter in microns and 50 represents those of 40 to 50 micron.

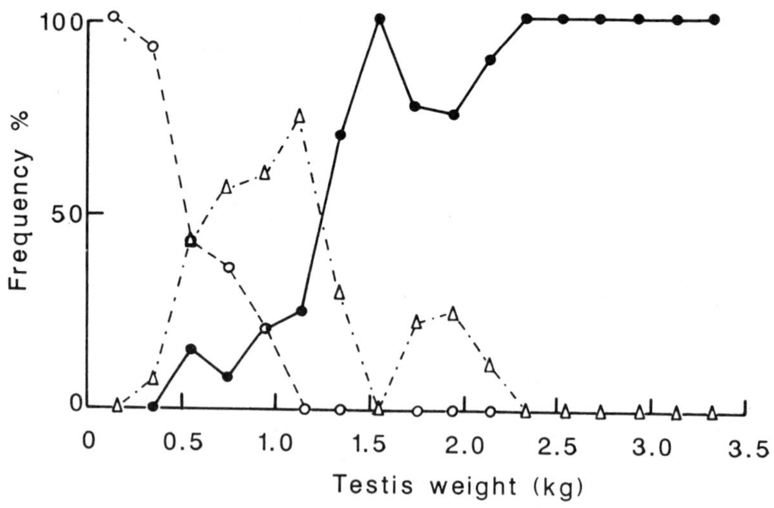

Figure 9.2: The proportion of mature (bold), immature (dashed) and pubertal (circles) males against single testis weight from the North Pacific.

length. At 12 m an immature whale had a tubule diameter of 60 micron and at 14.5 m it was 90 micron. After maturity a similar increase with body length was found but the mean diameter of the tubules was 60 micron larger. A similar trend was found in mature Antarctic samples but the few (10) immature samples showed no trend with length. In general the mean diameter of tubules from mature sei whales is over 90 micron. A similar critical size can be seen in samples from Canada (Mitchell and Kozicki 1974 ms), Iceland (Lockyer and Martin 1983) and the North Pacific (Masaki 1976a); the frequency distribution of tubule size for immature, pubertal and mature sei whales from North Pacific samples is shown in Figure 9.1, after Masaki (1976a).

The relationship of testis weight to stage of maturity in sei whales has been described by Gambell (1968), Mitchell and Kozicki (1974ms), Masaki (1976a), Jonsgard and Darling (1977) and Lockyer and Martin (1983), but in using the results of these studies care needs to be exercised in ensuring that relationships are described in terms of weights of single or both testes. In general, there is a slow steady rise in testis weight with length in immature sei whales, but a large increase is found at maturity. Gambell (1968) considered that a testis of 1.5 kg or over came from a pubertal or mature whale and this is confirmed by the other studies. Figure 9.2 shows the proportions of the different sexual classes against weight of a single testis from Masaki's North Pacific samples, and it can be seen that a few whales are mature with a testis of 0.5 kg. The mature sei whales have weights of single testes ranging from less than 0.5 kg to over 10 kg, with most immature sei whales having each testis less than 0.5 kg; consequently one can recognise a rapid and large increase in testis weight at length as the whales pass through puberty. The relationship of testis weight to body length has been described by the previous authors and in addition by Rice (1963, 1977) from Californian sei whales and by Omura (1950a) from Japanese coastal catches, and the relationship of testis weight to length from Norwegian samples is shown in Figure 9.3, after Jonsgard and Darling (1977). It is seen that the males mature at about 13 m but this is described in more detail later.

Seasonal cycles

Cycles of seasonal activity in spermatogenesis and

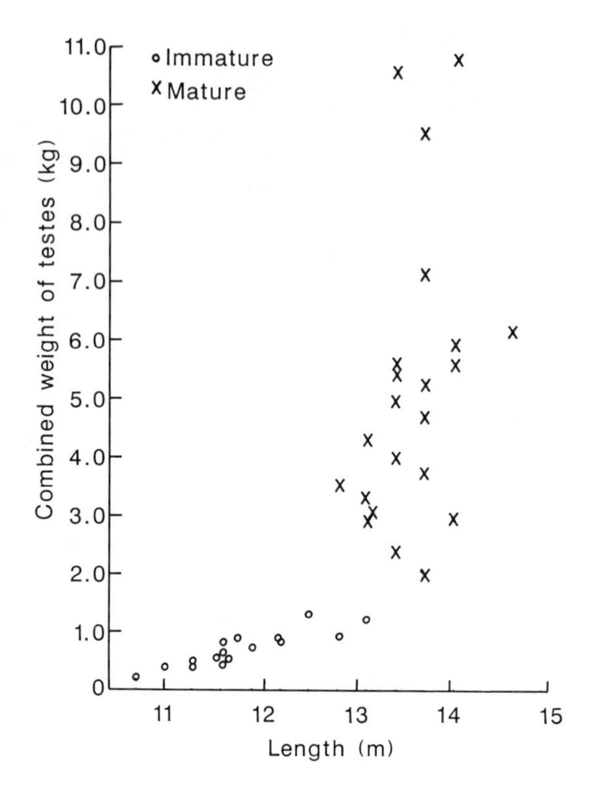

Figure 9.3: Weight of both testes against body length for immature and mature sei whales caught off Norway.

in testis weight have been found in dolphins and porpoises, and the subject of seasonality in male reproduction in Cetacea is briefly reviewed by Perrin and Donovan (1984). Its existence or otherwise is important not only in defining breeding periods, but in attempting to interpret information on stages of sexual maturity in possibly quiescent periods. An increase in testicular activity prior to winter and a decrease afterwards was described for fin whales by Laws (1961), and a decrease in the size of tubules was found in summer, and Laws concluded that a pronounced seasonal cycle existed; a similar result was found for humpback whales (Chittleborough 1955). Such a seasonal cycle is argued for in sei whales by Mitchell and Kozicki (1974ms). Their samples were collected from the north-west Atlantic in May to November 1966 to 1972 and the quantity of spermatozoa ranked on a scale 0 to 3. Generally groups 0 and 1 occurred before August and groups 2 to 3 from September onwards, indicating a pronounced increase at the end of the summer. Also, in the first part of the whaling season, the mean weight of both testes combined was 5.1 kg and in late summer it was 8.1 kg; they argued that as the mean length in the catch increased by only 0.15 m from early to late season then these changes represent seasonal cycles rather than the sampling of different sexual classes in the population. They also presented limited evidence for a small change in tubule diameter.

This interpretation is not supported by information from the Southern Hemisphere. From very limited Antarctic data of spermatozoa smears and testis sections, Matthews (1938) detected no seasonal difference. As already noted Gambell (1968) found that tubule diameter increased with length at the same rate in Antarctic and Durban samples of mature whales, but the difference in mean diameter size at the same length was only one micron. Similarly, Gambell considered that testis weights showed no marked seasonal variation, and they certainly do not show the 60% increase found from Canada. A limited amount of material was culled by Rice (1977) from Californian catches, and he found no changes in mean weight or tubule diameter from June to September although he thought his very few weight (4) and tubule (1) samples from October were suggestive of an increase.

Consequently we do not have unequivocal evidence for or against a seasonal cycle and additional material is needed to confirm the findings from the North Atlantic, but one must not exclude the possibility that the differences may be real physiological differences of whales from the different oceans.

Penis size

The length of the penis can provide a guide for determining maturity. From sei whales of the Antarctic and South Africa, Gambell (1968) classified males into immature, pubertal and mature classes from histology and testis size, and found that the longest penis of an immature whale was 96 cm and the smallest penis of a mature whale was 95 cm. The sample of 154 shows a clear relationship with sexual status independent of length. A similar exercise using 162 samples from the North Pacific was carried out by Masaki (1976a), but he shows a much greater overlap with the maximum length from an immature sei whale of 106 cm and the minimum size from a mature whale of 70 cm. The smaller mature lengths come from whales of a relatively small body size and most of the largest immature lengths come from large whales and it seems therefore that size of whales, independent of sexual status, additionally affects penis length. The large immature whales had large penises but they were still the smallest of their body-length group. This feature does mean that penis size can only be used as an approximate guide to determining maturity and a suitable critical length is about 95 cm.

FEMALES

Ovary

General description and size. The appearance and morphology of the ovary are similar to that found in fin and blue whales, which are described in detail by Mackintosh and Wheeler (1929) and Laws (1961), and a general review of the Cetacean ovary is given by Slijper (1966). In adults the organs are coloured a pinkish- grey and are usually described as looking like a bunch of grapes, for the development of Graafian follicles gives rounded protuberances which are added to by the accumulation of corpora albicantia and, during pregnancy, the corpora lutea.

The size of the ovaries generally increases with the size of the individual and in sei whales both ovaries are equally active (Ohsumi 1964b). Matthews (1938) and Gambell (1968) found that the combined weight of the ovaries in immature sei whales increased with length at a rate, given by Gambell, of 150 g m^{-1} of body length. However at Durban the mean paired (combined) weight of the ovaries of 14 m (46 ft) immature females was 569 g compared with 517 g from the Antarctic. This difference in absolute size is associated with the seasonal development of the Graafian follicles. Resting females are defined as those mature but neither pregnant nor lactating, and are those which have recently weaned a calf or have failed to conceive the previous year. From this group Gambell described paired ovary weights increasing with whale length at 197 g m^{-1}, with the weight from a 15.8 m (52 ft) whale being about 750 g from Antarctic samples. The paired ovary weight from resting females was described by Gambell (1968) in relation to the number of corpora, which is an index of age; a possibility of a decline in weight for the largest whales was suggested by Matthews (1938), but Gambell's more extensive study shows a steady increase to about 12 corpora, after which the weights stabilise and do not appear to decrease with increasing numbers of corpora. For pregnant whales Gambell found no difference between weights from samples from the Antarctic and Durban and concluded that, unlike fin whale ovaries, seasonal changes in weight do not continue after puberty. From the Antarctic the mean combined weight of the ovaries of the pregnant whales was 1188 g compared with 795 from the resting females; the difference is due mainly to the large corpora lutea. For comparison Laws (1961) reported the mean combined weight of pairs of ovaries from pregnant fin whales as 3.14 kg with a hypertrophic form in a blue whale weighing 59 kg; the largest record from the sei whale is about 2.5 kg.

Graafian follicles. In near-term humpback fetuses abundant primary follicles have been detected but in fin whales such follicles were not found in younger fetuses, and it is concluded that the genesis of the follicles begins in the fetal ovary, 2 to 3 months before birth (Chittleborough 1954, Laws 1961); no similar studies have been undertaken with sei whales. Minimum size limits for catches have meant that few immature specimens have been available for study, but in fin whales, which reach puberty at about 20 m, those of about 17 m had primary follicles of 45-70 micron in diameter, and by this stage they are separated from the germinal epithelium by a prominent tunica albuginea. In blue and fin whales approaching puberty follicles of about 30 mm are found and atretic folicles can exist. The size of follicles of 46 immature sei whales caught in the Antarctic and from Durban were reported by Gambell (1968) and, as mentioned earlier, the ovaries at Durban were heavier than those found in similar sized sei whales in the Antarctic, and from a 14.3 m whale the mean follicle diameter was 11 mm whereas from Durban the mean diameter was 22 mm; in addition the samples from Durban tended to have more primary follicles than those from the Antarctic.

In both immature and mature animals the mean size of the largest follicle increases with length, but in the mature whale there is variability involved associated with the phase of the sexual cycle; the mean size of the largest follicle in mature females ranges between 14 and 19 mm depending upon the length of the individual. Ovulation is thought to be spontaneous in the Cetacea (Slijper 1966), and has been shown to be so in some dolphins (Kirby and Ridgway 1984), and at ovulation the largest follicle ruptures to release the ovum and then forms the corpus luteum. At Durban a recently ruptured follicle from a sei whale measured about 83 mm in diameter, but in the Antarctic two similar measurements were 20 to 25 mm. Gambell (1968) considered that the size at which a follicle could be said to be maturing was 30 mm, as in humpback and fin whales.

In pregnant sei whales Gambell (1968) demonstrated a cycle of follicular activity somewhat different from that of the fin whale. The mean diameter of the largest follicle in the pregnant whale remains constant up to a time when the fetus reaches a length of 2 m, after which there is a pronounced reduction in size. At about the same period there is a substantial decrease in the abundance of follicles. Samples of lactating whales are rare as they have been protected, but some important ovarian data are available from the few catches from Saldanha Bay and the Antarctic, and from these Gambell estimated rates of post-partum ovulation of 11.1%, from South Africa, and of post-lactation ovulation of 9.1% from the Antarctic samples. (The 9.1% comes from one of eleven Antarctic whales sampled with a largest follicle exceeding 25 mm. The post-partum rate was obtained from one of twelve lactating whales sampled from Saldanha Bay with a follicle exceeding 25 mm, plus one of six with a large follicle and a fetus exceeding 4 m, to give two in eighteen.) Since post-partum ovulations occur at the same time of year as the main winter mating Gambell (1973) thought it probable that these ovulations would nearly always result in a pregnancy. From the sizes of fin whale fetuses Gambell (1973) deduced that end-of-lactation ovulations do not often result in pregnancy, and this is true also for sei whales.

The size distributions of the large follicles in resting whales was described by Matthews (1938) and Gambell (1968). In the Antarctic from January to April numerous follicles of over 25 mm are found with sizes up to 50 mm, indicating that they are very near to ovulation at that time, and one recently ovulated whale was found to have a corpus albicans of 53 mm, demonstrating that more than one ovulation can occur in these months. At Durban most females were not in a "resting" condition (75 out of 90) but of those resting most had large follicles and were nearing ovulation. Ivashin (1984) cites several examples of multiple fetuses and active corpora lutea as evidence of simultaneous ripening of two or more Graafian follicles in sei and other large whales.

The corpus luteum. A Corpus luteum is formed from a follicle that has ruptured to release its ovum. Initially these are small structures known as the corpora lutea of ovulation. Such corpora were described by Gambell (1968) from five sei whales caught in the Antarctic and 22 from Durban; of these two each had two corpora, and from the 29 samples the mean diameter ranged from 28 to 82 mm with a mean of 54.7 mm. In fin whales the mean diameter is about 830 mm. Laws (1961) considered that, as in most mammals, the corpus luteum of ovulation persists for 15 to 20 days and this has been assumed to be the case in sei whales.

If fertilisation occurs the ruptured follicle develops into a larger body, the corpus luteum of pregnancy, which persists throughout pregnancy. Laws (1961) described a range of morphological forms of the corpus luteum graviditatis, the main difference being between those possessing a central vesicle and those not. For fin whales, he arrived at the important conclusion that, as both forms occurred, the presence or absence of such a cavity in corpora lutea meant that such presence or absence could not be used to distinguish between corpora of ovulation and of pregnancy. Similar morphological forms occur in sei whales. From 223 samples Gambell (1968) found that the mean diameter of the corpus luteum of pregnancy was 76.2 mm, for whales with a fetus of about 0.1 m, increasing slowly to 83.8 mm for near term fetuses. From ovulation the increase in size of the corpora is rapid, the suggestion of a slower development by Matthews (1938) being a result of less extensive information, and the weight of the early corpus after ovulation is about 125 g and of near term corpus is about 390 g.

As described above more than one corpus luteum of ovulation can be found at the same time. In addition Gambell (1968) described eleven pregnant whales with more than one corpus luteum to give an estimate of multiple ovulations of 2.6% (13/493).

Twins and multiple fetuses also occur as described later.

The corpus albicans. If no pregnancy occurs the corpus luteum of ovulation regresses quickly to a smaller solid structure within the ovary which has no luteal cells. This structure is now commonly known as the corpus albicans although it is seldom unpigmented. A similar regression occurs at the end of pregnancy. The size of this body is about 36.5 mm during early lactaction, showing a reduction from 83.8 mm during late pregnancy. Samples from the Antarctic, assumed to be representative of late lactaction, have a mean diameter of 34.1 mm. From the North Pacific, Masaki (1976a) showed a reduction in size of the largest corpus with the reduction in thickness of the mammary gland during lactaction.

The corpora albicantia of Cetacea appear to exhibit a phenomenon unique in mammals in that they persist in the ovary and are not completely resorbed (Slijper 1962). If this is true then counts of corpora shed a substantial light on ageing and previous reproductive history. This feature was first established by Laws (1961), in his study on fin whales, and he classified the corpora albicantia into three groups of "young", "medium" and "old", depending upon the relative amounts of white connective tissue and brown, former luteal tissue. The "old" group formed a symmetrical distribution about a mean diameter of 20 mm and with very few less than 10 mm. A general increase in numbers of corpora with length and age, and the distribution of sizes strongly suggest a persistence of the reduced corpora albicantia. This study was repeated for sei whales by Gambell (1968), who plotted the size frequency distribution of the three classes from 6140 corpora. The oldest group had a mean size of 11.9 mm. Gambell noted that no sign of resorption of the small, old corpora albicantia was found in any sei whales examined; subsequent studies described later show linear relationships of corpora number with age and lend support to this interpretation.

Various authors have suggested that the corpus albicans from an unfertilised ovulation can be distinguished from that formed following a pregnancy (Robins 1954, Ivashin 1958, 1984). However Laws (1961) considered that the structure of the two types of corpora lutea was sufficiently similar that they would not give rise to corpora albicantia with different appearances. Gambell (1968) considered that this was also the case for sei whales. At present the two forms cannot be reliably distinguished although it is a subject of current debate.

Corpora aberrantia and atretica. These structures were described in fin whales by Laws (1961) as pigmented bodies distinct from corpora albicantia, and similar structures are described from sei whales by Gambell (1968). Gambell found that they occurred in 10.9% of the females examined, but they constituted only 2.5% of the total number of corpora. Usually they are excluded from counts of ovarian corpora, and different terminologies have been suggested by Best (1967b) and Harrison and Weir (1977).

Vaginal band

The presence of a vaginal band can give some indication of the sexual status of a whale. The vaginal band, probably analogous to the hymen in humans, was first described by Mackintosh and Wheeler (1929) but they found no incidence of this, or a tag from a ruptured band, in the few sei whales they examined. Subsequently the structure has been described in sei whales of the Southern Hemisphere by Matthews (1938) and Gambell (1968), and of the North Pacific by Ohsumi (1969). Gambell's data show that of immature whales only 7.4% had a band or tag and of mature whales none had a band and 5.9% had a tag. From the North Pacific Ohsumi's figures are higher with 28% of sei whales having a band or tag, and he considered that traces of the band or tag disappear with time after rupture. Consequently he discounted figures from multiparous whales to give a rate of 31% initial occurrence in the population. Ohsumi found that, when present, the vaginal band in immature sei whales is not ruptured but may be so in prepubertal stages (identified by no corpus luteum but developed Graafian follicles). Evidence of rupture probably indicates copulatory activity in prepubertal and pubertal stages of development. However, Ohsumi reported that, as found in fin whales, pregnancy can occur without rupture of the band, but the band is always broken by parturition.

Mammary gland

The mammary glands lie on either side of the mid-ventral line anterior to the genital groove and with the teats opening level with the genital groove. The size and state of the glands vary with the sexual status of the whale. From data from immature Southern Hemisphere sei whales Gambell (1968) showed that the depth of the mammary gland increased with the length of the whale; from Durban, samples showed an increase from about 12mm for whales less than 12m (40ft), to between 2.0 and 3.0 cm in the largest immature whales. Immature sei whales from the Antarctic had depths less than those from Durban to demonstrate a seasonal cycle. In the North Pacific Masaki (1976a) found few immature whales with mammary glands deeper than 2.0 cm. Cows with calves have usually been protected from capture and consequently data from early stages of lactation are rare. During the late stages the glands still have milk but the quantity is relatively small and it tends to be yellowish and viscous. From Saldanha Bay the mean depth of the gland in early lactation was found to be 13 cm and from the North Pacific the mean depth was 12 cm. In contrast the mean depth found from Antarctic samples was 9.6 cm, representing a later stage of lactation. In the North Pacific the mean depth during late lactation was 4.6 cm. The depths of the mammary gland of whales in pregnancy, anoestrus and at oestrus are all rather similar at about 4 cm, although Gambell suggests that there is a small shrinkage during early pregnancy.

Uterus

The relationship of distance across the collapsed uterine cornu with size of sei whale was described by Matthews (1938) and Gambell (1968). The increase tends to be exponential, even during the immature phase, with the diameter about 4 cm in whales less than 13 m, but quickly increasing to about 15 cm at 15 m. After parturition regression in size of the maternal uterine cornua is rapid.

The thickness and composition of the wall of the uterus varies substantially throughout the various phases of the sexual cycle (Matthews 1948). In many coastal whaling operations the belly of the whale is slit open before it is towed to the whaling station, in order to avoid overheating of the carcass, and with this process fetuses and often ovaries are washed out of the body cavity, and the sexual status of the whale is then difficult to determine. Developing the earlier work of Matthews,

Lockyer and Smellie (1985) used histological examinations of the uterine mucosa of Icelandic sei whales to determine sexual status.

Their initial examinations found no difference between fin and sei whales with respect to the thickness of the mammary gland, uterine cornua width and status of the uterine mucosa, and so they combined information from the two species. They demonstrated changes in size of the uterine cornua and the myometrium (the basal layer of the uterine wall) with state of the sexual cycle, along with the change in thickness and appearance of the stratum spongiosum, and the outer stratum compactum of the uterine wall. Cyclical changes were also found in the size of glands in the stratum spongiosum. Lockyer and Smellie also presented a key for distinguishing the sexual status of the whale on the basis of the uterine mucosa and mammary gland.

Conception and gestation

The size distributions of fetuses with month give good information on the approximate date of conception and approximate duration of gestation; however refining these times presents a substantial problem. As described earlier parturition occurs when the fetus is about 4.5 m. The mean date of conception appears to be easily estimated from the lower end of the fetal growth curve, but in the early months not all conceptions have happened and there will be a tendency to predict too early a mean date of conception. However, following Mackintosh and Wheeler (1929), a frequency distribution of apparent times at conception can be obtained by putting the fitted growth curve(s) through each datum to find the time of conception of the individual. Distributions based on such a technique are shown in Figure 9.4, from the Southern Hemisphere data of Gambell (1968), and the North Pacific data of Masaki (1976a). Because of the distribution of samples this relies heavily on fetuses found in summer catches and hence a substantial extrapolation is needed. In addition one must question whether the growth described over the first few months is representative of the individual; the general shape of the individual growth curve itself is assumed to follow that in the sampled fetal population. Notwithstanding these concerns the length data give a mean time of apparent conception in the Southern Hemisphere of June and in the North Pacific of December.

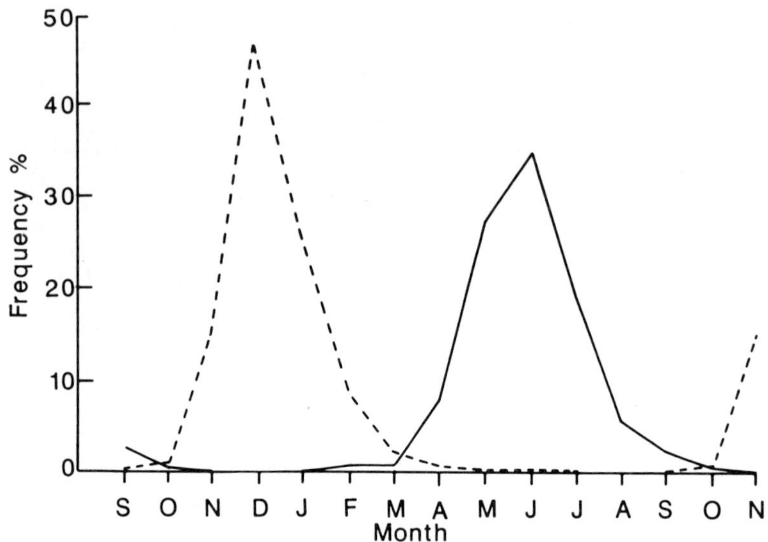

Figure 9.4: Frequency distribution of the time of conception in the Southern Hemisphere (bold) and North Pacific (dashed) from extrapolation back from the size of individual fetuses. The mean times are imprecisely determined, but the distributions about the means are better defined.

An alternative method of finding the mean date of conception relies upon the proposal of Huggett and Widdas (1951) that the constant t_o in the linear growth equation (see Chapter 7) is equal to about a fifth of the gestation period, for gestation times of 1-400 days. It is evident from the general picture of growth that the gestation time will be about 11 to 12 months and hence t_o will be 66 to 73 days after conception. Gambell's fetal growth data give a t_o at about 20 September to give a conception date of mid-July. Mizue and Jimbo (1950) gave an estimated date of January for conception for North Pacific sei whales. A similar exercise to that of Gambell was done by Masaki (1976a) for the North Pacific and he arrived at a mean conception data of late December, but using a period of 36 days prior to the "t_o" date. However, the value of 36 days, which was taken from the study by Laws (1959), was shown to be misleading by Rice and Wolman (1971) and consequently late November is probably a better estimate.

Table 9.1: Proportion of ovulating sei whales from Durban, excluding lactating whales.

Month	May	June	July	August	September
Sample size	6	11	23	29	21
% ovulating	50.0	54.6	30.4	10.3	9.5

Another approach was provided by Gambell (1968) who gave the frequency distribution of ovulating whales by month from Durban, Table 9.1, and peak ovulations occur in June. All these different data and analyses indicate that conception occurs over a range of months with a peak in June and July in the Southern Hemisphere and in November to December in the North Pacific; nevertheless the mean dates are not accurately defined. From Icelandic data Lockyer and Martin (1983) gave a time of conception of December to January.

Given these dates, or ranges, the gestation period can be calculated assuming a length at birth of 4.5 m. From the North Atlantic Lockyer and Martin (1983) give 10.7 months, from California Rice (1977) gives 12.7 months, from the North Pacific Masaki (1976a) gives 10.5 months. Gambell (1968) gives 12 months from the Southern Hemisphere and Lockyer (1977b) gives 11 to 11.5 months. For the purposes of describing the population dynamics of the populations it is essential to know the gestation period only to within a few weeks, and the range of 10.5 to 12.5 months is adequate for most exercises; nevertheless it is clear that the estimates of time of conception and birth could be much more refined.

Ovulation rates

The size of fetuses and occurrences of corpora lutea show that the main breeding season is in the winter months, April to August in the Southern Hemisphere and November to March in the Northern Hemisphere; however summer ovulations also occur. This section focuses on information gained from ovarian examinations and that from pregnancy rates is dealt with afterwards; not all ovulations result in a pregnancy.

As will be seen from the relatively high pregnancy rates most whales ovulating in winter become pregnant, and in order to estimate the average number of ovulations needed to result in a winter pregnancy Gambell (1968) examined the size frequencies of the first, second etc biggest corpora albicantia and found in each group a few large individuals. These he interpreted as coming from second, third etc ovulations that season. The cut-off points for size are somewhat arbitrary but on this basis he deduced that the average number of ovulations per pregnant whale was 1.117, with only 1% of the pregnant whales needing a sixth ovulation. A similar exercise was carried out on North Pacific material by Masaki (1976a) and he found a rate of 1.108 ovulations per pregnant whale.

To a lesser degree ovulations occur at other stages of the reproductive cycle, but Gambell's data are the only ones to provide information on post-partum ovulation rates, and he found a rate of 11.1% of lactating whales simultaneously pregnant. Summer pregnancies he found as 12.5% to give 1.236 ovulations per cycle. He assumed the rate of 1.117 ovulations per oestrous also applied during the summer as well as the winter to give 1.38 ovulations per cycle or 0.69 per year. These calculations were revised by Gambell (1973) particularly to include the fact that only a proportion of the population at the end of lactation and in winter should be attributed to the above rates, and he also modified the ovulation rate per oestrus to give an annual ovulation rate of 0.62. Masaki (1976a) repeated Gambell's earlier exercise to arrive at a rate of 0.666 per annum for the North Pacific, and if Gambell's later calculations are followed this figure would be a little reduced but both studies give similar estimates of the ovulation rate over the period of sampling.

Because of the persistence of corpora an examination of the number of corpora with age will also provide an estimate of total number of ovulations per year. This has been done in two similar ways, either by direct regression (normal or geometric) or by calculating the number of corpora per annum for the period of maturity of the whale (where the age at maturity is found from the transition layer of the ear plug). However first the relationship of ovulation rate with age should be considered. From the Southern Hemisphere the number of corpora with length were shown by Gambell (1968) and from Hokkaido and Sanriku by Omura (1950a). Plots of

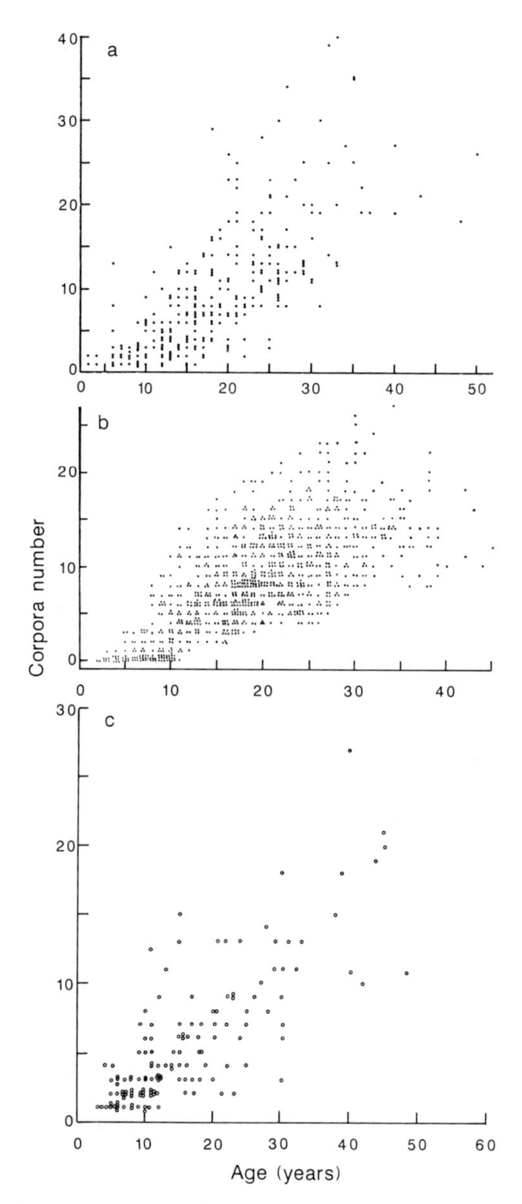

Figure 9.5: Number of corpora lutea and albicantia against age in years from a) the South Atlantic b) North Pacific and c) Iceland.

corpora number with age from the Southern Hemisphere are given by Lockyer (1974), Masaki (1978) and Best and Lockyer (1977 ms). From the North Pacific similar material is presented by Masaki (1976a), and Rice (1977), and from the North Atlantic by Mitchell and Kozicki (1974 ms), Lockyer (1978a) and Lockyer and Martin (1983). Examples from these areas are illustrated in Figure 9.5.

From a sample of 143 North Pacific sei whales Masaki (1976a) calculated the annual ovulation rate in those years after the age of maturity, as measured by the ear plug transition layer. His results are illustrated in Figure 9.6 and show a pronounced and continual decrease in average rate to the age shown. That is, the rate of 0.76 y^{-1} at age 20 years is the average accumulation rate from the age at maturity to age 20. Although this shows a drastic decline with age it represents a reduction in ovulation rate of about 4% per year to give the average cumulative reduction of absolute ovulation of 2% per year; by age 30 the ovulation rate would be some 40-50% of that at age 10 years. No explanation has been offered for this but it is possible that fewer ovulations are needed for conception in older and more experienced individuals, or there could be an underestimate of corpora with age. A trend such as this might also be generated if ovulation rates had increased with time independent of age. It seems unlikely that immediately after sexual maturity there should be a decline in ovulation rate. Evidence to support some decline in ovulation rate is presented from Masaki's data of corpora against age, Figure 9.5, where he suggests that the data are best described by two straight lines, that for the older individuals having a shallower slope. In addition he found that apparent pregnancy rate, measured by the presence of fetuses in the caught whales, was constant for whales with up to 20 corpora but declined rapidly after that, Figure 9.7, but this would not generate a decline as seen in Figure 9.6.

Data from other regions do not support the conclusion of a decline in ovulation rate with age. From the Southern Hemisphere the apparent pregnancy rates of sei whales caught over the period 1965 to 1976 showed a negligible decline of only 0.15-0.3% per years age (Masaki 1978) and importantly no decline is seen in North Pacific material. The plots of corpora against age, from the Southern Hemisphere over the same period, do not indicate anything other than a linear relationship, and from

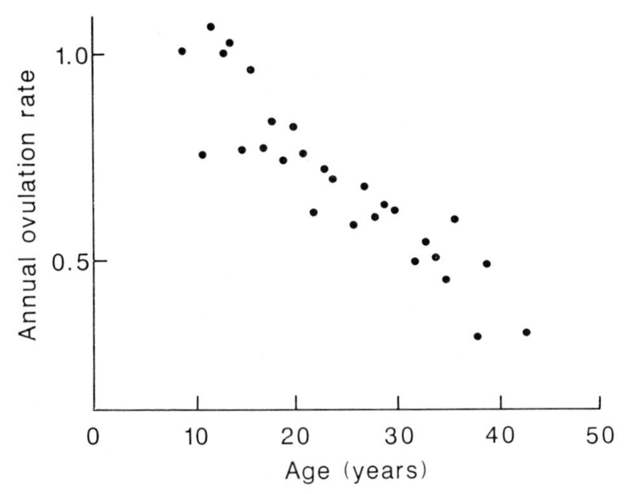

Figure 9.6: The decline in mean annual ovulation rate in North Pacific sei whales as calculated from number of corpora divided by the years after maturity. The rate is the average rate up to the age indicated.

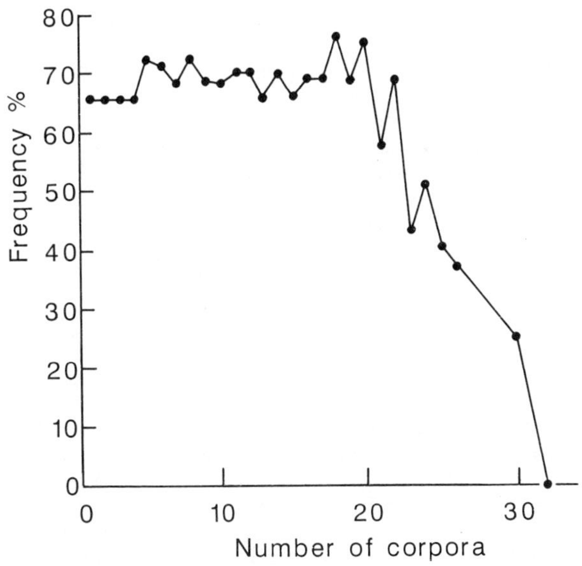

Figure 9.7: Apparent pregnancy rate of North Pacific sei whales against number of corpora in the ovaries of the whales.

Antarctic materials Gambell (1968) found no decline in pregnancy rate associated with whales having a large number of corpora. From the North Atlantic Lockyer and Martin (1983) state that they found no evidence to support a decline in pregnancy with age or corpora. Consequently the two sets of North Pacific data, that is those showing the decline in ovulation rates with age after maturity and those showing a reduced pregnancy rate for whales having a large number of corpora, appear different from material obtained in other studies.

It is evident that this issue is as yet unexplained but a linear relation of corpora with age provides a good fit in the majority of cases, and can be used to calculate average ovulation rates. For North Pacific sei whales under 24 years of age Masaki (1976a) gave a rate of 0.60 y^{-1} and Rice (1977) found a rate of 0.70 y^{-1}. From the Southern Hemisphere, Masaki (1978) gave rates of 0.42 to 0.57 y^{-1} depending upon the location of catches, Lockyer (1974) found a rate of 0.68 y^{-1} and Best and Lockyer (1977 ms) a rate of 0.63 y^{-1}. From the North Atlantic we have values of 0.26 y^{-1} from Canadian catches (Mitchell and Kozicki 1974 ms) and 0.59 y^{-1} from Iceland (Lockyer and Martin 1983). Except for the Canadian data, which indicate a three year cycle, the other data tend to indicate an average two year cycle. It is possible that ovulation rates can respond to the numerical status of the stock so trends with time may confound the interpretation with age; for instance the trend in the North Pacific could have been generated by low ovulation rates in the early years, and the differences in the rates amongst the different areas may be a real feature. Alternatively some of the apparent differences may reflect differences in interpretation of ear plug growth layers.

LENGTH AND AGE AT SEXUAL MATURITY FROM DIRECT OBSERVATIONS

Sexual maturity occurs over a range of lengths and ages and this feature is most easily shown by plots of proportion mature against age or length, which are usually smooth relationships if there are enough samples. For comparative purposes the median length or age is given as that for 50% mature after a smooth curve has been drawn, and this parameter is often used in simple population models as representing the length or age at which there are an equal number of immature whales at longer lengths or older ages as there are matures whales below these values, that is as a knife-edge length or age at maturity. Because of mortality over the range of lengths this parameter tends to be slightly biassed, and modifications to reduce the bias are reviewed by DeMaster (1978, 1984). It will be seen that there are clear differences in lengths and ages at maturity between male and female sei whales and data have been reviewed by ocean areas since different rates of exploitation may have generated changes in length and age at maturity.

North Atlantic

Males and females. Early information is reviewed by Tomilin (1967) and for males he cites Ruud as giving the size of the smallest mature male from Norwegian waters as 10.98 m and the largest immature male as 11.89 m: for females the respective lengths were 13.11 m and 13.72 m. Jonsgard and Darling (1977) give data from Norwegian captures from 1951 to 1957 and the median lengths at sexual maturity for males and females are 12.8 m and 13.4 m respectively. From Canadian catches from 1966 to 1972 Mitchell and Kozicki (1974 ms) give plots of testis maturity with length and ear plug layers. Taking their stages 3 and 4 as mature, 10% are mature at 11.9 m and 75% at 12.2 m; consequently 12 m appears to be about the median length at maturity. Mitchell and Kozicki consider that mature males can be recognised by having a combined testes weight exceeding 1000 g. On this basis also, the point of 50% maturity can be recognised as about 12 m, somewhat smaller than in other studies. Age at maturity of males from the Canadian catches is between 8 and 12 years, but there are too few data to be more precise. For females their figure 25 shows that 50% of the whales have one or more corpora by the time they have achieved lengths of 13.1 to 13.4 m (43 to 44 ft), and they show that although very few females are immature at 13.7 m or above, several immature whales were found with over 10 growth layers in the ear plug. Inspection of the proportion mature by age shows an erratic pattern from 7 to 20 years, presumably because of the limited data. The accumulation of corpora with time is described by a linear regression and this predicts an age of 6.15 years for the mean age when one corpus is found. This looks too low and is probably due to the scarcity of data, and low correlation, and the estimated age at maturity using corpora as the independent variable is about 16 years.

Lockyer and Martin (1983) described data on maturity of Icelandic sei whales from catches over the period 1967 to 1981. From histological examination of material only from 1977, they found both immature and mature males at a length of 13.1 m (43 ft); below that there were only immature males and above only mature individuals. From this study they considered that an individual testis weight of 1100 g could be used to separate mature and immature sei whales and they then applied this criterion to the greater time series of data from 1967 to 1981. This showed that at 12.5 m and below most males were immature and at 13.1 m and above most were mature. The mean length at sexual maturity is then about 12.7 m. Using the same criterion with age gave a mean age at maturity of about seven years. For females the rate of formation of corpora gave a predicted age at one corpus of 8.2 years. Finding the point of 50% maturity at age gave a value of 5.6 years (corrected for the small biasses discussed earlier) but they note that this statistic is likely to be biassed by selection for the larger whales at age, particularly in the younger ages. They found that maturity occurred over the lengths 12.8 to 14.0 m with the point of 50% mature at about 13.1 m.

Consequently the three studies for the North Atlantic gave estimates at 50% maturity for males of 12.8 m from Norway, 12.0 m from Canada and 12.7 m from Iceland. For females the lengths are 13.4 m, 13.1-13.4 m and 13.1 m respectively. The values are very similar for females but low for males from Canada; however the agreement would be much closer if testis weight at maturity was assessed at 2000 g combined, as in the Icelandic study. The median age at sexual maturity is less well defined and will be badly underestimated due to selection for the larger whales at age.

North Pacific

Males and females. For male sei whales, caught over the period 1948 and 1949 from Hokkaido and northeast Honshu, Omura (1950a) gave the proportions mature by length class. Fron Honshu the length at which 50% were sexually mature was 12.6 m and from Hokkaido 12.8 m. The transition from immaturity to maturity is very sharp from the Hokkaido data, but more extended from Honshu, and the distributions are not the same although the median lengths differ by only 0.2 m. Maturity was recognised by each testis exceeding 1.5 litres (calculated in the now standard way of multiplying length x width x thickness of the testis), or about 1 kg. This difference from the two localities is also shown for females, later, and one wonders if a few Bryde's whales may have affected the numbers. There seems no reason to believe in a segregation of small mature sei whales from similar sized immature ones between the two locations, which are essentially feeding sites. These samples were extended for the years 1950 and 1951, (Miyazaki 1952), and from Hokkaido a length at 50% maturity of 12.9 m can be recognised for males.

More recently Masaki's (1976a) study used material mainly from 1970 and west of 180°, but, in his more precise presentation, male length at 50% maturity is confused because of the explicit consideration of pubertal animals. Immature males constituted 50% of the catch at a length of 12.7 m and the mature at 13.0 to 13.2 m; a more precise figure depends on how the pubertal group is allocated. From this study he concluded that a single testis exceeding 900 g could be considered mature and this criterion was applied to samples from 7,178 male sei whales caught from 1963 to 1965 from coastal operations. The value of 900 g appears a good estimate for the purposes of finding a 50% mature statistic since this includes about half the pubertal group. Masaki gives plots of percentage mature with length, by longitudinal divisions, for males and females and it is evident that there is no difference with longitude; the overall length at 50% maturity is 12.9 m, and this is similar to the value of 12.8 m found by Rice (1963) from California. Masaki discusses the age at sexual maturity and, from 1924 specimens, the age at 50% maturity was precisely defined at 2.5 years; although he draws attention to the problems of ageing sei whales this very low value is largely associated with selection and the tendency for maturity to be determined by length rather than age.

From the presence or absence of corpora Omura (1950a) tabulated the proportions of mature females from Hokkaido and Honshu, and the lengths at 50% maturity are about 13.5 m and 13.1 m respectively, but as for males the distributions are different in the two areas. Miyazaki's (1952) data from Hokkaido in 1950 and 1951 indicate a length at 50% maturity of 13.4 m. From 5830 samples collected from 1952 to 1972 Masaki (1976a) found that 50% had one or more corpora at 13.4 m. Rice (1963) found a value of 13.7 m. Masaki's data show that no significant area effect exists. Age at sexual maturity from the catch is precisely defined at about 6.5 years but problems of selection may or may not exist. The minimum size limit for pelagic operations was 12.2 m (40 ft) and it is possible that the 6.5 years is not affected by selection. From regressions of ovulation rate against age, the estimate of age at one ovulation was found to be 5.2 years from all North Pacific samples.

Because of early problems of species identification, for the early studies it is best to rely only on the data from Hokkaido, and hence for males we have lengths at 50% maturity of 12.8 m (Omura), 12.9 m (Masaki) and 12.8 m (Rice), all relatively consistent. Age at maturity is ill defined. For females we have 13.5 m (Omura), 13.4 (Masaki) and 13.7 m (Rice). Again age at sexual maturity is ill defined but 6.5 years is likely to be a more accurate figure than the 5.2 years and comes from data collected from 1967 to 1972. As there is no difference with longitude the lengths at maturity from Masaki's greater sample size should be accepted.

Southern Hemisphere

Males. The volume of testes with body length was described by Matthews (1938) and if 1.5 litres is taken as an indication of maturity then his rather limited data show that maturity occurred between 13 and 14 m. Bannister and Gambell (1965) sampled sei whales from Durban in 1962 and 1963. They do not describe the testis weight or volume used in their study, but from examination of testes they found that 50% matured at a body length of 13.6 m. Gambell (1968) also gave data on mature, pubertal and immature whales with body length, caught between 1960 and 1966 from both Antarctic and Durban, and if pubertal whales with a testis over 1.5 litres are categorised as mature, then it can be seen that maturity occurred at between 13.1 to 13.4 m. By region Gambell gave estimates of 13.3 m from the Antarctic and 13.5 m from Durban. Nasu and Masaki (1970) reported on the maturity with length of sei whales caught by the Japanese Antarctic pelagic fleets from 1963 to 1969. Maturity was determined on the basis of a testis weight exceeding 1 kg and

Table 9.2: Length in metres at which 50% attain sexual maturity. Antarctic catches 1963 to 1969.

Area	III (0-70°E)	IV (70-130°E)	V (130°E-170°W)	VI (170-120°W)	All
male	13.2	12.9	13.3	13.3	13.3
female	13.9	13.8	14.3	14.4	14.0

Table 9.3: Length in metres at which 50% attain sexual maturity. Antarctic data 1956 to 1976.

Area	I	II	III	IV	V	VI
male	13.0	13.5	13.2	13.3	13.4	13.2
female	13.6	13.9	13.9	14.2	14.3	14.1

their extensive data allow the length at 50% mature to be determined precisely. The values for the whaling Areas III to VI are given in Table 9.2 and they range between 12.9 and 13.3 m, with the length from Area IV (70-130°E) somewhat smaller. This study was extended by Masaki (1978) using material from Antarctic catches from 1956 to 1976; although no criterion for male maturity is mentioned it can be assumed to be the same as in the previous Japanese study above. Except for Area I (60-120°W) the large amount of data means the lengths at 50% maturity are well defined and they are given in Table 9.3. No large differences are apparent with Area and values range between 13.0 and 13.5 m. Maturity was assessed in catches from Donkergat by Best and Lockyer (1977 ms) and if histological material was not used a combined testes weight of 3 kg was used to attribute maturity. They found a median length at maturity of 13.9 m which is higher than the other records and is almost certainly due to the different criteria for maturity. Lockyer (1974) found the proportions of sexually mature sei whales with age and then used the growth curve from the catch to predict a mean length at maturity of 13.6 m. However, as has been previously described, this is only valid so long as the growth curve is representative of the population.

The proportions mature with age were described by Lockyer (1974) for sei whales caught at Durban from 1960 to 1965, and by Best and Lockyer (1977 ms) for whales caught at Donkergat in 1962 and 1963. Lockyer's study used the criterion of combined testes weight exceeding 3 kg and showed that none up to and including age 5 years were mature. From age 9 over 75% were mature and the rapid change occurs over ages 6 to 8 years. Lockyer chose an age of 50% maturity of 7.47 years but the actual value could be anywhere between 6 and 8 years. Estimated age at maturity appears more variable than length at maturity. A similar criterion for maturity was used by Best and Lockyer (1977 ms) for the Donkergat samples. The spread of ages over which maturity occurs is about the same (5 to 10 years) and the median age is more precisely defined at 8 years. Before age 10 growth is rapid and selection of larger animals at age would tend to underestimate the median age at maturity in the population, but this bias would be less than two years. This is important in relation to trends in age at maturity described later.

Females. For female sei whales the numbers of corpora against length were plotted by Matthews (1938) and it can be seen that maturity occurs rapidly at about 14.5 m. The Durban data of Bannister and Gambell (1965) indicated a length at 50% maturity of 14.0 m, a length at which Matthews found none mature. Gambell (1968) regressed mean number of corpora on length to give a mean length at one corpus of 14.1 m, and a length at 50% maturity of 13.9 m from Durban and 14.1 m from the Antarctic. The lengths at 50% maturity from Japanese catches in the Antarctic over the period 1963 to 1969 are given in Table 9.2 by Antarctic Area and they range from 13.8 to 14.4 m. The differences do not appear to be due to statistical variability and they may be caused by a latitudinal segregation which is different with Area; sei whales are found further south in the Antarctic Areas V and

VI because of the more southerly position of the Antarctic Convergence. The lengths at 50% maturity from the more extensive data set of Masaki (1978) are given in Table 9.3. Larger median lengths are found in Areas IV, V and VI for females and these differences appear significant, and it would be of interest to inspect these data by latitude. Durban data from 1960 to 1965 are analysed by Lockyer (1974); by finding an age at 50% maturity and using a growth curve, a length at maturity of 14.0 m was calculated, the same as the overall average value given by Masaki. From Donkergat, Best and Lockyer (1977 ms) found a value of 13.9 m. Ages at 50% maturity from Lockyer (1974) are about 8.35 years, and certainly between 8 and 9 years. The similarity in the derived length at maturity with that of Masaki indicates a growth curve which is representative of the population over this range of lengths and ages. From Donkergat the corresponding age is 7.5 years.

From the Southern Hemisphere we see that 50% of the males are mature between 12.9 and 13.9 m. The lower values came from Antarctic samples and the larger ones from South Africa; however this is also confounded with slightly different criteria for maturity with a larger testis weight used in South African analyses. For females the range of lengths at 50% maturity are similar from the Antarctic and South Africa (13.6 to 14.3), but there is a significant difference in the samples from the different whaling Areas and a more detailed inspection of maturity with latitude is needed. Ages at 50% maturity are not as precisely defined as for length and the median age ranges from 6 to 8 years for males and 7.5 to 8.35 for females. These may be underestimates of mean age at maturity in the population as described above.

MATURITY FROM THE TRANSITION PHASE

General aspects

From sections of ear plugs it can be recognised that for the first few years the laminae are irregularly and widely spaced but later on a more regular distribution is found, and the earlier years of life show additional lighter laminations. For Southern fin whales Lockyer (1972a) found that the age at which the pattern changed, termed the transition phase or layer, was associated with the onset of sexual maturity. This is a most significant finding because it enables one to identify the age at which an old individual matured and hence develop a picture of changes in age at maturity with time and possible changes of population size. Consequently it is important to be convinced that the technique is valid. Lockyer presented a graphical portrayal of the spacing between adjacent laminae and showed the convergence to a more regular distribution; her interpretation of the position of the transition phase is also identified, and, from this random selection most do show a clear demarcation. Taking a testis size of 2.5 kg to represent maturity size she found that no immature fin whale had a transition phase, only 19% of pubertal males had one and 82% of mature males had a transition phase. For females no immatures had a transition phase, only 19% of pubertal females and 97% of older sexually mature females had one. Except for the problems of identifying a transition phase in recently mature fin whales and the few mature whales in which a transition layer was not found, it does appear that the presence or absence of a transition layer is associated with maturity. A second comparison that Lockyer used was to compare the age of a whale at capture, given by the total number of laminae, with the age predicted from the age given by the transition layer plus the number of corpora divided by the annual rate of corpora accumulation ($0.67\ y^{-1}$). A reasonable agreement was found and Lockyer says the two ages "agree closely". However a more pertinent question is whether or not they agree better than using a mean age at maturity and then adding on the age calculated from the corpora. From corpora data an age at one corpus was 6.7 years and a comparison can be made using the relationship: total age = 5.2 + corpora/0.67, where the 5.2 ensures an average age at one corpus of 6.7 years. Only eight specimens are used in the trial and the use of the transition phase gives only a slightly better relationship than does the use of the average age at maturity. However the deviants are dominated by one of the samples and this cannot be considered as a rigorous test or justification of the transition phase as an indicator of maturity. Nevertheless the previous findings do indicate that a basic association exists, and Lockyer showed that the growth changes of the ear plug are closely linked to body growth changes with which maturity may be associated. Using material from Southern Hemisphere minke whales Kato (1985) investigated the relationship between the number of corpora and the number of laminae after the transition phase. A linear regression predicted 1.03 corpora at zero layers after the transitions phase and this strongly supports the theory that the transition phase is formed at maturity.

The relationship of length and age at maturity in fin whales was also investigated by Lockyer (1972a), and, from the weight of testes, she found an age at 50% maturity of 6 to 7 years, for males and for females the presence of corpora gave an age of 6 to 7 years. From a regression of corpora data, an age at one corpus was found to be 6.7 years. Data from an extended period of time was used to construct a growth curve for the fin whales and this indicated a length at maturity of about 19.2 m for males and 20.0 m for females. The IWC minimum length for capture in the Antarctic was 17.4 m so it might be expected that these data are not seriously affected by selection. These lengths at 50% mature are similar to those reported earlier by Mackintosh (1942) and Laws (1961), for the same Areas of capture, of 19.2 m for males and 19.9 for females. For each of the six whaling Areas, from 1955 to 1963, Ichihara (1966b) found lengths at 50% maturity ranging from 18.6 to 19.1 m for males, and 19.7 to 20.0 for females. So within the Southern Hemisphere the lengths at maturity appear nearly constant and did not alter with time. However Ohsumi (1964a) reported on the age at 50% maturity of fin whales, caught in whaling areas IV and V (70° east to 170°W) from 1958 to 1960, as 9.4 years for males and 10.7 for females; both ages are precisely defined. Using corpora data gave an age at one ovulation of 10.8 years. These ages are very different from Lockyer's ages of 6 to 7 years, but subsequent analyses of a few ear plugs collected from Japanese expeditions in those areas in 1973 to 1975 do confirm this relatively higher age from the transition phase (Lockyer 1977d), and provide support for using the transition layer as an indicator of the age at maturity. In addition the fin whale material indicates that there may be differences in age at sexual maturity associated with time and location. Areas IV and V were subject to a later exploitation of fin whales than were Areas II and III. The tendency, in fin whales, of maturity to be at a constant length and for ages at maturity to have declined substantially with time and exploitation is clearly shown in the recent study by Ohsumi (1986) using materials from the North Pacific. One then must be sensitive to such possible changes with location and time when looking at similar materials from sei whales.

A transition phase has also been described from ear plugs of minke whales and the IWC ageing meeting of 1983 (Anon 1984) critically reviewed the aspects relating to identifiability of the transition phase in minke whales. The report and associated documents are especially important and those particularly interested in the transition phase material should read it. As originally noted by Lockyer (1972a) the transition phase in recently mature individuals is hard to recognise and it was found that, for females, the transition phase was only consistently recognised after four ovulations. For North Atlantic fin whales, Cooke and de la Mare (1984) show this effect persists for up to about six years after maturity. To show the trend in age at maturity with time, mean ages at maturity have been plotted against year of birth; however for whales maturing at, say about 10, and caught in 1980, the mean age at maturity, in the sample, will fall rapidly for the last 10 to 15 years of birth (1965-1980), as some potentially late maturing whales will have been caught before maturity and only early maturers will be represented in the later samples. This is clearly seen in Figure 9.9 and this feature is usually referred to as the "truncation effect". However the fact that recently matured animals are not recognised also exacerbates the problem. After a certain number of years back from the date of sampling this bias will have disappeared, but the fact that not all mature whales have a recognisable transition phase means that this time is uncertain. One way of partially, but by no means fully, overcoming this effect is to plot age at maturity against year of maturation (e.g. Best 1982). Cooke and de la Mare (1984) did this for Icelandic fin whale data and found recent median ages at maturity for both sexes of about nine years, and an earlier median age of about 12 years; however direct examinations of the catch showed ages at 50% maturity of six years for females and six to seven years for males. Lockyer and Butterworth (1984) presented maturity data for Icelandic female fin whales from 1978 to 1981 and showed that over this period 50% were mature at about 7.5 years which gave a mean age at maturity, using De Master's (1978) correction, of 8.6 years, which is more in agreement with the ear plug material. In addition, if selection is significant, the age at maturity from the biologically sampled specimens will underestimate that in the population. Consequently, it is not obvious that any discrepancy exists between the ear plug and biological material.

Comparative exercises with minke whale ear plugs at the IWC meetings showed up aspects of variability in ageing and recognition of the transition phase, and some large discrepancies between different ear plug readers were found. However not all readers were experts with minke whale ear plugs, and those who were, gave reasonable correlations. In general it is clear that there is an association of transition phase with sexual status and that this transition phase can be identified by different readers; however difficulties, and some substantial, can occur in interpreting trends from these data. Cook and de la Mare (1984) reviewed problems associated with possible biasses and variability, but the most perplexing aspect was presented empirically by Bengtson and Laws (1985), who illustrated declines in age at maturity of Antarctic crabeater seals (*Lobodon carcinophagus*), from the transition phase found in the teeth. Data from collections in 1972 to 1974 showed a steady decline from 4.5 years to 3.5 years in both sexes of ages over 7, from cohorts born in 1940 to 1965 - very similar to the trends seen in Antarctic whales. However samples from 1981 and 1982 showed a near identical trend for cohorts from 1950 to 1970, such that the lines showed similar slopes with one above the other. So far the only postulated mechanisms for this feature are either unidentified aspects in the interpretation of teeth laminae, or else a higher mortality rate of early maturing individuals. No one, as yet, has assessed the change in mortality rate required to generate such a trend but it would appear to be substantial, and hence would prohibit the reduction in age at maturity as a useful means of population homoeostasis; in the seals only one extra birth would ensue at some considerable later reproductive cost. Consequently, until these aspects are more fully understood one should interpret the trends observed from transition phase data with some caution.

Southern Hemisphere

The initial study of transition phase in fin whales was followed by one on southern sei whales caught in the South Atlantic, and from Durban, from 1960 to 1965 (Lockyer 1974). With mature males recognised on the basis of tubule size, sperm activity and

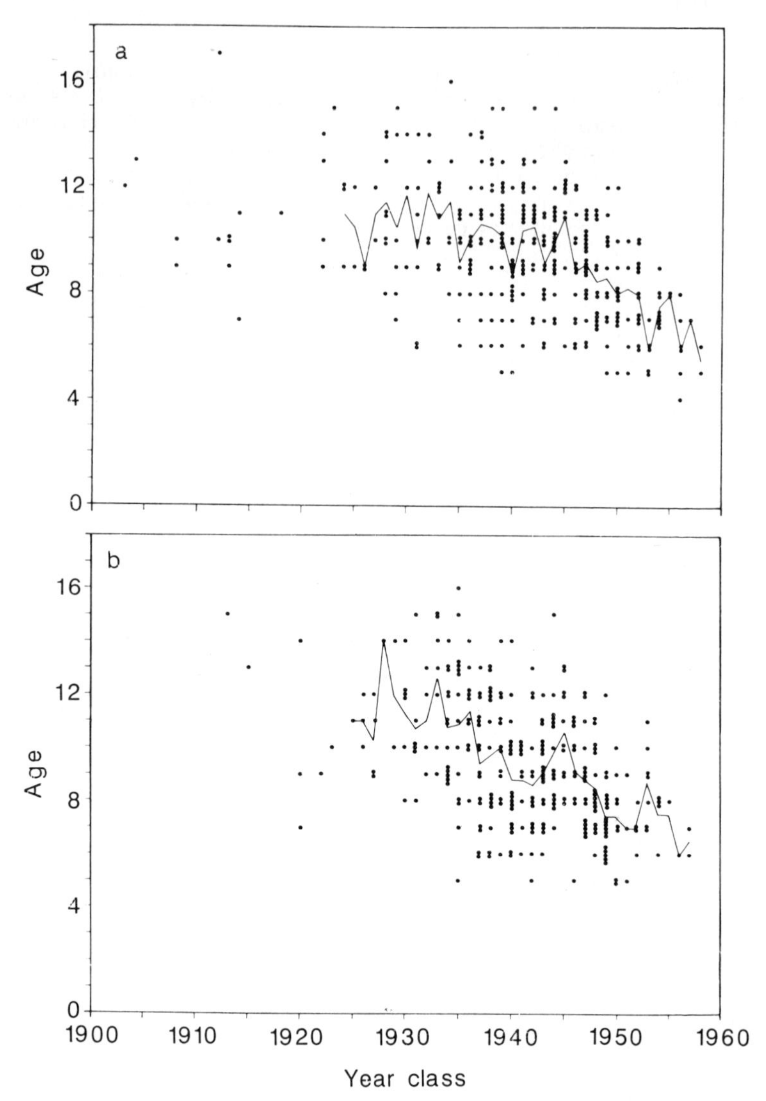

Figure 9.8: Age at maturity given by the number of growth layers up to and including the transition layers (with mean line) of sei whales from South Georgia and Durban from transition phase material. Data collected 1960-1965. (a) males, (b) females.

combined testes weight exceeding 3 kg, the percentage of sei whales having a recognised transition phase was calculated by sexual class; the percentage having a transition phase is given in Table 9.4. It can be recognised that the presence or absence of a transition phase is correlated with maturity, but as with fin whales, a transition phase was not recognised in most pubertal sei whales, and the lack of consistent recognition of a transition layer until a few years after maturity will account for most of the 11% missed in the mature group; from Lockyer's Figures 7 and 8 it can be interpreted that a transition layer was found in all old mature individuals. The number of growth layers up to and including the transition layer are plotted against year of birth in Figure 9.8. The first date of capture was 1960 and it can be seen that the oldest age at maturity was 17. This would indicate no, or negligible bias, due to truncation should exist for year classes before 1943. In addition, the fact that the transition phase might not be recognised for a few years after maturity means that this year of negligible potential bias should be reduced to 1940. In fact the selection need not be as strict as this for one could proceed in years until the first year class in which an ear plug with no transition phase is encountered. The female data show a statistically significant reduction in the mean age at maturity from 11.2 years in the pre-1930 year classes to 9.9 in the 1930-1940 year classes. For both males and females the mean ages at maturity are both highly variable and have declined over the period of apparently unbiassed data. Lockyer grouped the data by 1925 to 1934 year classes and 1940 to 1944 year classes and Figure 9.9 shows the percentage with a transition phase by age. The figure shows that between the two periods the mean age at maturity declined from 10.2 to 9.3 years for males and 10.6 to 8.4 years for females. The independent biological data from 1965 from Durban showed an age at 50% maturity of 7.47 for males and 8.35 for females; an approximate corrected value is 8.33 years for males and 9.72 years for females. The lower value for males tends to support a continued decline but the value for females is between the two. From the transition phase data, it seems that the main difference is caused by the fact that from the earlier years few whales matured before eight years but a significant proportion matured before age eight in later years.

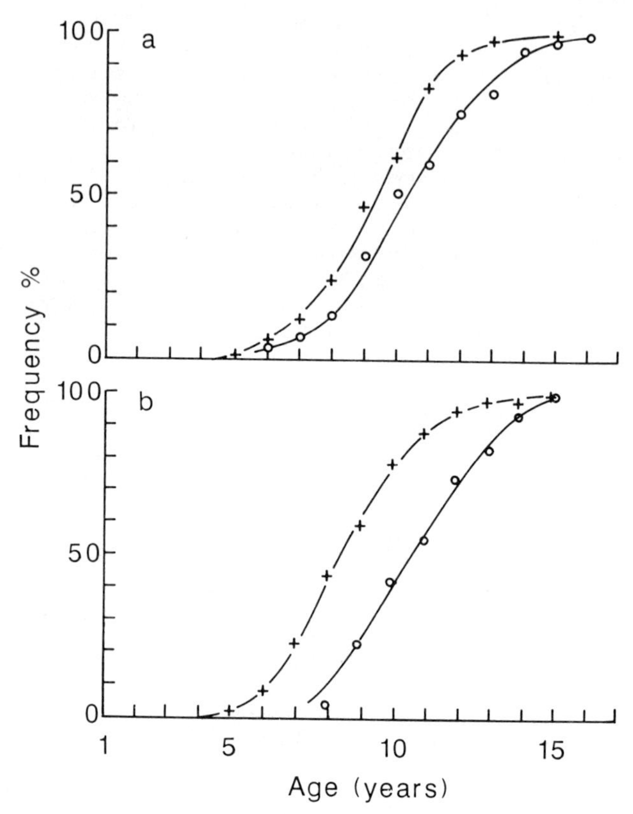

Figure 9.9: Percentage of whales maturing at age from transition phase data from the two periods 1925-1934 (o) and 1940-1944 (+). (a) males, and (b) females.

Table 9.4: Percentage of sei whales having a recognised transition pase by sexual class

Sexual class	Immature	Pubertal	Mature
male	5%	38%	89%
female	4%	20%	89%

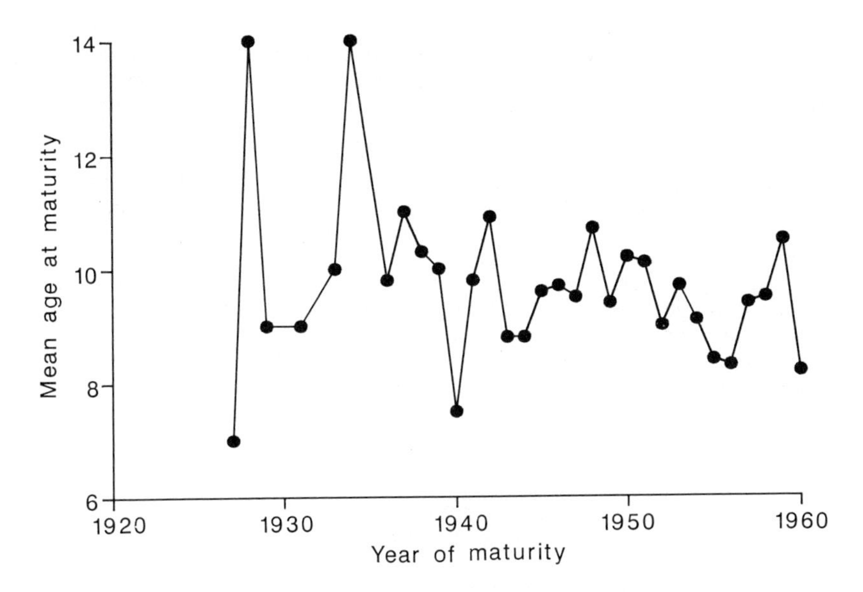

Figure 9.10: The mean ages at maturity against year at maturation for the female, data from Figure 9.8b.

Table 9.5: Relative declines (%) in mean ages at maturity from transition phase data relative to the pre-1920 year class by Antarctic whaling Area (1945-1950 values are approximate)

Area	II	III	IV	V	VI	I
1945-1950						
males	20	16	8	9	12	)11
females	26	22	0	4	12	)
1955-1960						
males	21	31	14	17	12	)26
females	26	28	6	17	17	)

An alternative presentation of these data is to inspect the age of maturity with year at maturity. This allows a seemingly unbiassed statistic up to very recent years, and the transformed data for females are illustrated in Figure 9.10. The data do not show a significant decline from 1930 to 1960 with a linear regression or geometric regression, although a negative slope is found. Grouping the transformed data by the years of maturation 1920-1940, 1941-1950, 1951-1960 gave mean ages at maturity of 10.15, 9.83 and 9.05 years respectively. The decline is not as pronounced as that implied by the previous analysis with a reduction of just over one year in the mean age at maturity of the females, but the mean values are statistically significantly different from each other. At the 5% significance level the 1951-1960 value is different from that of the previous two age groups. However the last few years at maturity do still suffer from the truncation effect, for although there is no upper limit to the ages at maturity there is a greater density of lower ages due to there being both more younger animals (because of mortality) and the fact that these cannot have or cannot be recognised as having a transition layer greater than a particular value. Consequently where possible only the unbiassed age groups should be considered with these analytical techniques, and the analysis based on survival rates of Harding and Horwood (in preparation) will allow the greater bulk of truncated data to be used. This data series was later extended.

For sei whales caught throughout the Antarctic, Lockyer (1979) described the relationship of age at maturity, determined by the transition phase, and year of birth. The analysis was presented for each of the six whaling Areas and was based upon ear plugs from whales caught in Japanese operations from 1973 to 1976, and the previously described material from 1960 to 1965 catches. The results are illustrated in Figure 9.11 as mean age at maturity for different blocks of years. Lockyer presented "only unbiassed means" and for the two Areas (II and III) from where earlier samples were taken only unbiassed year classes were used and added to the unbiassed data from the later years. (In this instance the only bias considered is that of the truncation effect). By presenting data to 1960 it can be recognised that for the later data a cut-off at about 12-15 years has been chosen and this may be too small to eliminate all bias. In addition recently mature whales are unlikely to be recognised

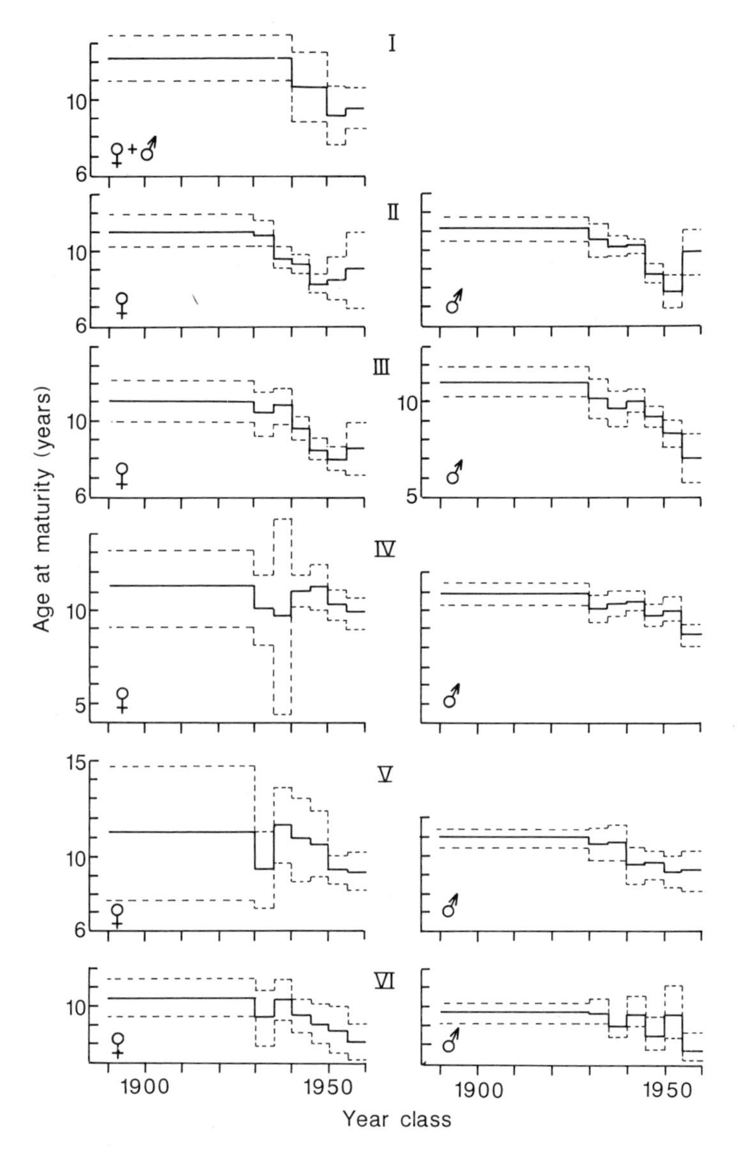

Figure 9.11: Mean age at sexual maturity from transition phase material by year class and Antarctic whaling Area with ± two standard errors.

as having a transition layer and so one needs to be looking at the data prior to 1950 for absolutely no potential bias from these sources, although bias in 1950-1955 year classes is likely to be small. If a similar cut-off was applied to the early data for Areas II and III it indicates that the estimates for the block of years 1940-1945 may be negatively biassed. From the results it can be seen that the initial ages at maturity were similar for both sexes with mean ages of 11 to 12 years. The relative declines in mean ages at maturity of the 1940-1945 and 1955-1960 year classes compared with those of pre-1920 year classes are given in Table 9.5. Declines are observed in all Areas with a 10% reduction representing a reduction of about 1 year in the age at maturity. The confidence intervals on the mean ages show that only in Areas II and III are there clear statistical differences amongst the unbiassed blocks of years, and the evidence for general declines is supported more by the consistent trend of decreases than by a series of statistically significant results. The greatest declines are found in Areas I, II and III and in these three Areas, catching was greater prior to 1930 than in the other three Areas; the declines in ages at maturity also occur earlier in these Areas. Total accumulated catches until 1970 were also greatest in Areas II and III and the implication is that the changes in mean age at maturity reflect a homoeostatic mechanism that is responding to a reduction in sei whales; however, as described later, interspecific effects may also have played a role.

North Pacific

From the North Pacific Masaki (1976a) presented information on age at sexual maturity from transition phase material and his results are shown in Figure 9.12. The data are from catches over the years 1968 to 1973 and consequently we might expect the data to be progressively biassed from 1950 onwards. Masaki noted that the age of maturity of those born prior to 1930 was about 10 years in both sexes and declined to six to seven years after 1960; however the later data are clearly strongly biassed. The data from the males was grouped by the year classes pre-1930, 1931-1935, 1936-1940, 1941-1945, and 1946-1950 to give mean ages at maturity steadily declining as 10.13, 10.04, 9.68, 9.49 and 8.77 years respectively. Only the final group of year classes showed a significant difference, but a continuous

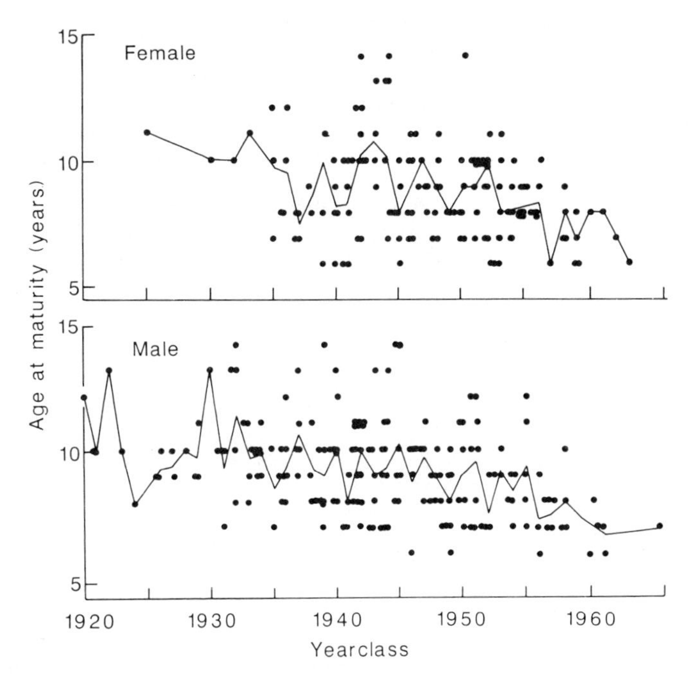

Figure 9.12: Age at sexual maturity by year class from transition phase from North Pacific catches in 1968 to 1973.

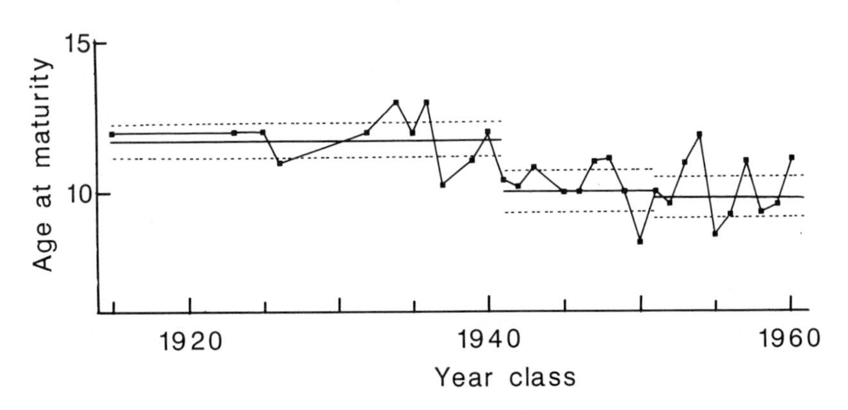

Figure 9.13: Age at sexual maturity from transition phase, data from Icelandic catches, both sexes combined. Dashed line represents 95% confidence limits on the mean.

trend can be recognised. For females the data were grouped as year classes from pre-1940, 1941-1945 and 1946-1950 to give mean ages at maturity of 8.88, 9.49 and 8.83 years respectively; the difference, between the last two groups is not significant. Consequently we might conclude that a trend was found for the males and not the females, but if this is correct it is somewhat surprising that the growth of the two sexes has responded differently with time, and the difference may be due to the low sample size; up to the 1950 year class this is 144 for males and 90 for females.

North Atlantic

Transition phase data from the North Atlantic were presented by Lockyer and Martin (1983) from Icelandic catches over the period 1967 to 1981. Their results are illustrated in Figure 9.13 for both sexes combined. From grouping various year classes, mean ages at maturity of about 11.7, 10.0 and 9.8 years can be recognised from the year classes pre-1940, 1941-1950 and 1950-1960. I assume that, as in the study of Lockyer (1979), unbiassed year classes from each year of capture were combined and Lockyer (1978a) reported that data were biassed from 1955 on, giving a cut-off value of some 12 or 13 years. This may be inadequate but in this case the post-1950 data are very similar to the 1940-1950 year class data which should not be affected by truncation bias. It can be seen that the early and later series are significantly different with a reduction in the mean age at maturity of some 1.7 years. A rapid decrease would not be expected when maturity is linked to growth which would take place over a 5 to 10 year period and with the eye of faith, a more steady decline can be perceived from 1934 to 1945 or 1950.

Summary

The previous description has shown that a transition phase can be identified in the sei whale ear plug some few years after it was established. It has been argued that this is associated with changes in growth with which maturation is closely associated, and a range of evidence provides strong support for the interpretation that the transition phase is formed in the year of maturity.

Trends in time of the age of the transition phase, from data free from truncation effects, have been found in all oceans. For sei whales significant declines have been found in Southern Hemisphere Areas II and III of up to 3 years, but more usually of about 1.5 years. Other Areas of the Southern Hemisphere indicate declines but not of a statistically significant nature. Significant declines of about 1.5 years were also found from Iceland and from North Pacific males. The largest declines and earliest declines are associated with the Southern Hemisphere Areas of greatest sei and baleen whale exploitation.

The interpretation of these trends was initially thought to be straightforward as a response in growth rate to declining stocks of sei or other baleen whales. However the empirical evidence from crabeater seals that a year-at-capture effect may explain all the trend, and supported somewhat by statistical analyses of whale data, indicates that the previous interpretation must be treated with caution until these issues are resolved. The historical material from the ear plugs is invaluable and the resolution of these statistical and biological problems deserves the greatest priority. Evidence from other species of whales is that there have indeed been changes in ages at maturity.

PREGNANCY RATES

The ovarian corpora provide a record of both successful and unsuccessful ovulations and are likely to give overestimates of population pregnancy rates; that is, the proportion per year of pregnant females of mature females. Estimates of pregnancy rates are obtained from examinations of caught females where the presence or absence of a fetus is recorded: this process is fraught with problems. On the summer feeding grounds the fetus size is usually large enough not to be missed if looked for, but in the winter whaling areas, conception has only just occurred and fetuses are easily missed without a thorough examination. As described earlier, there is a sexual segregation observed during migration with the pregnant females tending to arrive on the feeding grounds first; consequently early summer estimates of pregnancy rate are likely to be inflated. There is a latitudinal segregation with size in the Antarctic and this may influence the proportion pregnant. It is also possible that pregnancy rates can respond to fluctuations in the abundance of the stock or related species and the different stocks have a different history of exploitation. Consequently we might wish to describe pregnancy rates by stock, month, latitudinal zone, age and year. However, a complication is introduced by the operations of the whaling industry. From 1929 the Norwegian government had ordered the protection of calves, and cows with calves, and this was also adopted by the generally signed Geneva Convention of 1931, and has remained in all subsequent whaling conventions. The result is that a large proportion of those whales unlikely to be pregnant are excluded from the catch, and the apparent pregnancy rate is much greater than the population (true) rate. The term apparent rate is used throughout to mean the rate found in the sampled catch. In the early months of the summer season almost all calves will be near to their mothers but as the season progresses the calves become weaned, drift away from the mother, and allow the whales in late or post-lactation to be caught. Consequently a pronounced trend with month is introduced. All these aspects need to be considered to arrive at an estimated population pregnancy rate.

Trends with age

As we have just seen there is some argument as to whether ovulation rates decrease with age. Pregnancy rates with age have been described for sei whales by Gambell (1968) and Masaki (1976a), as pregnancy rate against corpora count, and by Masaki (1976a, 1978) as pregnancy rate against ear plug laminae. Masaki (1976a) shows a rapid decline in pregnancy rate with corpora (Figure 9.7); whereas Gambell (1968) does not; neither set should be affected by the problems described earlier.

From the North Pacific, data from whales caught over the period 1963 to 1973 shows that apparent pregnancy rate with age is nearly constant with a possible reduction after 40 years, Figure 9.14. From the Southern Hemisphere Masaki analysed the trend in apparent rate with age from the periods 1965/66 to 1970/71, 1971/72 to 1975/76 and 1965/66 to 1975/76, in order to avoid complications of trends by year. He found that over the ages 6 to 38, declines of 0.15 to 0.30% per year existed, equivalent to a decline of about 6% from age 10 to 40 years. Consequently no strong evidence of a decline in pregnancy rate exists from these data, and if the rates from 40 to 50 years are also

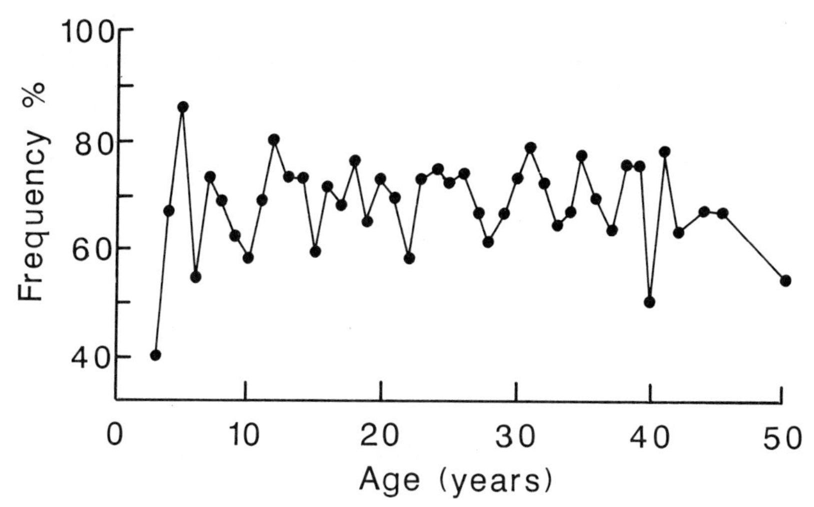

Figure 9.14: Apparent pregnancy rate with age in North Pacific sei whales.

considered the small, observed decline disappears; for practical purposes and subsequent analyses any trends with age are such that they need not be considered.

Trends with month and latitude

The trend of pregnancy rate with month in sei whales taken at South Georgia from 1927 to 1931 was described by Matthews (1938), with the three months of February to April giving apparent rates of 71, 47 and 14% respectively, and a similar trend can be found in the Antarctic and South Georgia data of Gambell (1968), Table 9.6, where the rates fall from 80-100% in October to January, to 38-50% in March and April. Doi, Ohsumi and Nemoto (1967) showed that apparent pregnancy rates in the pelagic Antarctic fell continuously from 70.5% in January to 45.3% in April. Clearly the protection of lactating females introduces a large trend with month, but there is also a trend with latitude. Ideally one might like to analyse the data stratified by month, season and location, to simultaneously describe all effects, and such BIWS data are tabulated by Mizroch (1980) from 1934/35 on, but let us inspect simpler summaries first. From Japanese data from the Southern Hemisphere from 1956/57 to 1975/76 Masaki (1978) extracted the apparent pregnancy rate by month and ten-degree latitudinal series. These are shown in Table 9.7. The trend with month is less

Table 9.6: Apparent pregnancy rate by month in the Antarctic. (1960-1966).

Month	Oct	Nov	Dec	Jan	Feb	Mar	Apr	Total
rate(%)	100	83	88	80	69	50	38	69

Table 9.7: Apparent pregnancy rate (%) from Southern Hemisphere sei whales by month and Series. (1956-1976).

	Series			
	40-50°S (D)	50-60°S (A)	60-70°S (B)	Total
Month				
Dec	65.5	51.9	-	65.4
Jan	58.0	66.5	69.2	59.9
Feb	52.0	61.4	67.8	58.5
Mar	45.5	52.0	57.1	50.4
Apr	50.0	75.0	-	62.5
Total	59.8	60.7	67.4	60.0

Table 9.8: Apparent pregnancy rate (%) of Southern Hemisphere sei whales by whaling Area. (1956-1976). I: 60-120°W, II: 0-60°W, III: 0-70°E, IV: 70-130°E, V: 130°E-170°W, VI: 120-170°W.

Area	I	II	III	IV	V	VI
rate	48.6	61.6	53.9	63.1	65.1	55.6

pronounced, but still exists, and it is seen that pregnancy rates increase towards the high latitudes, but with little difference between the rates in the D(40-50°S) and A(50-60°S) Series. Nasu and Masaki (1970) also found little difference north and south of 50°S, but Doi, Ohsumi and Nemoto (1967) described a strong change from January to April. Most catches are from north of 60°S and, if necessary, it seems permissible to neglect the series effect in analyses of other factors.

From the North Pacific similar trends in pregnancy rates were plotted by month and latitudinal band by Masaki (1976a) and he found a decrease in apparent rate from May to August, but there was no clear latitudinal effect.

Trends with stock or area

The discrimination of populations within the ocean regions is discussed in Chapter 10, but for management purposes the sei whales have been divided tentatively into sub-populations. Certainly different parts of the oceans have been subjected to varying degrees of exploitation and if there are homoeostatic mechanisms one might expect to find different pregnancy rates. In the North Pacific Masaki (1976a) described a decline in apparent rate with longitudinal sectors from 160°E to 130°W. In the Southern Hemisphere Masaki (1978) gave average apparent rates from the period 1956/57 to 1975/76, Table 9.8, and found a difference with whaling Area. Subsequent analyses of trends with time have tended to analyse the data by whaling region.

Trends with year

The numbers of sei whales have been reduced by exploitation since perhaps the beginning of the century, and certainly since 1950 in the North Pacific, and since 1955 in the Southern Hemisphere. It might be expected that, unless pregnancy rates were maximal before these times, increases in pregnancy rates would have occurred. It is also possible that reductions in the biomass of other baleen whales could lead to increases. From the basic two year breeding cycle and the few post-partum lactations it is unlikely that, in sei whales, true rates greater than 0.6 per year can exist, although it should be noted that it is believed that the minke whale has higher rates with a calving interval of about 15 months (a true pregnancy rate of 0.8 y^{-1}),

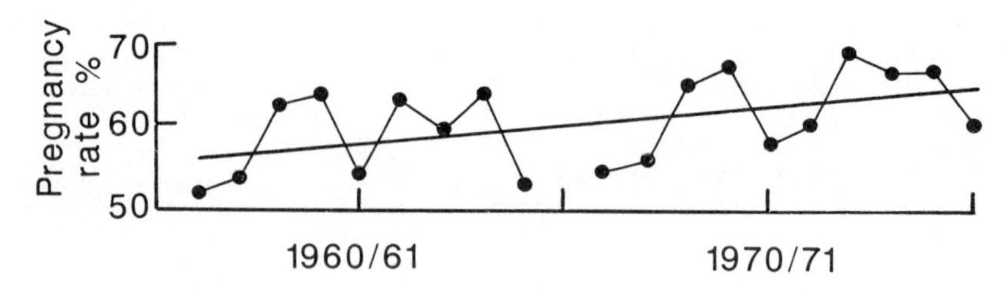

Figure 9.15: Variation of apparent pregnancy rate with year from the six Antarctic whaling Areas combined (Japanese data).

(Best 1982). However to assess the trends with season the basic measure used is the apparent rate and conversion to an absolute rate will be undertaken later.

Following on from the studies of Mackintosh (1942) and Laws (1961), that described large increases in the apparent pregnancy rates of Antarctic blue and fin whales from 1925 to 1941, Gambell (1973, 1975) investigated the time series of apparent pregnancy rates of Antarctic sei whales. His data, and that from the blue and fin whales are illustrated later in Figure 12.3, and it can be seen that in sei whales the pregnancy rate increased from about 25% in the immediate post-war years to rates of about 60% in 1969/70. This finding was of immense importance for it suggested a substantial degree of population homoeostasis and, as discussed in Chapter 12, significant interspecific interactions amongst the baleen whales. Analyses of trends with time over the later period of 1956/57 to 1975/76, from Japanese, Antarctic pelagic catches, are described by Masaki (1978) for each whaling Area, and Figure 9.15 shows the trend in apparent pregnancy rate for all Areas combined over this period. The data are grouped over months and latitudes. Masaki (1978) found increases in Areas IV to VI (70°E-120°W), and decreases in the remaining sectors. From all the Areas combined an increase of 0.45% per year was found. However it has already been shown that the period of intensive whaling for sei whales was associated with a northwards movement of the Antarctic whaling fleets and also, presumably, a change in the mean date of capture. To overcome these confounding factors Masaki (1978) investigated changes within the zone 40-50°S by month. For those whaling areas with sufficient data

Masaki concluded that, except for Area III in January, the remaining data sets showed an increase over this later period. However any increase is small and doubtful due to the large variation, and the more interesting consideration is whether pregnancy rates increased over the previous decades. Mizroch (1980) obtained an index of pregnancy rate using the BIWS data from 1930/31 to 1970/71, with factors of Area, month, latitudinal Series and year, and then looked for trends in the calculated year effect, but no significant trend with year was found; when the data were successively grouped, still no trends were detected. However the few data from the earlier years meant that the majority of the analyses were from the period of 1950, or onwards, and the sparse data prior to 1940 do indicate the possibility of low pregnancy rates.

Data from the earlier years are from catches from South Georgia, and these data were compared with later total pelagic data by Horwood et al (1980), extending the study of Gambell (1973). Gambell's earlier data were from the pelagic Antarctic and South Georgia catches combined, and these gave immediate post-war rates of 25%, but pre-war rates were not presented due to the small sample size. However over the period 1937/38 to 1945/46 the apparent rate was 30% from 444 mature females (13.7m). This is a much lower figure than the rate of 46% obtained from South Georgia, from 1926/27 to 1930/31 by Matthews (1938). Data from the BIWS are available from 1920/21 onwards but the catch was not recorded by sex or sexual status. These data and Antarctic pelagic and South Georgia catches are given in Table 9.9, and Horwood et al used them to compare with Matthews' data. It was assumed that from 1920 to 1936 both the proportion of females in the catch and the proportion of mature females was constant, since from 1937/38 to 1950/60 it was shown that these values were nearly constant, with 0.63 females of which 0.91 were mature, and the assumption gave a good prediction of observed rates until 1964/65. Applying these rates to the total catch of 1821, from 1920/21 to 1934/35, with 295 foetuses gives an estimated apparent rate of 28%. If parameters from the period 1937/38 to 1963/64 were used this predicted values is 33%. Both these are very similar to Gambell's figures and lower than those of Matthews, and Horwood et al argued that it was probable that the observed increase over the early years was real. One problem is that in the earlier years inspections were not always undertaken

Table 9.9: Apparent pregnancy rate data from sei whales caught at South Georgia and by Antarctic pelagic fleets. © Crown 1980.

Season	Total catch	No. foetuses	%	No. females	Sex ratio (%)	Mature females	% Mat.	Preg. rate
1920/21	36	1	3					
1921/22	103	1	1					
1922/23	10	1	10					
1923/24	191	44	23					
1924/25	1	0	-					
1925/26	13	0	-					
1926/27	365	31	8					
1927/28	95	17	18					
1928/29	396	63	16	No data		No data		
1929/30	216	48	22					
1930/31	144	47	33					
1931/32	16	-	-					
1932/33	2	-	-					
1933/34	0	-	-					
1934/35	265	42	16					
1935/36	2	-	-					
1936/37	490	18	4					
1937/38	161	24	15	103	64	99	96	0.25
1938/39	22	8	36	15	68	15	100	0.53
1939/40	81	15	19	51	63	48	94	0.31
1940/41	110	-	-	-	-	-	-	-
1941/42	52	2	4	27	52	24	89	0.08
1942/43	73	7	10	50	68	48	96	0.15
1943/44	197	32	16	120	61	96	80	0.33
1944/45	78	24	31	53	68	53	100	0.45
1945/46	85	24	28	62	73	61	98	0.39
1946/47	393	54	14	251	64	235	94	0.23
1947/48	621	94	15	354	57	346	98	0.27
1948/49	578	86	15	370	64	349	94	0.25
1949/50	1,284	310	24	796	62	773	97	0.40
1950/51	886	140	16	505	57	480	95	0.29
1951/52	530	127	24	302	57	293	97	0.43
1952/53	621	161	26	373	60	362	97	0.44
1953/54	1,029	220	21	566	55	521	92	0.42
1954/55	569	102	18	319	56	277	87	0.37
1955/56	560	171	31	358	64	348	97	0.49
1956/57	1,692	433	26	1,100	65	1,034	94	0.42
1957/58	3,309	901	27	2,151	65	2,093	97	0.43
1958/59	2,421	865	36	1,695	70	1,655	98	0.52
1959/60	4,309	1,225	28	2,758	64	2,669	97	0.46
1960/61	5,102	1,436	28	2,959	58	2,843	96	0.51
1961/62	5,196	1,501	29	3,221	62	3,068	95	0.49
1962/63	5,503	1,616	29	3,082	56	2,899	94	0.56
1963/64	8,695	2,084	24	4,260	49	4,075	96	0.51
1964/65	20,380	4,058	20	8,967	44	7,916	88	0.51
1965/66	17,587	2,741	16	7,387	42	5,644	76	0.49
1966/67	12,368	2,371	19	5,896	42	4,850	82	0.49
1967/68	10,357	2,047	20	4,701	45	3,950	84	0.52
1968/69	5,776	1,175	20	2,607	45	2,078	80	0.57
1969/70	5,857	1,312	22	2,725	47	2,188	80	0.60
1970/71	6,153	1,291	21	2,981	48	2,271	76	0.57
1971/72	5,456	995	18	2,530	46	1,808	71	0.55
1972/73	3,864	720	19	1,780	46	1,351	76	0.53
1973/74	4,392	784	18	1,976	45	1,633	83	0.48
1974/75	3,859	600	16	1,637	42	1,313	80	0.46
1975/76	1,821	381	21	829	46	655	79	0.58
1976/77	1,858	323	17	806	43			
1977/78	565	143	25	287	51			

by biologists and a trend may have been introduced through better reporting over the years. There does not seem to be a way of investigating this at present, unless some early expeditions did take biologists and then their records can be compared with others. The problem of whether the increase has occurred, from the values of about 30% in 1920 to 1944, to about 50% after 1964 is unresolved and more evaluation of the data and log-books is needed but my opinion is that the balance of evidence is in favour of such an increase.

From the North Pacific apparent pregnancy rates over the period 1952 to 1972 were given by Masaki (1976a), who found a decreasing trend with time, but with rates somewhat lower in the northern latitudes; he ascribed the decrease to an eastward movement of the whaling grounds. If this is so then it means that effects of time and longitude have not been disentangled. Apparent pregnancy rates were high at about 70%. From California, Rice (1977) gave pregnancy data for the period 1950 to 1970, and excluding lactating females a rate of 46.9% was found, which was much less than those from the more northern and western Pacific. Information on the pregnancy rates of early Japanese catches of sei whales was given by Omura (1950a). From the descriptions of Omura and Fujino (1954) the "sei" whales from Hokkaido (Kushiro and Kiritappu) and northeast Honshu (Kamaishi and Ayukawa) are virtually all true sei whales rather than Bryde's whales, and Omura presented graphical information on the number of whales from these regions with functional corpora lutea and the number with corpora albicantia. Reading from these rather difficult figures (his Figs. 37 and 38) gave a pregnancy rate of 20.8% from Honshu and 18.2% from Hokkaido; these are calculated as the number of whales with a functional corpus luteum divided by the number of whales with at least one corpus. The number of lactating whales recorded is very small and these rates can be considered an apparent pregnancy rate for the years 1948 and 1949. If both pregnant and ovulating whales are considered (Figs. 40 and 41) rates of 39.8% and 32.4% are found for the two areas. These are substantially lower than the later figures of Masaki (1976a); however I assume that as Masaki gave the mean length at maturity as 13.4 m (44 ft), his pregnancy rates are calulated as the number of pregnant whales divided by that of females of 13.4 m or over. Consequently I have repeated this exercise with Omura's data to give rates from Honshu of 28.7% (corpora data) and

51.9% (pregnant to resting), and from Hokkaido of 19.6% and 35.5%. These rates are still much lower than Masaki's and indicate a real increase in pregnancy rates from 1947 onwards.

Time series are not available from the North Atlantic but some estimates are available from the Canadian and Icelandic catches which indicate that low apparent pregnancy rates may be realistic; studies from both regions relied upon morphological or histological examinations to identify pregnancy. From Canada, Mitchell and Kozicki (1974ms) found that over the period 1969 to 1973 the rate was 45.7%. (Since 13 of 164 were found to be lactating and 69 pregnant I have calculated this rate as 69/(164-13)).

As demonstrated above there are indictions that pregnancy rates have increased over time. The lower apparent pregnancy rates, from early data, are found in the Southern Hemisphere and North Pacific and pregnancy rates from the North Atlantic do not appear as high as the present rates in the other two oceans. Evidence for such a change may be found in the reduced ovulation rates for older ages from the North Pacific studies, especially as pregnancy rates do not appear to decline with age; Masaki (1976a) found that ovulation rates of sei whales over 23 years was 0.44 per year and for the younger whales was 0.60. Lockyer (1974) found a rate of 0.69 from the South Atlantic, whereas the less exploited North Atlantic populations have an ovulation rate of 0.59 per year from Iceland, and 0.26 per year from Canada (Lockyer and Martin 1983, Mitchell and Kozicki 1974 ms).

Absolute pregnancy rates

Almost all pregnancy rates reported are apparent rates, due to the protection of lactating females, and assumptions need to be made to convert these rates to true rates representative of the population. Following Horwood et al (1980), assume the following model of the proportions of the different sexual classes in the population:

$$a + b + c + d = 1.0$$

where, a: is the proportion of the mature females simultaneously pregnant and lactating, b: is the proportion of mature females pregnant but not lactating, c: is the proportion of mature females resting and d: is the proportion of mature females lactating and not pregnant. It is assumed that the relatively few summer ovulations do not result in a pregnancy. If an equilibrium state exists then it can be assumed that the proportion pregnant $(b + a)$ will be the same as the proportion lactating $(d + a)$ and hence, $b = d$, and $a + 2b + c = 1.0$.

Most calves are weaned by February in the Antarctic and consequently, at that time, there is no protection of any component of the mature female population, and the apparent rate then may reflect the true population rate. Data from Masaki (1978), Table 9.7, indicated that in February and April apparent pregnancy rates were about 0.57 and therefore, $a + b = 0.57$.

It can then be found that for $c = 0.0$, $a = 0.14$ and for $c = 0.1$, $a = 0.24$, that is, even if no resting females existed in the population a post-partum pregnancy rate of 0.14 is predicted. Gambell's data indicate that, $a = 0.11b$, which is much lower than the model predicts. Clearly this model was wrong and Horwood et al (1980) suggested it was because of the differential migration of the sexual classes. In order to overcome this they focused attention on rates from the early part of the season when full protection was likely to exist. Data from Masaki (1978), indicated apparent pregnancy rates of 0.60-0.65 in December and January and then $b/(b + c) = 0.60\text{-}0.65$, and if $a = 0.11b$ then $b = 0.36\text{-}0.38$ and $c = 0.24\text{-}0.20$. Consequently the true population pregnancy rate $(a + b)$ is about 0.40-0.42 with about 22% of the females resting.

A modification of this model was employed at the IWC meeting on Southern Hemisphere sei whales in 1979 (Anon 1980). Firstly it was assumed that there was a 10% calf mortality between birth and arrival on the whaling grounds, although there is no specific justification for this figure, and thus we have $d = 0.9b$. In addition the post-partum ovulation rate of 0.11 was interpreted to give, $a = 0.055b$, which I consider erroneous. Retaining the first assumption only, and with p' defined as the apparent pregnancy rate determined early in the season, a true population rate, p, can be obtained from p $1.1\,p'/(1.0 + p')$. Apparent rates of 0.60-0.65 would give true rates of 0.41-0.43. This model was used to obtain population pregnancy rates of sei whales after correction for differences in monthly and latitudinal effects by Holt (1980) who presented tables of population pregnancy rate with season.

A somewhat similar exercise was carried out to obtain estimates of the population rate for Icelandic sei whales by Lockyer and Martin (1983), but as the sei whales are slit open at sea they used the presence of a large corpus luteum to indicate pregnancy. From 164 mature females caught in mid-season (13 August-11 September) 69 were determined as pregnant and 13 were lactating, and from a total of 15 lactating females none were pregnant. Assuming that there were no whales simultaneously pregnant and lactating, and that summer ovulations are few and can be neglected, they considered that true population pregnancy rates were between 0.37 and 0.46 per year, depending respectively upon whether lactation had ceased by 13 August or whether half of the calves were still suckling at that time.

FETAL SEX RATIO

The fetal sex ratio is obtained from examining the sex of fetuses from catches. From only 69 specimens Matthews (1938) found the proportion of male fetuses to be 0.65 but subsequent studies have shown the proportion to be near to 0.5. Data reported by Gambell (1968) gave exactly a rate of 0.5, and from BIWS records from 1934 to 1966, the ratio varied between 0.469 and 0.513. From Japanese Antarctic catches over the period 1966 to 1976 Masaki (1978) reported a rate of 0.509 males. These last two studies involved over 18,000 and 7,000 fetuses respectively. From the North Pacific over 3,000 fetuses from 1967 to 1972 gave a proportion of males of 0.506 (Masaki, 1976a). For sei whales the sex ratio is very near to 1:1; for other large whales there may be a small tendency towards an excess of males (Kato and Shimadzu 1983).

MULTIPLE FETUSES

In blue whales seven fetuses were found in a specimen from South Georgia, all of a reasonable size and state of development (Paulsen 1939), six have been found in fin whales (Jonsgard 1953) and four in sei whales (Masaki 1978). As demonstrated by the presence of only one, or more than one, corpus luteum and by the sex of the multiplets, they can arise both from a single ovulation or by polyovuly. Paulsen (1939) presented information of sei whale fetuses from the IWS and from 1921 to 1934 three sets of twins were found from 448 pregnancies, to give a rate of 0.7%. Ohsumi quotes Matsuura (1940) as also finding a rate of 0.7% in sei whales (Kimura

1957), and over the years 1933 to 1953 Ohsumi calculated that the BIWS statistics gave a rate of 2.3%, much higher than for the other large whales. However Gambell (1968) noted that from 452 pregnancies examined by biologists a rate of 1.1 to 1.3% was found. From Donkergat, Best and Lockyer (1977 ms) reported that 0.67% of pregnancies were twins in 1962 and 1963 and that from 1926 to 1967 0.85% had twins, and from Durban samples a rate of 0.97% was found. From the BIWS, for the seasons 1959 to 1966, 1.6% had multiple fetuses. Masaki (1978) described the incidences of multiplets from Japanese Antarctic catches from 1966 to 1976; from 7556 fetuses there were 95 twins, five triplets and one quadruplet to give an average litter size of 1.014. From the North Pacific, fetal information is given by Risting (1928) and Masaki (1976a); the latter found that 19 of 3686 pregnant sei whales had twins to give a litter size of 1.005. Ohsumi (ibid) discussed twinning in fin whales and found that fetal growth of twins was similar to that found in single pregnancies and that twinning might increase with age.

SUMMARY

The process of maturation of the testes is described and maturity can be determined by the diameter of the seminiferous tubules and more approximately by testis weight or volume. Maturity has been determined by mean tubule diameter exceeding 90 micron or single testis weight over about 1 kg; however different authors have used a range of 0.5 to 1.5 kg, and this has resulted in different sizes or ages at maturity reported in various studies. In addition, the statistics needed for most population purposes are the proportions mature at age, or in the population, of those mature plus half the pubertal group, rather than the proportions of certainly mature and for this a figure of 1 kg appears appropriate, although real differences may exist between different populations. Penis length over 95 cm is also an indicator of maturity. The evidence for seasonal cycles in the testes is contradictory.

The gross histology of the ovary is described and it is shown that corpora albicantia persist to reveal the reproductive history of the individual. As well as the presence and type of corpora, the uterus and mammary glands can be used to define sexual status. Gestation periods are not precisely defined with a range of estimates of 10.5 to 12.5 months. Conception occurs mainly in June and July in the Southern Hemisphere, and in November and December in the Northern Hemisphere, with the variation known more accurately than the mean dates. Most mature sei whales become pregnant from the first ovulation of the season, and there are about 1.1 ovulations per pregnancy. Pregnancy rates appear constant with age, but contradictory results are found for pregnancy rates against number of corpora. From most populations, rate of corpora formation is about 0.6 per year, but rates as low as 0.26 have been found, and older individuals from the North Pacific showed a lower rate than younger whales; if pregnancy is constant with age this can be interpreted as reflecting ovulation and pregnancy rates lower in earlier years, and the lower ovulation rates are associated with the less depleted populations. Most recorded pregnancy rates are artificially high as lactating females are protected from capture. The weaning in the summer and a differential migration of the sexual classes means that the apparent pregnancy rates vary considerably with month over the summer whaling season, and models to convert apparent rates to population rates are discussed. In the Antarctic there have been substantial differences with season, and although post-war apparent rates have been over 50%, rates from 1920 to 1940 were nearer 30%. Similar low rates were found in early North Pacific samples and from recent Canadian material. These would indicate real differences, although possible changes in efficiency of reporting of fetuses from the Antarctic need further investigation. Fetal sex ratios are equal and the average litter size is negligibly over one.

From direct observations over a large number of years the length at which 50% mature has been relatively constant. From males of the Northern Hemisphere this length has ranged from 12.7 to 12.9 m, with the only different statistic being the 12.0 m from Canada, and a much lower testis weight was used to determine maturity in the Canadian study. In the Southern Hemisphere the lengths are all much bigger ranging between 13.0 and 13.9 m. A similar difference was found in females with lengths at 50% maturity ranging from 13.1 to 13.7 m in the Northern Hemisphere, and 13.6 to 14.5 m in the Southern Hemisphere. It would appear that lengths at sexual maturity are different in the two hemispheres but have remained unchanged with time. Ages at sexual maturity are much affected by selection for larger individuals and this tends to underestimate age at maturity. For males it is not obvious how useful these data are, but for females regressions of corpora with age give estimates of age at one ovulation of 5.5 to 6.5 years from both hemispheres; however there may have been changes in age at maturity and ovulation rate with time.

From the ear plug a transition phase can be recognised by a change in the spacing and regularity of the laminae, and this is associated with a change in growth with which maturity is also associated. The transition phase provides historical information on possible growth rates and times to maturity of past year classes. A particular problem is the "truncation" effect which gives distributions of ages at maturity unrepresentative of the year class for year classes younger than the oldest possible age at maturity. In addition, a few years after the formation of the transition phase are needed before the transition phase can usually be identified. This means that data only from year classes older than 15 to 20 years usually can be used to describe trends with traditional methods; however it is anticipated that new methods of analysis will be developed so that the entire data set can be utilised. Using data free from this source of bias significant declines in the age to the transition phase have been found with time in the Southern Hemisphere, North Pacific and North Atlantic. The initial mean ages at maturity were about 10 to 12 years for both sexes with the maximum mean reduction about three years, but most of about 1.5 years. The largest and earliest declines are in those Areas in which whaling for sei and other baleen whales was greatest and earliest. However recent evidence both empirical and statistical has found that declines in ages at maturity from transition phase data can best be explained by an "age-at-capture" effect, and consequently this throws doubt on the interpretation of the observed declines being a real phenomenon. Nevertheless declines in mean age at maturity have been described in other rorquals, with data other than that using transition layers. Given the vast importance of this historical record immediate priority should be given to resolving the outstanding problems of interpreting the transition phase material.

As a source for recognising mechanisms of population homoeostasis we must look to reproduction, for estimates of mortality are very imprecise and detecting changes in mortality is not likely to be possible with any present techniques. It has been shown from pregnancy and ovulation data that calf production may have doubled over the period of exploitation and it is possible that the whales may have matured one to three years earlier. Consequently both these processes can provide homoeostatic mechanisms, although there are problems in proving that each has occurred. As to whether these changes are associated with interspecific or intraspecific effects will be discussed later.

Chapter Ten

STOCK SEPARATION

INTRODUCTION

Within species or subspecies, genetically different groups can frequently be identified and are referred to as distinct populations or stocks, and since relatively small mixing rates would, over several generations, cause these differences to disappear, the groups of animals are considered to be isolated breeding units. However other stocks that are evidently geographically isolated do not show any genetic differences and this reveals that the techniques of genetic separation can still be improved. In many cetacean studies, genetic techniques have failed to separate populations which phenotypic clues suggest are different.

Ideally a population or stock is a group of animals which are freely interbreeding with no immigration or emigration. Such a situation cannot be easily identified for whales, which show extensive movements, and physically and genetically indistinct boundaries. In the case of large whales, a starting point for the identification of stocks is to identify the breeding locations and to follow the movement of whales from these sites. If these whales are largely isolated then the essential dynamics of the populations can be described with reference only to their own numbers and environment; if there are interactions with other breeding groups then more sophisticated models may have to be used. Time scales play a significant role, for over short enough time scales mixing between populations can sometimes be neglected. Consequently we are led to a working definition of a population or stock, as that of a group of interbreeding individuals, whose dynamics can be described essentially in relation to themselves. If mixing is such that two breeding units need to be considered, in order to describe their mutual dynamics, then this definition would have to be expanded. In fact the management of the large whales has followed a more pragmatic approach to the definition of stocks. It has been assumed that true stocks do exist but that their delineation is very difficult, and consequently the ocean areas have divided into a series of management stocks or areas, on the implicit understanding that this gives greater flexibility and safety in management. One definition of a stock used by the Scientific Committee of the IWC (Anon 1977) is taken from the North West Atlantic Fisheries Commission as, "a relatively homogeneous and self contained population whose losses by emigration and accessions by immigration, if any, are negligible in relation to the rates of growth and mortality". This can be recognised as a definition adequate for management purposes. Jamieson (1973) reviewed the concept of "the stock" in relation to fisheries biology and management and discusses the concepts of genetic stocks, and the problems and principles discussed are identical to those confronting the Cetacean biologist. From the above three sorts of "stocks" have been described. The first a genetic stock, the second a stock that allows its variation in numbers to be adequately described with reference to itself and the third a management unit devised for practical or cautious reasons. Research into intraspecific variability, for population studies, strives to reconcile the three definitions. To illustrate the techniques used, the extent of current knowledge, and the difficulties in defining stocks of large whales, several examples are presented before the specific details of the sei whales are described.

In the Southern Hemisphere Hjort et al (1932) divided the Antarctic into five management areas on the basis of the distribution of mainly blue, but also some fin, whale catches. Later Mackintosh (1942, 1965) reviewed data on the distribution of catches of blue, fin and particularly humpback whales and divided the Southern Hemisphere into six Areas: I: 120°W-60°W, II: 60°W-0°, III: 0°-70°E, IV: 70°E-130°E, V: 130°E-170°W and VI: 170°W-120°W. The humpback whales showed particularly discrete units with the breeding areas just north of 40°S and concentrations of catches in the feeding areas due south. Mackintosh (1965) says "there is no reasonable doubt of the independence of these humpback groups" and cites as evidence the different responses of these groups to exploitation. In addition there are differences in the colour patterns of the humpbacks from the different groups (Lillie 1915, Chittleborough 1965), and marking returns indicated little mixing. These studies are of further practical significance for these six Areas have been used as a basis for management for all Southern Hemisphere, large, baleen whales. In contrast, in the North Atlantic the humpback whales presented a particular problem for management. The sub-committee of the IWC Scientific Committee (IWC 34, 1984; IWC 35, 1985), reviewing these humpbacks, noted that feeding aggregations existed in the Gulf of Maine, off Newfoundland, Greenland and off Iceland. From the identification of individuals, from photographs of tail flukes, no mixing between these feeding groups has been detected; however whales from all of these groups breed in the same areas of the Caribbean, and as yet the relationship of the different groups is undefined.

From the above it can be seen that data on distributions of catches, sightings or indices per unit effort can be useful in indicating stock boundaries, but the largest problem is that most of these data are from the feeding and not the breeding areas. The distributions sometimes give only little useful information, for example catches of Southern Hemisphere minke whales do not support boundaries between Areas III, IV and V, but this is not surprising as Southern Hemisphere minke whaling has only seriously taken place since 1970 and, as quotas have been small in relation to the stock size, the fleets tended to operate near to a boundary to save fuel (Shimadzu and Kasamatsu 1983). Phenotypic patterns can also be used to distinguish stocks but tend to be useful in the more patterned forms, such as humpbacks and killer whales (Evans and Yablokov 1978, Evans et al 1982), and similar identification studies have been carried out on the dialects of sounds produced by humpback and killer whales (R Payne 1979, Winn et al 1981, Ford and Fisher 1982). Marking studies, or the recognition of individuals, allow the integrity of groups to be evaluated. In a series of biochemical studies Lund (1938, 1951) found that the iodine levels in various tissues differed with location of capture, and that these differences were consistent from year to year. He argued that these values indicated that stocks of whales existed. At least it indicated slow rates of mixing amongst the feeding areas. These techniques all yield some information about stock units but frequently, although not always, more insight is afforded by investigations of blood groups and protein structure.

As described in Chapter 1 the studies by Fujino on blood groups of whales has advanced our knowledge of Cetacean genetics. Fujino (1962, 1964a) described the Ju blood groups in Southern Hemisphere fin whales from about 6000 specimens. Two homozygous and one heterozygous forms were found from agglutination reactions with rabbits' and fowls' sera. Four distinct groups were identified one in Area II, one in Area IV and two in Area III, north and south of 50°S. This last separation is unexpected and possibly indicates two forms of fin whale as there are two forms of blue whale. The blood grouping allowed Fujino to estimate rate of mixing amongst the stocks. Perhaps of even greater significance are his studies of the maternal-fetal incompatibility of some blood groups (Fujino 1963b, 1964a), which will assist in the perpetuation of a stock even if mating takes place between whales from different stocks. Significant differences in the frequency of proteins from various stocks of a species have been described by Wada. In comparing minke whales caught by the Republic of Korea in the Yellow Sea and Sea of Japan, with those caught by Japan off north and east Hokkaido and off northeast Honshu, Wada (1984) examined electrophoretically 15 loci; five were polymorphic. The Adh-1 (alcohol dehydrogenase) locus had four alleles and seven phenotypes were detected, and clear differences found between the two groups of minke whales. Wada (1983) also described differences between North Pacific and Southern Hemisphere minke whales. The existence of polymorphic protein systems in sei whales has been described by various authors, see Chapter 1, and although this indicates that the techniques could be used to identify sei whale stocks, the limited catching of sei whales after the refining of these techniques has meant that little progress has been made. Sharp (1981) discussed some of the statistical difficulties likely to arise without large samples and a controlled programme.

The techniques and examples given above illustrate some of the difficulties in identifying large whale stocks, but also what can be achieved. These techniques are now used to separate stocks of sei whales, but often the data are inadequate for the task and then lines have been drawn for management purposes, on the assumption that it is safer to segregate. A general review of cetacean speciation and population differentiation is given by Gaskin

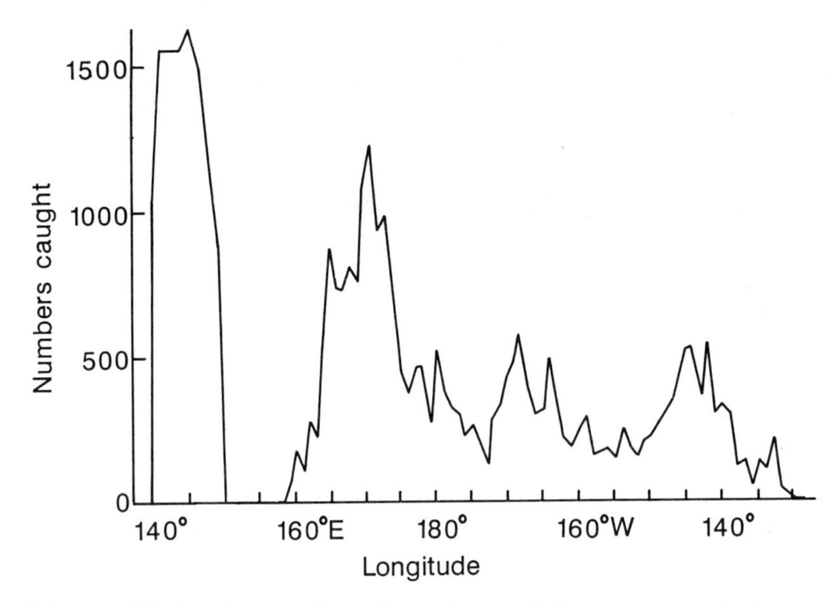

Figure 10.1: Accumulated number of Japanese catches of sei whales by longitude from the North Pacific 1952 to 1972.

(1982). It has been seen earlier that sei whales seldom if ever cross the equator and that there are differences between southern and northern forms; consequently the sei whale stocks are described from the North Pacific, North Atlantic and Southern Hemisphere.

NORTH PACIFIC

Indices of abundance

The distribution of Japanese catches of sei whales accumulated over the years 1952 to 1972 is illustrated in Figure 2.1; the high density of catches to the west is due to the coastal operations, otherwise little structure is obvious in the data. Masaki (1976a, 1977a) summed these catches by longitude, as shown in Figure 10.1, and peaks of catches can be seen from 165°-170°W and 140°-145°W. The large hiatus to the west is caused by Japanese domestic legislation restricting Japanese coastal whaling to the west of 150°E, and pelagic whaling to the east of 159°E. The Japanese sighting surveys are described by Ohsumi and Yamamura (1982) and results from the North Pacific are illustrated in Figure 2.2. As with the catch data Masaki (1976a) summed the sightings data by longitude, but there does not appear much structure in the information. Masaki considered that high density areas could be recognised west of 180°, between 150°W-180° and east of 150°W. These distributions could well be reflecting preferred feeding locations, and in themselves do not give adequate information on the existence or otherwise of stocks.

Morphometrics

The relationship of the length to breadth of the baleen was used by Omura and Fujino (1954) to distinguish between sei and Bryde's whales. Masaki (1976a, 1977a) used their material, from sei whales, and additional plates collected in 1973, to look for differences in the ratio of length to breadth, from sei whales from different longitudes. The length was measured along the outer arcing side from tip to the gum line and the breadth along the gum line of the largest baleen plate. Following the finding by Omura and Fujino, Masaki found no significant difference with sex, of the length to breadth ratio. There was no relationship between body length and this ratio. The following groupings were considered coastal, 160°E-170°E, 170°E-170°W, 170°W-150°W and 150°W-130°W and the results are given in Table 10.1. Masaki interpreted the results as indicating that the 170°W longitude may provide one boundary; however Table 10.1 shows a more complex pattern, and without more details it is difficult to say more than that several significant differences were found. Values of the ratios are published by Masaki (1977a), from the pelagic catches, and these show the ratio to be somewhat higher to the east of 170°W, at 2.52-2.57, compared with 2.42-2.47 to the west. However the quoted standard errors are a minimum of 0.13 increasing to 0.45, and it is difficult to see that these differences are statistically significant.

The angle at the outer base of the baleen plate was also calculated by Masaki (1976a, 1977a), and a comparison of this angle amongst the different longitudinal groupings is shown in Table 10.2. As with the previous results several significant differences are given, but again the specific mean values with the quoted standard errors makes these differences difficult to accept. Unlike the case of the baleen ratio, the value of the mean angle fluc-

Table 10.1: Comparison of the means of length to breadth ratio from the largest baleen in the North Pacific. A: coastal Japan. B: 160°E-170°E. C: 170°E-170°W. D: 170°W-150°W. E: 150°W-130°W. Asterisks show means different at $p < 0.95$; dash = $p > 0.05$.

males

	B	C	D	E
A	-	*	-	-
B		-	-	-
C			*	*
D				-

females

	B	C	D	E
A	-	*	-	-
B		-	-	*
C			-	*
D				-

Table 10.2: Comparison of the means of the angle of the outer corner of the baleen. Symbols as in table 10.1.

males

	C	D	E
B	-	-	*
C		-	*
D			*

females

	C	D	E
B	-	-	-
C		*	-
D			*

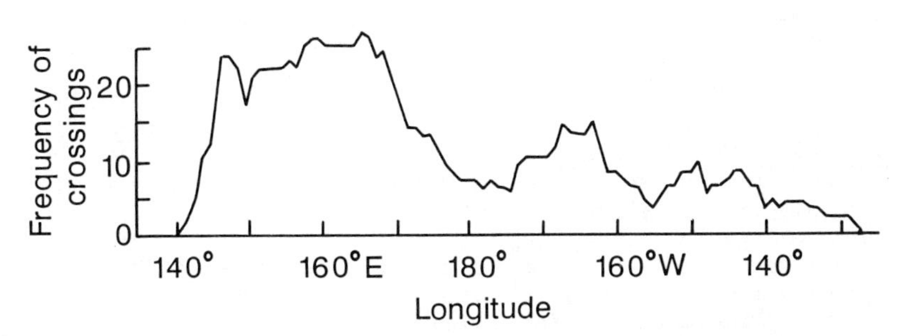

Figure 10.2: Frequency distribution of longitudes crossed by sei whales in the North Pacific marked and recovered in Japanese operations.

tuates with longitude, and probably does not provide any evidence for different populations. If there were prior reasons for believing in the longitudinal segregation, then the baleen ratio results would support such grouping; by themselves they are difficult to interpret, and one would wish to see if the results were repeatable from year to year.

Marking

The whale marking and recoveries from the North Pacific are described in Chapter 3 and we know the annual migrations are of a north-south direction. It is then instructive to consider the east-west movement of whales since this could show whether whales from the different longitudes mix. This exercise was attempted by Masaki (1976a), using data from Japanese marking and recovery, over the period 1949 to 1973. Mixing across sixty degrees of longitude was found but his Figure 63 illustrates that mixing is not random and most recoveries were from locally marked whales. However these also include recoveries of whales marked only a short time beforehand. The traditional way of showing movements of whales by a line joining the positions of marking and recapture is unsatisfactory when large numbers of recoveries are involved, and the tendency is to give more visual weighting to those that moved large distances. Masaki presented his marking data as a frequency distribution of the number of marked whales crossing a particular longitude; this is shown in Figure 10.2. There are only a few first year recoveries and the data therefore represent longer term, longitudinal movements. Although somewhat sensitive to the number of whales marked on a particular longitude, the plot shows two hiatuses in the longitudinal movement of the marked whales, at 155°W and 174°W.

Summary - North Pacific

The indices of abundance and marking data indicate possible segregations about 155°W and 175°W. Examination of blood types, by Fujino (1964b, c -cited by Masaki, 1976a, 1977a), indicated different populations in the inner part of the Gulf of Alaska, and off Vancouver. The existence of several populations of North Pacific fin whales, identified by blood typing (Fujino 1960), indicate that different populations of sei whales probably exist. The baleen material, and the differences in reproductive activity described earlier between Californian and Japanese catches, provide further support for differences. Masaki (1976a, 1977a) argued for three populations with boundaries at 174°W and 155°W. Ivashin and Rovnin (1967) also proposed three stocks but did not give precise boundaries. The International Whaling Commission has set quotas for the whole North Pacific rather than by any finer breakdown. This is because whaling for North Pacific sei whales ceased after 1975, and the more thorough studies on stock separations occurred after this time, along with the setting of regional quotas by the IWC. Nevertheless assessments were sometimes undertaken by dividing the North Pacific into American and Asian components at 180° (Ohsumi et al 1971). Rice (1977) described the parasite burden of Californian sei whales. Different forms and species of parasites are frequently a characteristic of different isolated host populations, and Rice described the species of the parasitic crustacean *Penella* from Californian sei whales as different from *P. balaenopterae* found in Californian fin whales and Japanese coastal sei whales.

NORTH ATLANTIC

Knowledge of the structure of stocks in the western North Atlantic comes mainly from the distributional information of Mitchell and Kozicki (1974ms), and they suggest that two stocks exist, one summering off eastern Nova Scotia and the other in the Labrador Sea. Sei whales are found in the Labrador Sea from the first week of June and, at the same time, sei whales are found moving northwards along the continental shelf of the United States. The latter group migrates south, from mid-September to mid-November, having gone as far north as the south coast of Newfoundland. In fact Mitchell and Kozicki considered that there may be either two stocks or one that is widely dispersed, but they favoured the hypothesis of two because of the returns from marked whales, albeit only three (see Table 3.8). Table 3.2 shows 72 sei whales marked in the North Atlantic (including a few that were possibly Bryde's whales) and Table 3.8 gives fourteen returns. These limited data indicated a segregation of the whales caught at Iceland from those caught and marked off Canada and the United States. The summer distribution of these animals is not known and, as with the humpback whales described earlier, and possibly minke whales (Dorsey 1981), these aggregations could be social and feeding groups that may interbreed in the winter. The very limited comparative reproductive data described in Chapter 9 do, however, suggest a difference.

In considering research proposals for the North Atlantic the Scientific Committee of the IWC considered that, "The indentification of the various

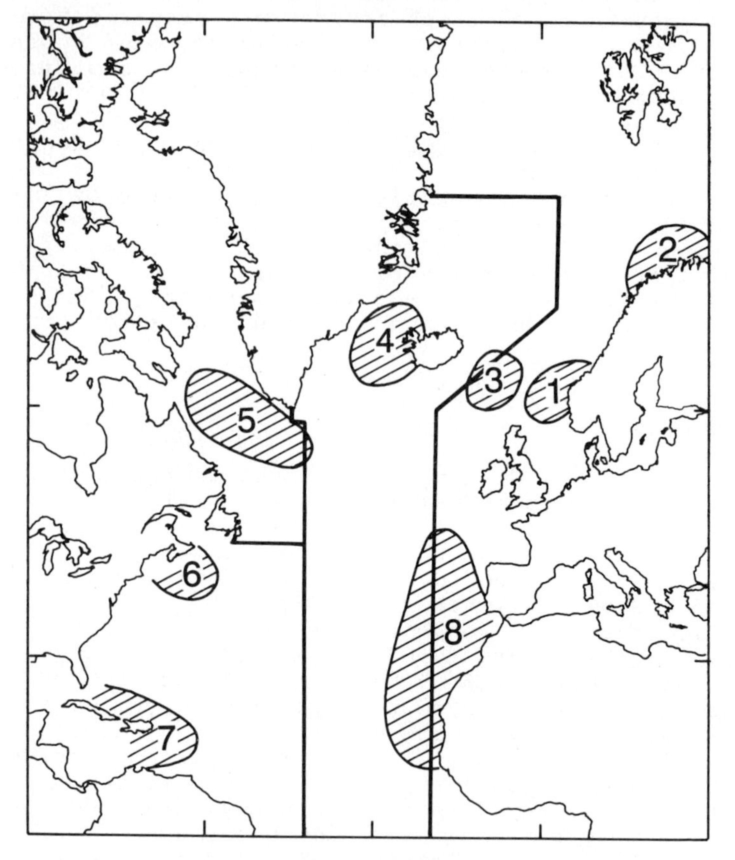

Figure 10.3: Presumptive stock units of North Atlantic sei whales as described in the text. 1: SW Norway, 2: N Norway, 3: Faroe Is, 4: Denmark St, 5: Labrador Sea, 6: Nova Scotia, 7: Caribbean Sea, 8: SE North Atlantic. Bold lines show current IWC management areas.

stocks is a major research problem." (Mitchell 1976). It still is, but at that time they divided the North Atlantic into eight stock units and these were called, South-West Norway, North Norway, Faeroe Islands area, Denmark Strait, Labrador Sea, Nova Scotia, Gulf of Mexico and Caribbean Sea (including Bryde's whales) and South-East North Atlantic (including Bryde's whales). No explanation for these divisions is given but I assume it was based on the previous study of Mitchell and Kozicki (1974ms) and the known distribution of catches and sightings (Chapter 2). These divisions are given in Figure 10.3. Of particular interest is the unit in the Caribbean Sea, for Mitchell and Kozicki noted that the sei and Bryde's whales seen in the Caribbean were relatively larger than those seen off Canada. Also none of the, ten, marks, including some possibly put into Bryde's whales, were recovered. From this they concluded that they were "probably Southern hemisphere whales". This is surely an over enthusiastic conclusion at this time, although Schmidly (1981) also speculated that a Caribbean and Gulf of Mexico stock might exist.

Compared with catches from other oceans the numbers of sei whales caught annually since 1930, in the North Atlantic, have not been great, and annual quotas for sei whales were first set for the 1977 season, by the International Whaling Commission. At that time quotas were set for the two areas of catching, those off Nova Scotia and of the Iceland-Denmark Strait. As early as 1972 the Scientific Committee had considered the Canadian sei whales to differ from other stocks (IWC 23, p32). In June 1980 the Scientific Committee noted that some sei whales were being taken off Spain and the Faeroe Islands and referred to these as being from a separate stock (IWC 31 p64). This was adopted from the 1981 coastal whaling season. The three IWC management areas are illustrated in Figure 10.3 and are defined by;

Nova Scotia: to the south and west of a line through, 47°N 54°W, 46°N 54°30'W, 46°N 42'W, 20°N 42°W.

Eastern: east of a line through, 20°N 18°W, 60°N 18°W, 68°N 3°E, 74°N 3°E and north of a line through, 74°N 3°E, 74°N 22°W.

Iceland-Denmark Strait: east of a line through Kap Farvel (South Greenland), 59°N 44°W, 59°N 42'W, 20°N 42°W and to the south west of the eastern boundary.

It can be seen that no area has been recognised in the north west North Atlantic but in this region fin whales have been divided into two groups.

Initial studies of protein polymorphisms were reported by Arnason and Sigurosson (1983) from Icelandic sei whales, but the information is insufficient to allow discussion of any intraspecific differences.

SOUTHERN HEMISPHERE

Indices of Abundance

Based upon the general knowledge gained whilst working with the Soviet whaling fleets, Budylenko (1978a) considered that eight populations, or "herds", of sei whales could be recognised through the different concentrations and timings of occurrence. These he identified as West and East Atlantic, West and East Indian, West and East Pacific and Central Indian and Pacific groups. No quantitative information is provided.

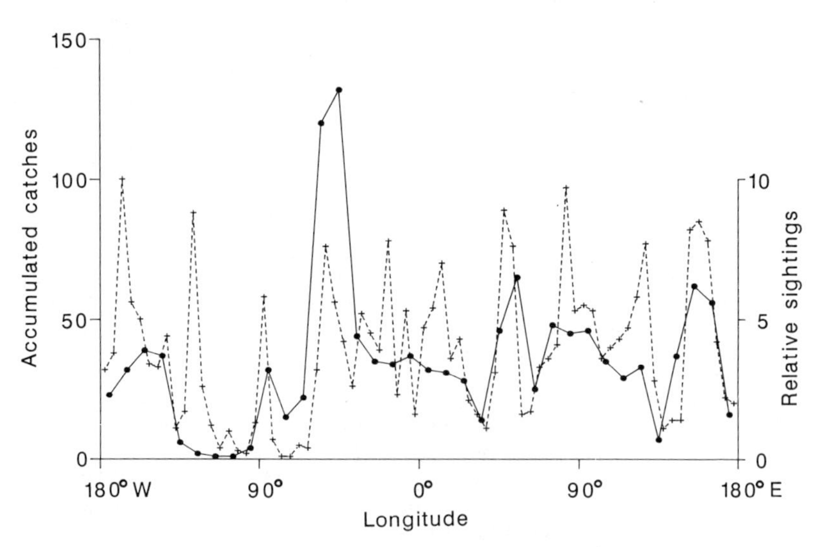

Figure 10.4: Accumulated catches by 10° longitudes, from 40°S southwards, 1956/57 to 1977/78 in thousands (bold line and dots) and of relative sightings (crosses and dashed line).

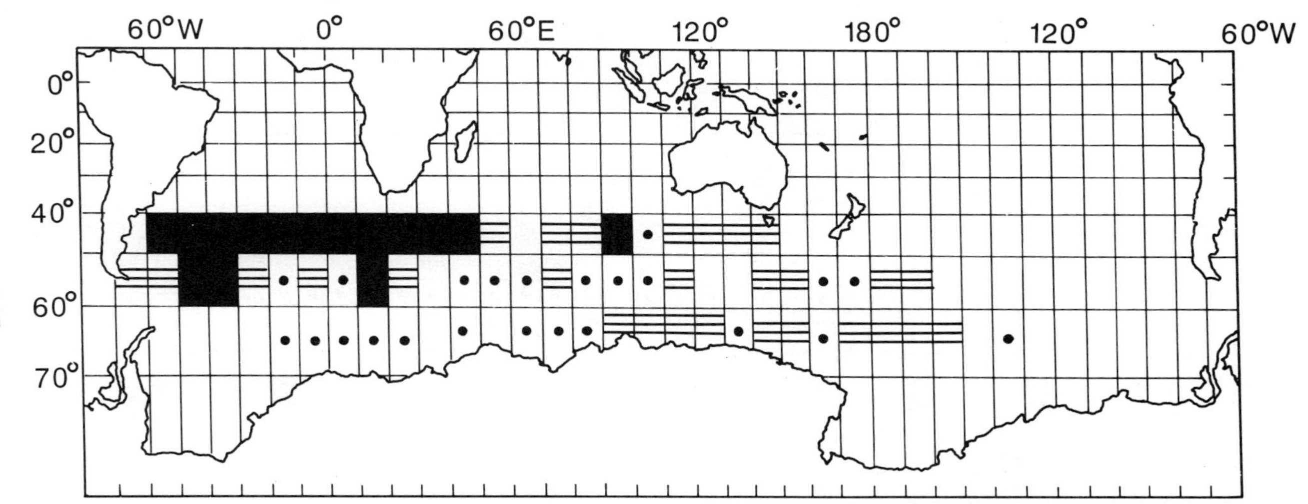

Figure 10.5: Distributions of catch of sei whales per catcher days worked in the seasons 1964/65 and 1965/66. Solid: > 1.0, hatched 0.5-1.0, dot < 0.5, blank: unvisited.

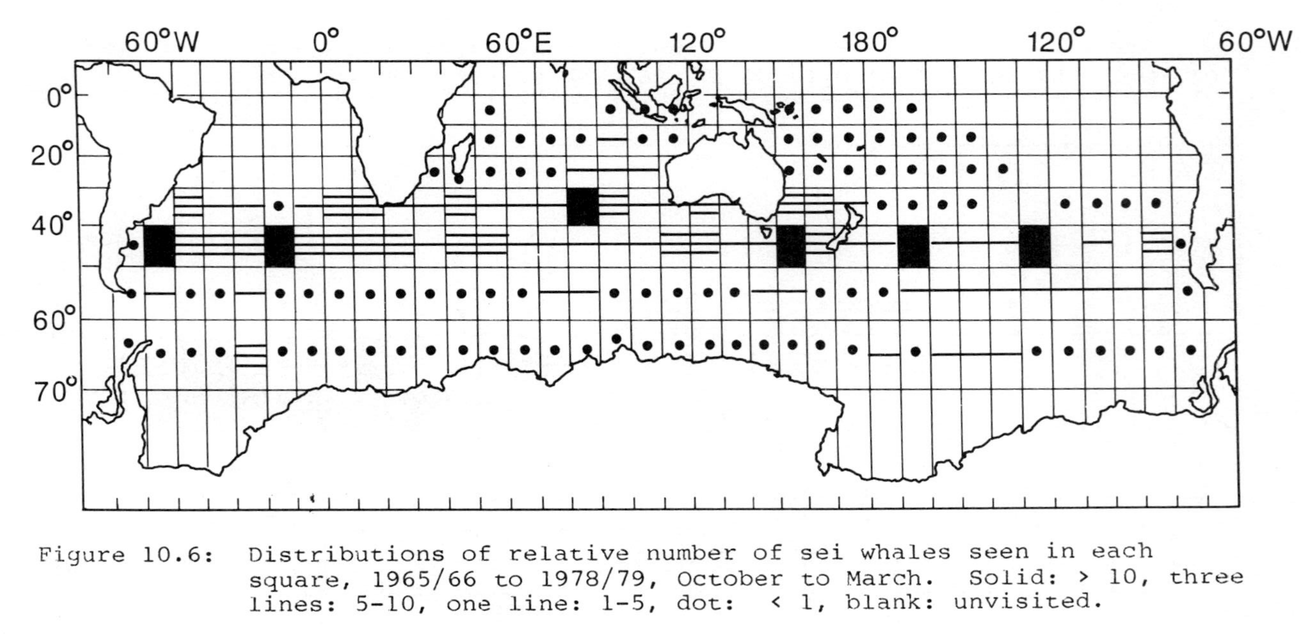

Figure 10.6: Distributions of relative number of sei whales seen in each square, 1965/66 to 1978/79, October to March. Solid: > 10, three lines: 5-10, one line: 1-5, dot: < 1, blank: unvisited.

Catching by the pelagic fleets was mainly restricted to south of 40°S and the accumulated catches from 1956/57 to 1977/78 are illustrated in Figure 2.3 by ten degree squares, and summed over latitudes from 40°S in Figure 10.4. As can be seen there are substantial variations with longitude, and since major exploitation did not begin on sei whales until after 1956 and then severely depleted the populations, this distribution probably reflects the distribution of the sei whales independent of whaling effort. High numbers were found from 20°W to 80°W, from 20°E to 100°E, from 140°E to 170°E and from 140°W to 180°. Particularly low catches were found from 70°W to 140°W, at 35°E, 135°E and 175°E.

The density of sei whales was described by Doi, Ohsumi and Nemoto (1967) using an index of average catch divided by the number of working days of the catcher boats. This is a crude index, affected by various operational considerations and the catches of other species. However the data are from the seasons 1964/65 and 1965/66 when the sei whales were the most important baleen catch and biasses might be similar longitudinally. Their distributions are illustrated in Figure 10.5, and they claimed that at least three stocks could be identified, between 70°W and 60°E, 70°E and 160°E, and 160°E to 130°W.

As described earlier, the Japanese programme of sightings from scouting vessels has provided an index of abundance, largely independent of whaling operations, and the indices of abundance from this programme were provided by squares of five degrees of latitude and longitude (IWC 30 p499); these are illustrated in Figures 10.4 and 10.6. Although the sightings data are much more variable than the CPUE data, hiatuses are found at the same longitudes, at 70°W to 140°W, at 35°W, 65°W, 135°W and 180°. In the lower latitudes some Bryde's whales may have been included.

In all, these data do not provide a clear picture of isolated aggregations and all that can be concluded is that the overall longitudinal pattern does not differ substantially from that found earlier for blue and fin whales.

Body scars

Some sei whales have numerous oval, whitish scars, of some 4 x 3 cm dimension, and these are also common on other large whales. The wounds from which the scars develop tend to be fresh on whales caught in warmer waters. The number of scars tends to

Table 10.3: Number of white scars formed during the previous winter.

Region	numbers examined	number of scars min	max	mean	s.e.
South-east Atlantic (0-15°E)	37	171	395	242.3	2.6
Prince Edward Is (30°-42°E)	33	11	88	24.7	0.2
Crozet and Kergeulen Is	42	8	64	25.6	0.2

Table 10.4: Numbers examined (n) and percentage of sei whales with scars on one side of the caudal peduncle

Region	n	new scars	number of total scars (%) <30	30-60	60-200	>200	marbled
South East Atlantic	41	51.2	0	0	4.9	75.6	19.5
Prince Edward Is	31	5.7	32.3	41.8	28.6	3.2	0
Crozet and Kergeulen Is	47	6.1	36.2	36.2	27.6	0	0

increase with age, giving further indirect evidence of regular migrations. The origin of the scars has produced some debate. Pike (1951) concluded that scars found on North Pacific whales were caused by the sea lamprey, *Entospherius tridentatus*. Nemoto (1955) also investigated these scars, and found some that clearly matched the array of denticles around the sucking disc of the lamprey. However he concluded that the regularity and persistence of other scars indicated an additional source of attack. Subsequently both Shevchenko

(1970) and Jones (1971) suggested that the damage in warm water was caused by the small shark, *Isistius brasiliensis*. However Ivashin and Golubovsky (1978) managed to record some interesting observations from Bryde's whales, caught at about 30°S, in that they saw the parasitic Penella falling off the whales, and leaving the wound that they claimed healed to the characteristic white scar.

Irrespective of their exact cause the relative numbers of scars have been used by Shevchenko (1977) to differentiate stocks of Southern Hemisphere sei whales, from catches from between 40°S to 50°S in 1971/72. Table 10.3 gives the number of fresh scars and scars less than one year old (recognised as bright white scars), on the entire body of whales caught in the three areas of South East Atlantic (0-15°E), Prince Edward Island (30°-42°E) and from the Crozet and Kerguelen Islands. As can be seen the number of scars are ten times as many on the western individuals, and there is a possibility that the other two groups also may be different, (In Table 10.3 I have interpreted Shevchenko's sigma as a standard error).

The total number of scars is difficult to count, so Shevchenko also concentrated on counting the scars only on one side of the cordal peduncle, from the posterior of the dorsal fin to the beginning of the tail fluke. The degree of scarring on the peduncle was also judged into five categories of "very little", 0-30 total scars, "little", 30-60, "moderate", 60-200, "many", too numerous to count and "very many", a marble-like pattern. These results and the counts of bright white scars are given in Table 10.4 for males of 13.5 (or 14.0 m) to 14.5 m. A similar result is found to that above. Based on these data, it can be concluded that there are at least two groups that do not mix in either the warmer or colder waters. Shevchenko used the frequency of scarring to determine whales from the South Atlantic group and a boundary was found at about 20°E, although the exact criteria for determination were not described.

Marking

As described in Chapter 4 the number of sei whales marked in the Southern Hemisphere since 1954 is 848 and there have been 81 recoveries. The pattern of recoveries is such that it does not lead to any postulated groupings but the data can be used to investigate the immigration and emigration out of the six Areas defined for fin, blue and humpback whales. Recoveries from the Soviet programme over the period 1954/55 to 1977/78 are given in Table 3.11 and the distribution of marks was described by Ivashin (1980). The results are summarised in Table 10.5. The results show that for whales caught in the same season of marking or after (with eleven years the maximum elapsed time between marking and recapture), the large majority were recaptured in the same Area of marking. For recoveries after six months this was 86%. The other few were recovered from adjacent Areas. A similar exercise is shown in Table 10.6 for marks and recoveries from the International scheme over the years 1955/56 to 1975/76. Over this period there were 377 sei whales marked and 51 recoveries. As shown in Tables 3.3 and 3.10 there were 160 whales subsequently marked but only one extra return, these data have not been used as the numbers marked have not been given by Area of marking. The one extra mark was placed in and returned from Area IV. Table 10.6 shows that of those marks returned, over six months after marking, 74% were returned from the same Area that they were marked in.

These results are influenced by the relative whaling effort in the different Areas, for instance if there had been no whaling in Area IV then marks from Area IV could only have been found from other Areas. However whaling was relatively intensive in all Areas over these periods and the data indicate a relatively high degree of adherence to the same feeding grounds. Brown (1977a) remarked that the movement of sei whales within the whaling Areas may be more restricted than in the case of fin whales, and he concluded that there was no evidence, from marking data, to suggest that any modification of the boundaries of the six whaling Area was necessary, to reflect more accurately the distribution of sei whale populations in the Antarctic. The subsequent recoveries support Brown's conclusion.

Morphometrics

No thorough morphometric investigation has been carried out on Southern Hemisphere sei whales, but some limited information is available. Jacobson's Organ is a shallow depression found on the ventral side of the tip of the rostrum and differs in shape with species; in sei whales the convex parts oppose each other. It is thought to have some sensory function. Mikhaliev (1979) and his colleagues measured the distances from the centre of the organ

Table 10.5: The number of whale marks recovered by Area of marking and recovery from the Soviet programme, 1945/55-1977/78. The number marked is given in brackets. The first number recovered is from marks found after six months and the second before six months.

Area	recovered					
marked	I	II	III	IV	V	VI
I(69)	3/2					
II(63)	1/0	7/4	1/0			
III(29)			1/0			
IV(34)				6/0		
V(62)					0/0	1/*
VI(37)					1/0	0/1

* plus two recoveries somewhere in Areas III-V

Table 10.6: The number of whale marks recovered by Area of marking and recovery from the International programme, 1955/56-1975/76. The number marked is given in brackets. The first number recovered is from marks found after six months and the second before six months.

Area	recovered					
marked	I	II	III	IV	V	VI
I(16)	2/0					
II(70)		5/11	2/1			
III(56)			3/1	3/0		
IV(132)				7/3	2/0	
V(89)				2/0	8/1	
VI(14)						0/0

to the anterior end of the rostrum (l_1), and to the beginning of the row of baleen (l_2). The total length of the rostral lip, L, is then $l_1 + l_2$. Measurements were made of 317 sei whales, from catches from off the coast of Chile, off Graham Land and about the Islands of Gough, Prince Edward, Crozet and Kerguelen. For both sei and fin whales the relative size of L to the body length was substantially greater in females than males. The information, presented graphically, shows that the relative sizes of l_1, l_2 and L to body length, for each sex, are similar from sei whales taken from off Chile and the other group of islands excluding Gough Is. From Gough Is the relative sizes appear significantly smaller. It is argued that the relative position of the organ has proved to be useful for population differentiation, but in fact the rostral lip length (L) gives the same information and until analyses are undertaken to find what additional information is held by l_1 or l_2, in addition to L, then one can only conclude that the use of Jacobson's Organ as a means of stock separation has not yet been established.

The most obvious measurement, and most easily available since IWC regulations have required its collection, is body length. This has been measured as a straight line from the tip of the upper jaw to the apex of the notch between the tail flukes. However, in an analysis of length measurements of fin whales, Clark (1983) found differences that indicated lengths measured from Japanese operations in the 1950s were some two to three feet (in about 70ft) smaller than those measured by UK and Norwegian operations; this is claimed to be a measurement difference rather than a real difference. The absolute length of the whales in the population depends upon its level of depletion following exploitation, and in the catch it also depends upon the average or minimum size at capture and location of the catch. Woolner and Horwood (1980) described the mean lengths of sei whales in the Antarctic catches by season and by ten degree latitudinal zones. As with the marking exercise these data can be used to investigate the validity of the six IWC management Areas. Analyses of variance or covariance were used on the mean length data, after prior selection for appropriate trends. The data used are those of Mizroch (1980), but only series D (40-50°S) had sufficient information for a study of Area differences. The results indicated that mean lengths of males were different in Area II compared with Areas III and IV over the period 1961/62 to 1975/76. Over the reduced time period of 1961/62 to 1965/66 the analyses showed that in zone A (50-60°S) mean lengths of females in Area II were higher than those from III and IV. From zone B (60-70°S) mean lengths of females were significantly higher in Areas II and V compared to areas III and IV. These results again provide some support for the existing management Areas, but Horwood (1986) has shown the importance of the Antarctic Convergence in determining the size distribution of the sei whales and analyses by ten degree zone do not compensate fully for the longitudinal changes of the Antarctic Convergence.

Summary - Southern Hemisphere

So far as I am aware these are the only studies that give any information on possible stocks of sei whales in the Southern Hemisphere. The rostral lip lengths and degree of scarring show localised groupings on the feeding grounds, and the marking data show that the sei whales tend to return to the same large feeding areas. The distribution data shown patterns of high and low concentrations of sei whales around the Antarctic. Chapter 4 has shown that there is an annual migration north to south. The six management Areas of the IWC were first used for stock assessments of sei whales by Doi, Ohsumi and Nemoto (1967), although no argument was posed for their use, but only since the season of 1974/75 were they used for the allocation of catch limits. The above studies plus the differences found in the timing and extent of declines in age at maturity (Chapter 9) indicate some degree of isolation of sei whales within these six Areas.

SUMMARY

The amount and extent of the studies on differentiating populations of sei whales is depressing. Electrophoretic studies were developed to a useful degree after most, but by no means all, sei whaling had ceased and although blood-typing and immunological studies are very time consuming such studies would have been highly desirable. As for the morphometric studies, one might have imagined that if work was easily possible it might have been done; the above shows this not to be true. Not to mention a whole variety of techniques scarcely mentioned above such as comparative chemistry and parisitology. The identification of populations must be the starting point, and probably the most important single aspect, of studies designed to further our knowledge of why and how numbers of animals vary, and if such future studies are possible on the sei whale they are likely to contribute to re-analyses of the population dynamics.

The previous studies have shown that within oceans it is highly probable that different stocks of sei whales exist, as has been demonstrated genetically for fin and minke whales. These are described under the previous headings. For management, the oceans have been divided into regions for practical purposes and to help in avoiding overdepletion. What information there is, indicates that there is a reasonable degree of isolation of whales within these areas, and even the suspicion that the areas may be too large.

Chapter Eleven

HISTORY OF SEI WHALE EXPLOITATION

INTRODUCTION

The hunting of sei whales has only been associated with what is known as "modern whaling". Pre-modern whaling was characterised by harpooning from rowing boats that operated from sailing ships, although other techniques were used, such as the netting off Japan. In the North Atlantic there was a large, early whaling industry, that operated in the Arctic waters from the opening of Spitzbergen in 1610, to its gradual decline in the early nineteenth century. The whale caught was the Greenland right whale or bowhead, and was a suitable quarry for several reasons. It was originally found near coastal locations and could be caught and towed to land (or ice) for processing. It was fat, gave a good oil yield and had a very large baleen, and the value of the baleen was often more than that of the rest of the animal. But the most important features were that the whale was slow swimming and usually did not sink when killed. From 1712 sperm whales were killed from the Atlantic coasts of the United States, and by 1792 whaling vessels had reached the Pacific to start the southern whale fishery for sperm, right and humpback whales.

From whaling vessels based in Peterhead, Scotland, the number of Arctic whales processed per voyage declined from 16.8 about the year 1800 to 2.1 in the 1840s (Gray 1932). De Jong (1983) showed that the number of whales caught and yields per whale steadily declined from 1660 to 1800 in the German and Dutch fleets and it is clear that the industry for the Greenland right whale could not be sustained. The blue, fin and sei whales had not been caught. Early experiences had shown that the whales were fast, strong and consequently too dangerous to catch. They also sank. If they were to be caught a new technology was needed and it was the introduction of such techniques that gave rise to modern whaling. A variety of propelled harpoons, some exploding on impact, had been developed by the early 1800s and the next two innovations were to place such a harpoon gun on a fast moving motorised vessel and to secure the harpoon line so that the whales would not escape and sink.

The British Vice-Consul to Iceland reported that in about 1860 an American bought a 40 ton steamer and used to tow to sea a whaling boat. To kill the whale a harpoon was shot from a rocket-like apparatus, from the steamer or boat, and the harpoon exploded on entering the whale. However, he does say that when the whale was killed a rope was attached to it and, if it subsequently sank, the whale was retrieved later. From the descriptions of Tonnessen and Johnsen (1982) this was Thomas Roys who, with his colleague Lilliendahl, caught over 100 rorquals between 1865 and 1867. Over the same period the now famous Norwegian Svend Foyn was experimenting with similar techniques and visited Iceland. In 1866 Roys patented a sprung compressor to take the strain off the harpoon rope when attached to both whale and vessel, and this must have helped considerably. Eventually the Icelandic venture failed but Foyn's contribution was to improve every aspect of the capturing process so that an efficient and economic system was developed. The whale catchers still took the whales to a shore station for processing but before the turn of the century the Norwegians had developed a form of pelagic whaling, wherein whales could be processed at a factory ship at sea, and the final phase of development came with the rear slipway, up which whales were dragged for processing.

The regulation of whaling, that is the restriction of the number of companies operating or numbers of whales caught, was essentially in the hands of the coastal state governments; the open seas had no regulation. Following the decline of the right whale fisheries, and the catching of rorquals off Finnmark as early as 1911, it was mooted that international agreements might be wise. The first such agreement followed the 1931 meeting in Geneva which gave rise to the Convention for the Regulation of Whaling, and which was based upon Norwegian and British domestic legislation. It took several years for the different countries to ratify this agreement and Japan and Germany were not involved. Even though the Convention was not fully operative, production agreements were achieved for most years, and for the 1932/33 Antarctic season a quota was set which was the equivalent of about 18600 BWU (blue whale units). Oil was the main interest of the industry, and on this basis it was considered that 1 BWU could be obtained from 1 blue whale, 2 fin, 2.5 humpbacks or 6 sei whales. Additionally the 1931 agreement gave protection to right whales, calves and females with calves. From 1937 on, a series of international agreements were formulated which introduced minimum size limits for blue, fin and sperm whales, protected humpbacks south of 40°S, regulated the duration of the Antarctic season, banned pelagic whaling from the North Atlantic and generally between 20°N to 40°S, and closed off an area of 70°-160°W, south of 40°S which is generally known as the Sanctuary. Several nations were not included in these agreements. However, at this time, sei whales were a negligible catch in the Southern Hemisphere and the Second War then stopped operations.

The most important step in international regulation was taken in 1946 when interested governments participated in the Washington meeting that gave rise to the 1946 International Convention for the Regulation of Whaling (HMSO 1947- Cmd. 7043). The history, principles and procedures of the IWC are reviewed by Birnie (1985), particularly in respect of decision making within an international legal framework. Like most international agreements this Convention took some time to be ratified by various countries, but by the first meeting in May 1949, 15 countries had ratified the treaty; Japan and Brazil joined soon afterwards. The Convention consists of 2 main parts, the Convention itself, which is still unchanged and only changeable by unanimity, and the Schedule, which can be changed with a lesser agreement. The aims of the Convention are implicit in the preamble, cited below:

"The Governments whose duly authorised representatives have subscribed hereto,

Recognising the interest of the nations of the world in safeguarding for future generations the great natural resources represented by the whale stocks;

Considering that the history of whaling has seen over-fishing of one area after another and of one species of whale after another to such a degree that it is essential to protect all species of whales from further over-fishing;

Recognising that the whale stocks are susceptible of natural increases if whaling is properly regulated, and that increases in the size of whale stocks will permit increases in the numbers of whales which may be captured without endangering these natural resources;
Recognising that it is in the common interest to achieve the optimum level of whale stocks as rapidly as possible without causing wide-spread economic and nutritional distress;
Recognising that in the course of achieving these objectives, whaling operations should be confined to those species best able to sustain exploitation in order to give an interval for recovery to certain species of whales now depleted in numbers;
Desiring to establish a system of international regulation for the whale fisheries to ensure proper and effective conservation and development of whale stocks on the basis of the principles embodied in the provisions of (various previous international agreements) and;
Having decided to conclude a convention to provide for the proper conservation of whale stocks and thus make possible the orderly development of the whaling industry;"

Following the preamble The Articles of the Convention are cited. The Convention includes principles and rules of working, but the Schedule includes operational regulations which can be changed each year. Those changes associated with catch limits, sizes, times of operation or other direct operational matters need a three-fourths majority; each member government having one vote. The regulations of the IWC have greatly affected whaling on sei whales, not only because of direct quotas but because of indirect effects such as the protection of other species, but it should be noted that often regulation tended to follow industrial action with a response that was both late and of inadequate strength. Initially, Chile, Ecuador and Peru did not join the IWC and operated through their own Commission for the South Pacific, but Chile and Peru joined for 1979 and by the annual meeting in 1984 there were 39 member governments. The restrictions of the industry that affected sei whales through the IWC schedule are given below. It will be noted that initially little attention was paid to sei whales and then only in the Southern waters.

IWC REGULATIONS

The Schedule of 1946 incorporated the following restrictions. Calves and females with calves were protected along with non-aboriginal killing of gray and right whales. Humpbacks were protected south of 40°S. The use of factory ships for taking and processing baleen whales was prohibited in the following areas: North of 66°N, (except between 150°E to 140°W and 66° to 72°N in the North Pacific), in the Atlantic Ocean and the Indian Ocean, North of 40°S, in the Pacific Ocean East of 150°W between 40°S and 35°N, and in the Pacific Ocean West of 150°W between 40°S and 20°N. The Sanctuary area (70°-160°W, South of 40°S) was maintained. Land stations could only operate for 6 months in each 12. South of 40°S baleen whales could only be taken from 15 December to 1 April and South of 40°S a quota was set of 16000 BWU, and 6 sei whales were recognised as equivalent of 1 BWU. Minimum size limits were given for 5 species and that for sei whales was 40 ft (12.2 m) or 35 ft (10.7 m) for sei whales delivered to land stations for local consumption. In the Schedule sei whales were deemed to be sei or Bryde's whales.

For the 1953/54 Antarctic season a 4 day season for humpback whaling was allowed, with a baleen season of 2 January to 7 April (for blue whales from 16 January); the quota was 15500 BWU. For the 1954/55 and northern 1955 seasons, blue whales were protected for 5 years in the North Atlantic and in the Eastern North Pacific. Those countries most concerned all objected and were thus not bound by the agreement. The objection procedure allows any country to say within a certain time period that it will not comply with any IWC schedule amendment. It was encouraged by the USA as a safety-valve which would allow the IWC to persist in the face of difficult decisions. On balance one must conclude this was correct. For the 1955/56 and 1956 seasons the Sanctuary area was opened. Humpbacks were protected in the North Atlantic and in a limited part of the Antarctic. The Antarctic baleen season was from 7 January (1 February for blue whales) to 7 April, and the quota was reduced to 15,000 BWU and 14,500 for years afterwards. Most whaling countries objected to the latter part, but for 1956/57 a quota of 14500 was agreed. For the season 1956/57 the quota remained the same but was increased to 15000 BWU for 1957/58 to 1962/63, but with greater protection given to humpback whales. For the 1963/64 season and after humpbacks were protected in the Southern Hemisphere, and blue whales in the Antarctic, except for an area in which pygmy blue whales were found. In addition the quota was reduced to 10000 BWU. This quota remained for 1964/65, but was then reduced to 4500 BWU, with the agreement to further reductions aimed at ensuring sustainable yields. Blue and humpback whales were protected in the North Pacific. For the 1966/67 season the Antarctic quota was 3500 BWU, and for 1967/68 it was 3200 BWU. Blue whales were protected in the Southern Hemisphere. For 1969/70 the quota was further reduced to 2700 BWU. For the Antarctic season of 1970/71 the quota remained at 2700 BWU, but importantly, in the North Pacific, quotas were set by species. Fin whale catches (excluding those of the East China Sea) were not to exceed 1308, and sei and Bryde's whales catches should not exceed 4710; some trade off between fin and sei or Bryde's was allowed. For the 1971/72 season the Antarctic quota was set at 2300 BWU. In the North Pacific the sei and Bryde quota was reduced to 3768, the fin whale quota was cut to 1046 and sperm whale quotas were introduced.

Over the previous decade concern both within the IWC and outside of it was being expressed over the state of whale stocks and the ability to manage exploitation. The severe depletion of the blue whale in all oceans, notwithstanding the cautions urged by scientists advising the IWC, was a focus for this concern and is well expressed by Small (1971). A consequence was that the United Nations Conference on the Human Environment, held in Stockholm in 1972, called for a moratorium on commercial whaling. This was considered by the IWC in 1972 and was rejected. Some support for the rejection was afforded by the Scientific Committee of the IWC, which generally argued that if quotas were set by species and area such a blanket moratorium was biologically unjustified. This opinion is generally held to date. Nevertheless many nations were pressing for a moratorium or else substantial change. For the Antarctic season of 1972/73 quotas were set by species and a quota for the small minke whale was introduced; these were 1950 fin, 5000 sei and Bryde's, and 5000 minke whales. In the North Pacific quotas were again reduced, to 650 fin and 3000 sei and Bryde's whales. For the following Antarctic season of 1973/74, and elsewhere in 1974, blue whales were totally protected. In the Antarctic the quota for fin whales was 1450, for sei and Bryde's whales it was 4500 and for minke whales it was 5000. In the North Pacific the quota for fin whales was 550 and for sei and Bryde's whales it was 3000.

The next year the Antarctic quotas were set by the whaling Areas (I to VI) with total quotas not to exceed 1000 fin, 4000 sei plus Bryde's and 7000 minke whales. By Area the catch of sei and Bryde's whales were not to exceed 1275 for Areas I plus II, 1503 for III plus IV and 1664 for V plus VI, but with the catch not to exceed 810 and 495 in Areas II and III respectively. In the North Pacific the quota for fin whales was reduced again to 300 and that of sei and Bryde's whales to 2000. In 1974 the IWC agreed in principle to a substantial change in the way quotas were set and this became operative, as the so-called New Management Procedure, for the seasons including and after 1975/76. It relied upon the Scientific Committee assessing the status of each whale species and stock and judging its relative depletion. Based on the relative abundance of the present stock size to its initial stock size the population would be classified as either an Initial Management Stock (IMS), a Sustained Management Stock (SMS) or a Protection Stock (PS). The Scientific Committee then estimated the maximum sustainable yield (MSY) for each stock, and the algorithm provided an automatic means of then arriving at a quota, which was less than the sustainable yield. The three important elements involved were a relative level below which stocks were afforded protection, the fact that quotas were less than the sustainable yields and hence stability of stock and industry were to be assured, and that the quota was arrived at through this calculation rather than by alternative arguments. The principles were good but they placed a too exact demand on the science. The critical levels were made in relation to the level of the stock compared with the level at which maximum sustainable yield was found (MSYL), such that the categories were:

IMS: stock size greater than 120% of MSYL,
PS : " " less than 90% of MSYL,
SMS: " " between the two.

For the IMS category the catch limit was calculated as either 90% of MSY or 5% of the initial stock size. Limits were zero for the PS category. For the SMS category the catch limits were 90% of MSY for stocks above MSYL and this was reduced by 10% for every 1% that the stock was less than MSYL. The scheme was introduced as a compromise to a moratorium and the break up of the IWC, but great problems arose through the difficulties in identifying both the MSY and MSYL, although estimates of relative stock sizes were somewhat easier to calculate. For baleen whales the Scientific Committee has generally assumed that MSYL is at 60% of initial stock size, although evidence for any one value is extremely meagre; comments about the general character of MSY and MSYL are given by Fowler (1980) and Holt (1981).

Notwithstanding the difficulties in arriving at the specific figures, catch limits were set for 1975/76 according to the new procedure. The six Antarctic Areas were extended to the equator for sei and minke whales, regulating the coastal whaling. In the Southern Hemisphere Area III sei and Bryde's whales were classified as a Protection Stock with a zero catch limit, and in the other Areas I-VI catch limits were 198, 567, 671, 693 and 297 respectively, but the total was not to exceed 2230. Total catch limits for fin and minke whales were 200 and 6810 respectively. In the North Pacific sei whales were classified as a Protection Stock with a zero catch limit, but a catch of 1363 Bryde's whales was allowed. In the North Atlantic catches of sei whales were not regulated but quotas were set by area for fin and minke whales.

In the Southern Hemisphere a catch limit of 1863 sei whales, separately from Bryde's whales, was imposed for 1976/77 and 1977, with those of Area III designated a Protected Stock, and the others as Sustained Management stocks. In 1977/78 Areas II, III, V and VI sei whales were protected, with a total catch limit of 771 for the remaining Areas, and in 1978/79 and 1979 onwards sei whales were protected in all southern waters. In the North Pacific sei whales remained protected. In the North Atlantic the stock of sei whales off Nova Scotia was classified as a Protected Stock from 1977 onwards. The stock caught in the Iceland-Denmark Strait was classified as a Sustained Management Stock from 1977 to 1984 with a catch limit of 132 in 1977, 84 in 1978 and in 1979, and from 1980 to 1984 the limit was of 100 per year with a maximum of 504 in 6 years. The last limit was given in the form of a block quota as the sei whales appeared erratically off Iceland, and in some years the catch limit might not be reached. From 1981 an Eastern stock was recognised and given a precautionary, zero catch limit.

Even though it had been shown that reduced quotas could be agreed, and set by species and area, and that some species could be counted in hundreds of thousands of animals, the IWC imposed a cessation on all commercial whaling from the 1985/86 and 1986 seasons. At the time of writing sei whales may still be taken from Iceland under a Scientific Permit which is allowed through the 1946 Convention. Several countries have also objected to the moratorium and the future pattern of whaling is somewhat uncertain.

NORTH ATLANTIC

Norway, Murmansk and Spitzbergen

The North Atlantic experienced the start of modern whaling with rorquals caught from the mid 1860s. In Norway, Foyn obtained the right to operate a monopoly over the years 1872 to 1882 and was sufficiently successful that competitors were ready to start whaling in 1883 near to his Finnmark operations. At the same time limited whaling was pursued from the Murmansk coast. Blue, fin, humpback and sei whales were caught, but statistics are incomplete from those early years. From about 1902 whaling was prohibited in Murmansk and, following conflicts with the fishing industry, whaling from North Norway was banned from 1905. In the last few years Tonnessen and Johnsen (1982) claimed that stocks were depleted and coastal whaling gave way to a pelagic whaling. The catches of sei whales from 1885 to 1971 from North Norway are given in Table 11.1, from Tonnessen (1967), Jonsgard (1977) and Tonnessen and Johnsen (1982). In some years not all of the catch was identified by species and in a few years there are minor differences between the catches in Table 11.1 and those of the IWS. In 1885 Collett (1886) recorded an additional 47 sei whales taken from the Murman coast. It can be seen that catches were extremely variable but the years of high and low catches are not well correlated with catches of other species, and the variability is probably due to availability rather than selection. Collett noted that the managers of the whaling stations in Finnmark agreed that about an equal number of males and females were caught.

Catching did restart in North Nòrway in 1948, but from then the Skjelnan whaling station caught only two or three sei whales (Jonsgard 1977, Jonsgard and Darling, 1977); the station closed in 1971. Off West Norway catches are described by Hjort and Ruud (1929) and Jonsgard and Darling, and sei whales were taken from 1913 to 1969; from 1946 three stations operated in West Norway but the last closed in 1969. Records of catches from Jonsgard (1977) are given in Table 11.1 and it can be seen that catches were between 85 and 305, from 1918 to 1929 but after 1940 catches were small. There are minor differences compared with the figures in the IWS, and the IWS give catch by sex. In 1910 the IWS records 7 sei whales caught from Spitzbergen (Svalbard).

Scotland and Ireland

Associated with the possible depletion of stocks off North Norway and the banning of whaling in these waters from 1904, (although operations were allowed in that year), whaling stations were introduced into Scotland and Ireland. In 1903 two Norwegian stations operated from the Shetland Islands, and two other companies started in 1904. By 1907 ten catcher boats were in use and operations continued until stopped by the Great War in 1915. Brown (1976c) records that the season was generally from April to September. Whaling restarted in 1920, was absent in 1921, and then continued until 1929. Catches of sei whales are given in Table 11.1 and are taken from Brown (1976c). His sources are various and generally agree with records of the Fishery Board for Scotland. Only in one year did the catch of sei whales exceed that of fin whales and over the period 1903 to 1929, for those years when species were identified, over twice as many fin whales than sei whales were caught. In terms of oil yield this is a difference of over seven times and consequently the sei whale can be considered as a minor component of Shetland whaling. Information from Haldane (1907-1910) shows that over the seasons 1906 to 1909, 53% of the catch was of males. From the Hebrides whaling started in 1904 at Bunaveneader, and operated each year until 1915. Whaling restarted in 1920 but then ceased for two years before the period of 1923 to 1928. There were only two more years of operation, 1950 and 1951, and the catches of sei whales (Brown 1976c) are given in Table 11.1. Sei whales were caught in every year but constitute a smaller component of the catch than at Shetland. From Haldane (1907-1910) it can be seen that over the seasons 1906 to 1909, 65% of the catch was of males.

Whaling was not extensive from Ireland and is reviewed by Went (1968). The Arranmore company started in 1908 and closed in 1913 and the Blacksod company operated from 1910 to 1922, the first from Inishkea Island and the second from near Belmullet; catch statistics are given in Table 11.1. Over the seasons six times as many fin whales were caught than sei whales, and more blue whales than sei whales. Statistics for 1924 and 1925 are for Scotland and are taken from the IWS (Vol 2); catches of sei whales in this volume for the years 1910 to 1920 appear to be for Scotland only, although they claim to include the Irish catches.

Faroe Islands

In 1894 the first whaling station was started by a Norwegian company, and from 1906 to 1911 six stations were in operation. Tonnessen and Johnsen (1982) reported that over the period 1894 to 1916 a total of 6682 whales of all species were caught. Blue and fin whales were taken but sei whales were again reported as being erratic in their annual occurrence. From 1917 to 1919 there was no whaling but it resumed in 1920. Although some profits were made the whales were generally caught on their northwards migration when they were in a poor condition and hence oil yields per whale were low. From 1935 to the cessation of whaling due to the War, only one station operated. Whaling restarted with two stations in 1946, there was none in 1959 or 1961 and only spasmodic attempts after this. Limited catching of large whales has gone on in recent years under a scientific permit arrangement, but which to date has yielded little scientific material. Catches of sei whales are available from 1910 onwards from Jonsgard (1977) and the IWS, and these are given in Table 11.1 Catches by sex are given in the IWS.

Iceland

As described earlier operations at Iceland helped in the development of modern whaling techniques, and records of catches of sei whales exist from 1893 to the present. From 1883 to 1915 there were 14 shore stations, eight on the West coast and six on the East, and known catches of sei whales are given in Table 11.1 for the years 1893 onwards, from Jonsgard (1977) and the IWS; catches by sex are given in the IWS. As an early conservation measure whaling was banned from 1916 and did not recommence until 1935 when there was one west coast whaling station.

Table 11.1: Catches of sei whales in the North Atlantic by region

Year	Norway West	Norway North	Iceland	Shetland	Hebrides	Ireland	Faroe	Canada	Iberia	Others
1885		724								47
1886		61								
1887		202								4
1888		144								
1889		22								
1890		213								
1891		244								
1892		449								
1893		422	3							
1894		156	2							
1895		93	4							
1896		133	1							
1897		513	2							
1898		547	3							
1899		117	2							
1900		39	20							
1901		22								
1902		34								
1903		59		5						
1904		24		5	4			39		
1905				32	2			2		
1906				262	64			2		
1907				140	11					
1908				198	34	31				
1909				168	55	9				
1910				171	19	39	103			6
1911			4	86	44	2				
1912				73	35	4				
1913	13		9	149	10	1	21	1		
1914				203	45	2	40			
1915										
1916							6			

1917											
1918	154	8									
1919	305	6									
1920	173	24		228		31	3	75			
1921	85							6			
1922	99			44				16			
1923	237			9		1		8			
1924	131				(57)			28			2
1925	248				(18)				4	20	
1926	188			16		5		9	3		
1927	121			21		12		16	9		
1928	140			25		3		9	23		
1929	121			4				14	3		
1930	60							10	1		1
1931	52										8
1932	59										24
1933	22							7			
1934	172							13			
1935	108		1					3	13		
1936	154		1					1	2		
1937	55							11	7		100
1938	94		5					6			
1939	46		3					8	2		
1940											
1941	149								2		
1942	48								4		
1943	45										
1944	31								1		
1945	55								5		
1946	11							1			
1947	12							2	4		2
1948	34	1	5					15	4	10	1
1949	6		12					21	23	18	
1950								2	16	11	
1951	11		2				3	126	39	91	
1952	24		25					2		37	
1953	48		70					27		12	
1954	20		93					3		7	

Table 11.1 (cont'd)

Year	Norway West	Norway North	Iceland	Shetland	Hebrides	Ireland	Faroe	Canada	Iberia	Others
1955	10		134				11			3
1956	16		72				3	2		1
1957	1		78					5	2	1
1958	6		91				1		1	
1959	4		67					5	18	
1960			42						11	
1961	6		58					1	15	
1962			44						10	
1963			20						10	
1964			89					1	13	
1965			74					2	19	
1966		1	41					8	12	
1967			48					62	11	
1968			3					104	12	
1969	1		69					152	33	
1970			44					94	11	
1971			240					235	15	
1972			132					183	13	
1973			139						15	
1974			9						5	
1975			138						26	
1976			3						10	
1977			132						5	
1978			14						15	
1979			84						13	
1980			100						2	
1981			100							
1982			71							
1983			100							
1984			95							

Early catches were dominated by blue and fin whales, and oil yields were much greater per whale than off North Norway, but from 1901 to 1912 most of the whales were not identified by species, and in four of those years the unidentified animals exceeded 1000 per year. It seems likely that sei whales were a significant, but minor part of this group. Post-war whaling has been described by Jonsson (1965) and sei whales from the West coast are generally caught in August and September in the shelf or deep waters of the Denmark Strait. During 1986 sei whales were caught under the IWC scientific permit scheme.

East Canada and West Greenland

Modern whaling was introduced to Canada with the establishment of a whaling station in Notre Dame Bay in North Newfoundland in 1898, and a few sei whales were taken in the earliest years. Whaling was initially successful, mainly because of blue and fin whales, and by 1904 there were 14 stations in operation, but by 1915 only three were left. In 1903 True (1903) described the first authenticated sei whale from the Western North Atlantic, and this was caught from Placentia Bay in the South of Newfoundland. Whaling continued from Newfoundland and the coast of Labrador until 1972 with small intervals of no whaling, but the catching of sei whales was negligible compared to the fin whale catch. Highest sei whale catches were recorded in 1904 and 1951 of 39 in each year compared with about 500 fin whales. Whaling took place in the Gulf of St Lawrence from 1911 to 1915 but no sei whales were taken. Sei whales were more important in the catches from Nova Scotia and whaling operated from 1964 to 1972. Catches of sei whales from Labrador, Newfoundland and Nova Scotia are given in Table 11.1 from Mitchell (1974a) and Jonsgard (1977).

The catch of sei whales from West Greenland was reviewed by Kapel (1985), and the area was operated in by early Norwegian pelagic expeditions, and by a limited number of Danish small whaling vessels post-war; a few whales were caught by local fishermen. Eight were taken by pelagic whaling in 1931 and two in 1924, although these are recorded in the IWS as coming from the Davis Strait. Otherwise there are eight authenticated catches since 1947 and 17 fin or sei whales. These are recorded under "others" in Table 11.1.

Spain, Portugal and Morocco

Over the period 1921 to 1929 whaling was conducted from land stations, moored factory ships and pelagic factories. The first station, at Getares, on the coast of the Straits of Gibraltar was particularly successful due to high catches of fin whales. Over this period the IWS and Jonsgard (1977) reported 66 sei whales caught, but Sanpera and Aguilar (1984) reported only 20. Three land stations operated during the period 1946 to 1954, one at Benzu on the coast of North Africa, one at Getares and the last from Portugal. As explained earlier sei and Bryde's whales were not clearly identified at the southern stations. From 1957 catches continued from land based operations near Vigo, mainly for fin whales, but until 1982 there was a small but continuous catch of sei whales. Catch statistics for these areas, Table 11.1, are mainly taken from Aguilar and Sanpera (1982) and Sanpera and Aguilar (1984), as there are discrepancies in the data from the IWS, and the sei and or Bryde's whales have been included.

Pelagic Whaling

From 1903 to 1912 Norwegian pelagic and shore whaling operated from Spitzbergen to Bear Island taking mainly blue, fin and humpback whales, although Ingebrigtsen (1929) reported that sei whales are found occasionally in these waters. Jonsgard (1977) reports that sei whales were caught only in four years, 1930-1932 and in 1937 with catches of 1, 8, 24 and 100 respectively. The 100, caught in 1937, were mainly from the Davis Strait. The catching of right whales from the Hebrides encouraged southern pelagic expeditions in 1908 to 1911, and although no sei or right whales were recorded as caught rorquals were found about the Straits of Gibralter, and this gave rise to the Norwegian whaling interests in Spain and Portugal.

NORTH PACIFIC

Eastern

Modern whaling was introduced to the North East Pacific in 1905 with a whaling station on the west coast of Vancouver Island, and subsequently expanded to the islands about Alaska. From Canada, whaling stations operated from 1905 to 1943 with breaks only in 1921, 1931, 1932 and 1939; from 1948 to 1967 one station only operated, from Coal Harbour. Pike and

MacAskie (1969) gave catches from the Canadian stations from 1919 to 1967, from 1948 these data are the same as those in the IWS, and are given in Table 11.2. Sei whales were slightly more important at the Canadian operations than at other Eastern locations but fin whale catches were usually greater. Shore stations were started in Alaska and adjacent islands as early as 1911, and the evolution of the industry is described by Tonnessen and Johnsen (1982) and Kellogg (1931). The former say that catch statistics prior to 1930 are unreliable but Reeves et al (1985) compiled detailed catches for the stations at Akutan, which operated from 1912 to 1939, and at Port Hebron, which operated from 1926 to 1937. From both these stations over all the years not more than five sei whales were landed compared with a total of over 8000 other whales. These are included under "others" in Table 11.2.

Further south, shore stations operated from Washington from 1911 to 1925 and from over 2500 whales caught less than 1% were sei whales. From 1961 to 1965 a station operated from Oregon but no records of sei whale catches are known. From 1918 whaling took place from California with a station opening at Monterey, and in 1920 one opened just north of San Francisco. In 1921 and over the winter of 1926/27 pelagic whaling took place from California to Alaska but the numbers of sei whales taken, if any, would have been small. Kellogg (1931) gives the catches from California and Mexico by species for the years 1919 to 1929 but he admits that sei whales could possibly have been confused with Bryde's whales; nevertheless the numbers are small. The IWS records a few whales taken from California from 1937 to 1947 but whaling was revitalised when stations were opened in California in 1956 and 1958; this last phase ceased in 1967. Statistics for California are given in Table 11.2 from Kellogg (1931), Rice (1974) and the IWS.

Western

From coastal Japan fin whales have been caught for two centuries and it can be assumed that some sei whales were also taken, but modern whaling was introduced to the Western Pacific by Soviet operations from 1864 to 1885. These were in the Sea of Okhotsk and along the coast of Kamchatka. From 1895 to the Soviet-Japanese war of 1904 whaling took place from Hajamak to Sakhalin, and in 1903 a floating factory, the MICHAIL operated. In Japan

Table 11.2: Catches of sei whales in the North Pacific

Year	Japan coast	Canada	Kamchatka	Kuril Is	California	Pelagic	Others
1910							
1911	261						
1912							
1913							
1914	221						
1915	718						
1916	383						
1917	436						
1918	653						
1919	433	74					2
1920	357	121			1		16
1921	421						
1922	361	1					
1923	454	53					1
1924	618	100					2
1925	432	68					45
1926	516	25			25		
1927	487	7					48
1928	275	13			12		4
1929	352	67			9		2
1930	397	89					
1931	347						
1932	295						
1933	313	1					
1934	252		1				2
1935	307						6
1936	306	2					
1937	332		1		12		1
1938	445						
1939	598						
1940						3	
1941	578		11		1	7	
1942	235				2		
1943	269				2		
1944	485						
1945	70						
1946	503						
1947	373				3		
1948	530	2	3	36			
1949	728	3	21	60			
1950	299	24	7	49			
1951	419	5	16	52			
1952	666	22	13	188		14	
1953	585	14	26	86		98	
1954	646	134	22	126		129	
1955	488	139	28	128		21	
1956	782	37	16	171		48	
1957	478	93	36	108	1	166	
1958	823	39	19	336	2	330	
1959	1340	185	93	131	37	32	
1960	788		59	140	47	203	
1961	782		54	52	51	4	
1962	1229	340	303	79	22	260	

Table 11.2 (cont'd)

Year	Japan coast	Canada	Kamchatka	Kuril Is	California	Pelagic	Others
1963	855	154	514	16	96	945	
1964	854	613	595	35	13	1533	
1965	466	604	695		22	1398	
1966	291	354	1545		62	2208	
1967	536	89	1997		3	3474	
1968	806		1105		14	3819	
1969	466				10	4682	
1970	484				4	4017	
1971	276				2	2727	
1972	215					2112	
1973	43					1813	
1974	48					1232	
1975	30					478	

modern whaling is recognised as starting in 1898 with Oka the pioneer. The Soviets leased stations in Korea and were joined by the Japanese in 1900, and from 1905 the Japanese industry started a rewarding expansion with dividends of over 50% declared. Competition was fierce, and from 1908 governmental and self regulation occurred with the formation of the Japan Whaling Association. In terms of numbers of whales caught and duration of operations, Japanese coastal whaling has been the most significant in the world. Kasahara's (1950) review of Japanese whaling shows over 65 land stations in Japan and the adjacent islands. After 1945 Japan operated 10 land stations, and those in the Kuril Islands that were used from 1913 to 1943 were utilised by the Soviets from 1948 to 1964. Other whaling areas were lost around Taiwan, Korea and Kamchatka, but in recent years whaling, for Bryde's whales, recommenced in the Bonin Islands. From 1956 a limited whaling was undertaken by the Peoples Republic of China, in the Yellow Sea, and by Korea from 1965 to the present.

Catch data prior to 1910 are not available. An extensive series from 1910 is given in the IWS but Tonnessen and Johnsen (1982) point out that at least from 1920 to 1930 these data differ from those provided by the Whales Research Institute, Tokyo. In addition sei and Bryde's whales were not differentiated until the mid 1950s and not incorporated into the IWS as two species until about 1964. Consequently one must pay attention to where the "sei" whales were caught. Those from the Kurils and Kamchatka will be true sei whales, whereas those recorded in the IWS as coming from Japan, or Japan and Korea, will be mixed. For the next section catches by Japanese pelagic operations near the Bonin Islands from 1946 to 1952 are assumed to be all or nearly all Bryde's whales, as shown by Omura, Nishimoto and Fujino (1952). However Omura and Fujino (1954) found that Bonin Island coastal catches in 1935-36 were all true sei whales; the differences being due to the timing of the whaling seasons. At the northern localities Bryde's whales are rare from Hokkaido and uncommon from Sanriku. Consequently in Table 11.2 I have only included those from Kashara's regions I to V, or approximately North of 35°N. This includes most of the sei or Bryde's whales caught as the large majority of the two species were caught from the Kurils, Hokkaido and particularly North East Honshu.

Pelagic

A Norweigan pelagic expedition explored the North Pacific in 1925 and 1926, but was generally unsuccessful, and continuous pelagic exploitation only began in 1932 with the Soviet factory ship the ALEUT. This operated as the only Soviet pelagic factory until 1961, but in 1940 it was joined by a Japanese factory ship. From 1946 to 1952 a Japanese fleet operated near the Bonin Islands but took Bryde's whales. From 1961 to 1965 the Soviet Union introduced another three factories and in 1954 a second Japanese fleet operated, but nevertheless catches of sei whales were small until 1963. Following this build up of catching effort there were arguments within the IWC about regulation and catch limits, but the catch of sei whales rose rapidly from 1963 to 1967. The following year was the last year of Soviet coastal whaling but a peak pelagic catch of 4682 was taken the following year, and eventually the protection of sei whales was established from the end of the 1975 season. Pelagic catches of over 1000 per year only lasted for one decade, whereas Japanese coastal whaling of several hundred per year lasted for 60 years.

SOUTHERN HEMISPHERE

Coastal and North of 40°S

Substantial catches of sei whales have been taken from South Georgia, Brazil, Natal and the Cape Province. Catches from Peru have been small and those from Australia and New Zealand negligible. Catches, as recorded in the IWS, from Chile are of moderate size but include some Bryde's whales. From the Antarctic, catches were taken from 1906 and factory ships, which were of the type of mobile shore stations, operated about the Falkland Islands and South Shetland Islands; catches from these operations are included in Table 11.3 under "pelagic". From South Georgia several bases operated, initially taking sei whales late in the season, and the stations caught sei whales from the season of 1913/14 to 1965/66. The numbers of sei whales caught are given in Table 11.3 and for the Southern Hemisphere are from Ohsumi and Yamamura (1978a), unless otherwise stated; they also give the catch by sex.

In Brazil two stations operated, one at Costinha opened in 1911 and has continued to the present day, the second, at Cabo Frio, operated for a short period from 1960 to 1963. At both, sei and Bryde's whales were caught, but catches were predominantly of sei whales (Paiva and Grangeiro 1965). These whales were taken from 1947 on, only humpback whales being caught before then, and da Rocha (1983) gives the catches of sei and Bryde's whales combined from Costinha. From 1967 they are separated by species. In Table 11.3 the figures from 1946/47 to 1965/66 are from Ohsumi and Yamamura the only difference being that da Rocha has a catch of 198 in 1955. In the years after 1967 90.1% of the catch was sei whales and the proportion of sei to Bryde's whales was likely to have been higher in earlier years when the sei whales were more abundant.

From the west coast of South Africa whaling started in 1910 just north of Cape Town at Saldanha Bay and continued until the closure of Donkergat in 1967; however there was no whaling from 1931 to 1934, from 1938 to 1946 and from 1954 to 1956. Cumulative catches of sei whales exceeded 9000. Catches are given in Table 11.3 and are unaltered from those given by Oshumi and Yamamura from 1929 onwards, but from 1917 to 1928 they have been modified following the study by Best and Lockyer (1977ms) who also gave additional catches from 1910 to 1914. Sei and Bryde's whales are combined in some of these statistics, and Best considered that only from 1962 onwards were the two species well identified, but that the magnitude of the error would not be great as sei whales predominated. In 1963 the catch statistics show 721 sei caught but Best (1967a) explains that this includes 89 Bryde's

Table 11.3: Catches of sei whales in the Southern Hemisphere by whaling season. 1910 refers to the pelagic season of 1909/1910 and the coastal season of 1910.

	Area I			Area II			Area III				Area IV	Area V	Area VI
	Peru	Chile	Pelagic	Brazil	South Georgia	Pelagic	Cape Province	Natal	Pelagic	Pelagic N of 40S	Pelagic (total)	Total	Pelagic (total)
1910			173			173							
1911			97			98							
1912							90	11				2	
1913			22			21	45	1					
1914	4		52		86	53	92	3					
1915								7					
1916							39	10					
1917							55	5					
1918					49		95	4					
1919			1		7		190	3					
1920					71		127	15					
1921					36		34	49					
1922					103		76	48					
1923					10		128	60					
1924			1		191	1	364	57					
1925	13				1		21	112					
1926	32		91		13	91	263	97					
1927			206		365	207	62	89					
1928			394		95	394	342	51					
1929			206		396	205	193	42					
1930					216		159	52					
1931					144			29					
1932					16			23					
1933					2			11					
1934								30					
1935					125	140		90	1				
1936	10					2	214	68					
1937	3	13			471	19	49	64	1				
1938		44			155	6		64					
1939		15			19	3		42				1	
1940					80			25					

1941					88			5					
1942					52			13					
1943					73			34					
1944					197			24					
1945		1			76			34					
1946		13			82	3		75					
1947		2		14	391	2	39	119					
1948	1	6		10	609		83	109			2	10	
1949				18	562	15	119	101				1	
1950				98	1183	101	324	101					
1951	1	2		151	519	350	237	247			11	1	
1952	17	10		153	498	27	711	155	3			1	
1953	36	27		161	498	13	295	96	10		3	89	
1954	2	26		183	778	158		71	11		1	1	
1955	1	33		198	423	120		176	18		4	3	
1956		47	16	196	284	242		101	6		1	15	8
1957		39	275	115	980	284	263	60	15		1	4	132
1958		16	540	118	924	180	405	172	16		276	397	968
1959	1	17		294	1019	53	525	379	79		250	481	531
1960		13	159	750	1075	423	498	183	230		526	1649	232
1961		13	102	958	792	1146	228	475	336		103	564	2030
1962		9	1629	610	447	1249	388	319	427		633	411	369
1963		6	807	346		1812	721	369	1457		631	433	345
1964	1	47	28	256	409	4150	673	282	1984		274	1824	
1965	2	487	40	149	506	15584	764	459	443		1564	2207	
1966		210	35	72	4	12718	417	273	2756		442	1008	599
1967		191	3	49		1553	152	66	6860		2825	718	402
1968	355	83		58		194		24	2352		2271	2653	2880
1969	784	31	73	56		188		40	1771	318	1030	2156	552
1970	413	17		23		1278		8	1997	455	1925	474	156
1971	415	1		18		640		10	1065	380	3967	285	194
1972	337	15	22	5		269		4	1172	156	2608	1384	
1973	19	14	460	6		98		4	600	490	1685	943	77
1974	63	34	1126	2		3		4	391	449	1537	519	814
1975	130	58	1027	3		4		1	93	275	1071	507	1157
1976	12	2	198	3		515				23	466	393	249
1977	5	15	298	5		83					383	556	536
1978	3	98	311								254		
1979		65											
1980													

whales taken mainly prior to the main season (Table 11.3 excludes these), and over the normal operating season in 1963 the Bryde's whales constituted 7% of the catch (see Chapter 2). Further north along the coast of Africa a shore station operated from the Congo but catches of sei whales were small or nonexistent. On the east coast, whaling started in 1908 near Durban in Natal, and the success of the companies was such that seven factories operated in 1913 between Durban and Cape Town. This initial success did not last, but catching of sei whales has been virtually continuous since then. A very few Bryde's whales have been caught off Durban and they constituted a very small percentage of the catch of sei whales (see also Other Operations).

From Peru the IWS show limited erratic whaling in the pre-war years and this appears to be from factory ship operations. In 1951 a shore station opened but concentrated on sperm whales. From 1968 baleen whales provided a direct interest and these were mainly sei and Bryde's whales with a few fin whales. Table 11.3 gives the recorded catches from Valdivia et al (1981, 1984). From 1973 sei whales were differentiated from Bryde's whales and Bryde's whales constituted 90% of the catch. Data from 1968 to 1972 are of both species combined, and clearly overestimate the numbers of sei whales taken; however, these figures are very different from those of Ohsumi and Yamamura (1978a) and the IWS. In Chile shore based whaling started in 1906 and developed erratically, but whales have been caught continuously from 1929 with a break from 1940 to 1942; catching of sei whales ceased in 1979. Catch data are given by the IWS, Ohsumi and Yamamura (1978a), Aguayo (1974) and Maturana (1981): they agree in some years. There are only minor differences to and including 1964 and those from Aguayo are given in Table 11.3 (In 1964 the catch is 48 greater than that given in the IWS). Catches were small until 1965 and Ohsumi and Yamamura show that 38% were males. From 1967 the figures given by Ohsumi and Yamamura are incomplete and Table 11.3 gives the catches from Maturana (1981) and the IWS, which are the same. Aguayo only gives catches to 1970 but his figures for 1967 to 1970 are 330, 70, 81 and 19. All these catches from Chile could include some Bryde's whales, and Gallardo and Pasterne (1983) suggest that 90% may be Bryde's whales based on the data from Peru. The whaling station in Peru is at about 5°S whereas the stations from Chile operated between 30°-40°S, and these are latitudes in which one might expect to find many sei whales and the assumption of 90% sei whales is unreasonable. It may be preferable to assume that they were all sei whales until proved otherwise. For some years operations were under Japanese control and the opinions of their operatives would be useful. Catches of sei whales from Australia and New Zealand are taken from Ohsumi and Yamamura (1978a), and there is some confusion with Bryde's whales but catching of sei or Bryde's whales was insignificant.

Pelagic

True pelagic whaling did not begin until the season of 1923/24 but before that mobile processing vessels operated near to safe anchorages. Consequently this allowed sei whales to be caught from the Islands of the Falklands, South Shetland and South Orkney and these catches are considered here. From 1910 to 1930 the IWS records no catches of sei whales from South Orkney and six from South Shetland. Table 11.3 gives the pelagic catches, by Area, from Ohsumi and Yamamura (1978a), and for these Islands they have taken the IWS catch data and apportioned them half each to Areas I and II. As described in Chapter 2 sei whales were taken from the spring of 1906 when 97 sei whales were caught, either mainly or exclusively from the Falkland Islands (Tonnessen and Johnsen 1982), (The figure of 122 in the British Government report (Anon 1920 p55) appears erronious). Ohsumi and Yamamura have a catch of zero for 1911/12 and a blank for 1914/15 but in these two years only the total catch of all species is known for the Falkland Islands from the IWS; from 1909/10 to 1913/14 sei whales were numerically the most abundant species in the Falkland catch and so the total catches for the two years, of 103 and 255, are likely to have been mainly sei whales (they are not included in Table 11.3).

In the season of 1928/29 one sei whale was caught by a true pelagic expedition to the Ross Sea, but as the pelagic fleets operated near to the ice edge, hunting mainly blue, fin and humpback whales, the catches of sei whales remained small until the season of 1956/57. Until this season coastal catching took many more sei whales than did the pelagic operations. From then catches of sei whales rocketed from less than a thousand per year to 20,000, and in the Southern Hemisphere, from the season of 1909/1910 to the closure of the fishery in

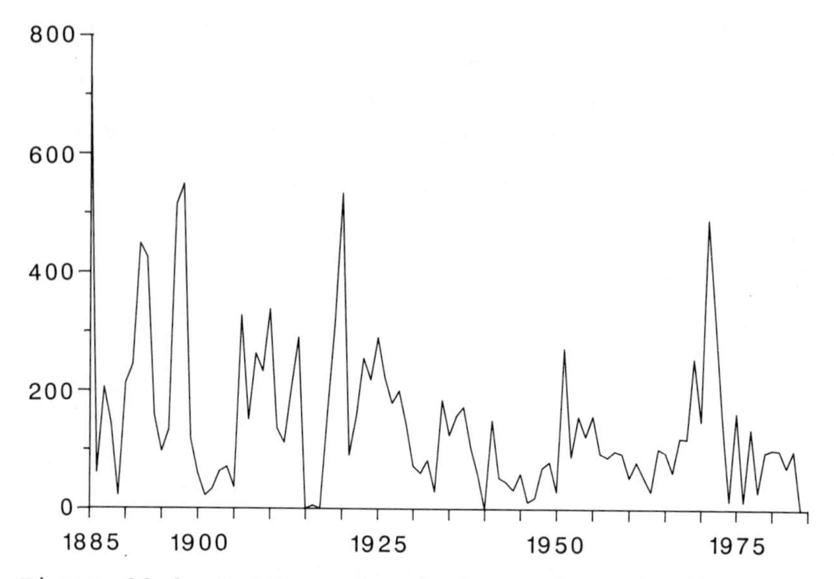

Figure 11.1: Catches of sei whales from the North Atlantic by year.

1979, over 132,000 sei whales were taken by the pelagic fleets compared with 37,000 from coastal operations. The move to sei whales followed restrictions on the catches of blue, fin and humpback whales and the whaling fleets operated more to the north than in earlier operations. For a few seasons there was limited pelagic whaling South of Madagascar and these catches are given in Table 11.3 although, in total, the IWS give 13 whereas Ohsumi and Yamamura give only one sei whale.

Other Operations

The IWS records catches of sei whales from other operations, particularly about the coasts of Africa, and, following Ohsumi and Yamamura, these catches have not been considered as they are likely to be mainly or totally Bryde's whales. These include catches of less than 1000 in total from off Angola and smaller numbers from the Congo. Nevertheless sei whales have been reported from these locations.

Other operations are those of the "pirate" whalers. These are operations which are either illegal or else operating outside of the jurisdiction of international agreements. One such example was the OLYMPIC CHALLENGER, which operated from

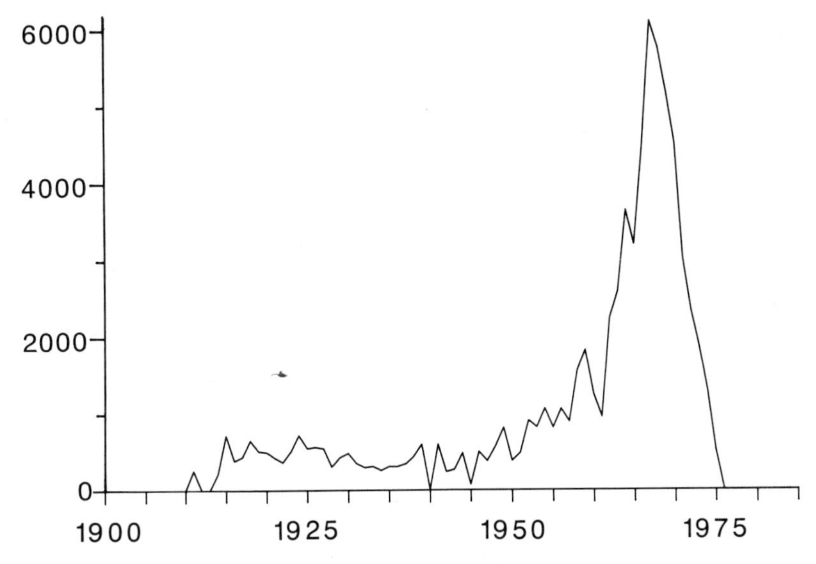

Figure 11.2: Catches of sei whales from the North Pacific by year. There are no data prior to 1910.

1950/51 to 1955/56 catching whales protected by the IWC, and which presented falsified catch statistics. The operation was very unpopular with the governments of the whaling countries, and bombing by the Peruvian armed forces and Norwegian legal action stopped the whaling. However we do not know what the catches were, but Small (1971) recorded 21 sei whales taken from off Peru in 1954. As described in Chapter 2 a similar operation existed with the vessel SIERRA which worked in the South Atlantic, and probably also just North of the Equator, from 1969 to 1976. It is thought that the catches forwarded to the BIWS and reported to the IWS are accurate and are given as 2536 sei whales, however this figure will include some unknown proportion of Bryde's whales (Table 11.3, Area III, N of 40°S, including some whales taken by Antarctic pelagic fleets). The log books of this operation are particularly important as few rorquals have been caught in this region and it may be hoped that they could be studied.

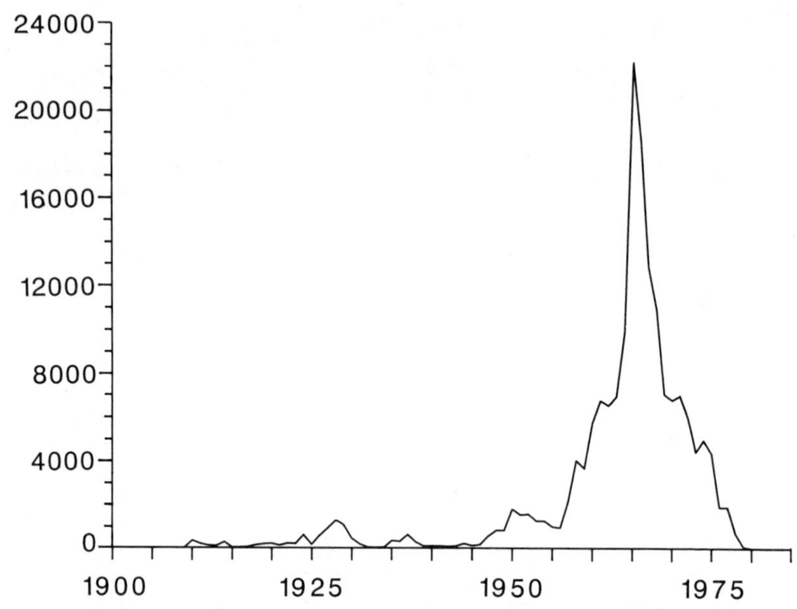

Figure 11.3: Catches of sei whales from the Southern Hemisphere by whaling season. 1900 refers to catches from the summer season of 1899/1900 plus the winter season of 1900.

SUMMARY

Information has been provided on the history of catching of sei whales and the numbers caught. Their speed and pelagic existence meant that they could only be caught easily with modern whaling techniques, and so have only been caught regularly for the last 100 years. This is only a few generations and so catches this far back may well be significant in describing the dynamics of populations. Figures 11.1-11.3 give the catches of sei whales by year, and they show the very large burst of catching from 1955 to 1975, but also the very long period of continuous exploitation. The high catches were a consequence of reduced quotas of the other rorquals and eventually led to the protection of sei whales in all areas.

Exploitation

The numbers in the tables are the best data currently available on the catches of sei whales, and it has been explained where the data were obtained and discrepancies with other data have been mentioned. These data are probably good for the purpose of describing the dynamics of sei whales, and the catch by sex is given in the references indicated. However for the sei whale and other species a more complete and definitive set of catch statistics is needed based, where possible, on original log books. For sei whales there is always likely to be the confusion with Bryde's whales, but from many localities this is seasonal and the approximate disaggregation of the catch is also best done from original log books.

Chapter Twelve

INTERSPECIFIC AND INTRASPECIFIC ASSOCIATIONS

INTRODUCTION

From the last three chapters it can be seen that there have been substantial differences in the estimated rates of reproduction from sei whales caught from various locations and from a range of years, and that these differences may be associated with separate populations which have experienced different histories of exploitation. Pregnancy rates from samples from the South Atlantic increased from 0.3 to 0.5 between 1920 and 1960, and greater increases were reported from Japanese catches in the North Pacific. Ovulation rates were different from Canadian and Icelandic samples and ovulation rates possibly increased with time in samples from the North Pacific. Ages at maturity, estimated by examination for a transition layer in the ear plug, have shown declines in most locations. As explained previously, sampling has been totally inadequate to detect changes, if any, in mortality rates. Although the estimates and trends in demographic parameters are subject to criticism, and there are certainly unresolved problems in interpretation, it is to be expected that some demographic parameters will have altered due to exploitation or changes in the environment. Consequently, it is accepted here that the trends described in Chapter 9, for both pregnancy rates and ages at maturity, are real and that we may then enquire as to how these changes were generated; in particular the relative roles of inter- and intraspecific competition should be evaluated.

It has been shown that in most oceans large numbers of sei whales have been caught and in Chapter 13 it will be shown that the relative depletion in many cases has been very severe. Such reductions in population numbers might be expected to give rise to better survival or birth rates; however this need not be the case. It is not thought that the sei whale has a particularly sophisticated social behaviour but, for instance, sperm whales do, (Best 1979), and it is quite probable that disruption of schools of sperm whales by the unnatural process of whaling could depress pregnancy rates, even if the population was being thinned. In addition modulation of the environment has had significant effects on many marine populations (Cushing 1982). Notwithstanding these possible causes of changes of the demographic parameters, Gambell (1973, 1975) reviewed the exploitation of the baleen whales of the Southern Hemisphere in relation to trends in pregnancy rates and ages at maturity and concluded that there were substantial interspecific effects due to the severe depletion of stocks of blue and fin whales. In this chapter Gambell's hypothesis is examined in relation to exploitation in the Southern Hemisphere, since in this area only limited whaling for sei whales occurred before the late 1950s and sampling has been undertaken since 1930.

EXPLOITATION IN THE SOUTHERN HEMISPHERE

The most intensive exploitation of the large baleen whales of the Southern Hemisphere occurred this century, with the introduction of modern whaling into the Antarctic. The numbers caught in the Antarctic, including those from South Georgia, are illustrated in Figure 12.1 for blue, fin, sei and minke whales for the years 1909-1977 and it can be seen that catches of blue and fin whales steadily increased until the time of the Second World War, during which few catches were taken. After the War quota restrictions and a much depleted blue whale population reduced catches of blue whales and consequently consistently large numbers of fin whales were taken until 1960. Sei whale catches rose dramatically from 1955 to 1965 and fell in a similar fashion. The last increase is of minke whales.

The catches of humpback whales are not included in Figure 12.1, but they were the dominant component of the Antarctic catches at the start of the fishery until the season of 1912-13; in the season of 1910-11 over 8000 were caught in the Antarctic. Although the catches were variable in number they were an important component of the catch from African land

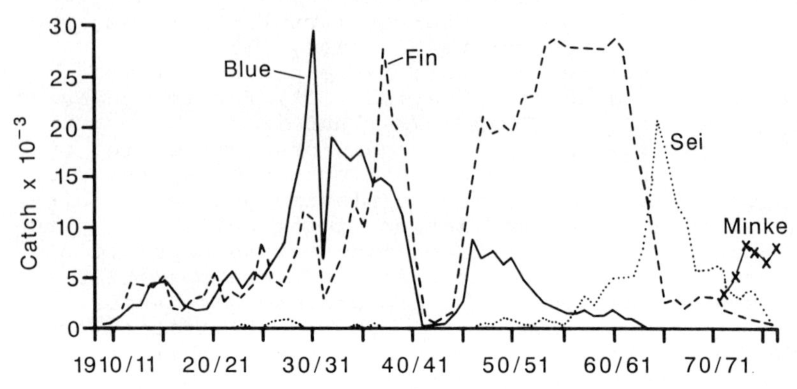

Figure 12.1: Catches of large whales, by species and whaling season, from the Southern Hemisphere. © Crown 1981

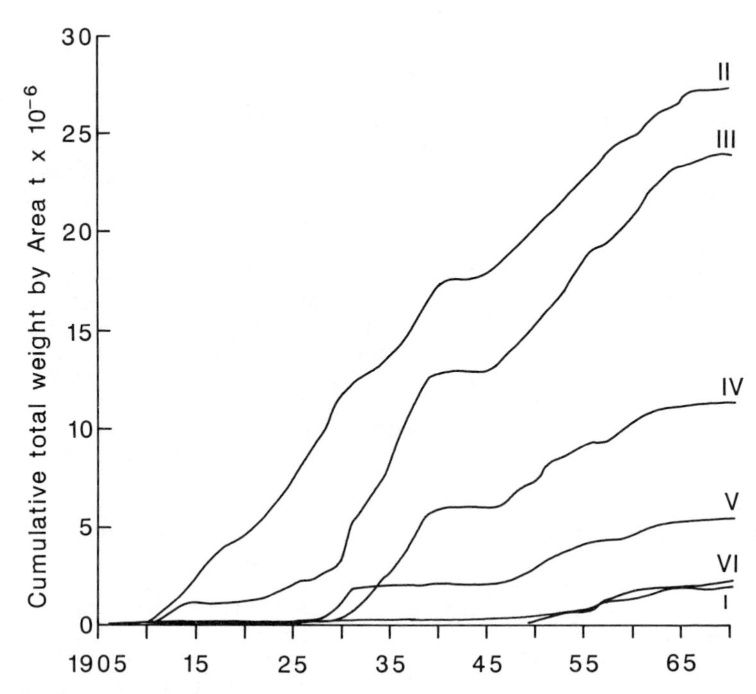

Figure 12.2: Cumulative catches of large whales by whaling Area. © Crown 1980

stations until 1940, but nevertheless the catches of humpbacks by weight were dwarfed by the catches of blue and fin whales. However, humpback whales were taken during the earlier era of open-boat whaling. They were usually not pursued at sea for they were too fast and sank when dead, but they were caught in the bays and on banks during the winters; a review of humpback whaling is given by Mitchell and Reeves (1983). The southern right whale was also subject to early whaling but of a much more intensive character and only remnants of the populations existed at the turn of this century (Best 1970, Townsend 1935).

If there are interspecific interactions it is likely that they are associated with competition for food, especially since in the Antarctic the plankton tends to be dominated by the one species of krill; it is then more instructive to look at the catches by weight than by numbers. In addition there are different populations of the different species in the Antarctic and one should consider catches by region and Figure 12.2 shows the cumulative catches by weight and IWC management Area for baleen whales over the period 1905-1970. It is clear that from each Area catches are of different magnitudes and started at different times. Removals have been at a constant rate from Area II since 1910 and by 1970 the cumulative catch exceeded 25 million tonnes. Catches from Area III also started in 1910 but from 1915 to 1930 catches were few, and an accelerated exploitation then took place to give a final removal of about 25 million tonnes. Exploitation in Areas IV and V, (70°E-170°W) started much later at about 1930, when larger catches were also taken from Area III, but in Area IV cumulative catches only reached 10 million tonnes and in Area V it reached 5 million tonnes. Whaling from Area I started early but continued at a very low level and from Area VI, the sanctuary, catches increasing from 1948, and a cumulative catch in 1970 of only 2 million tonnes is found from both Areas. Consequently exploitation has differed in its extent and timing from Area to Area and any pattern of interspecific effects should reflect these differences.

CHANGES IN PREGNANCY RATES OF SEI WHALES

Changes in the apparent pregnancy rates of blue and fin whales were noted first by Mackintosh (1942) when he compared data over the period 1925-1931, and

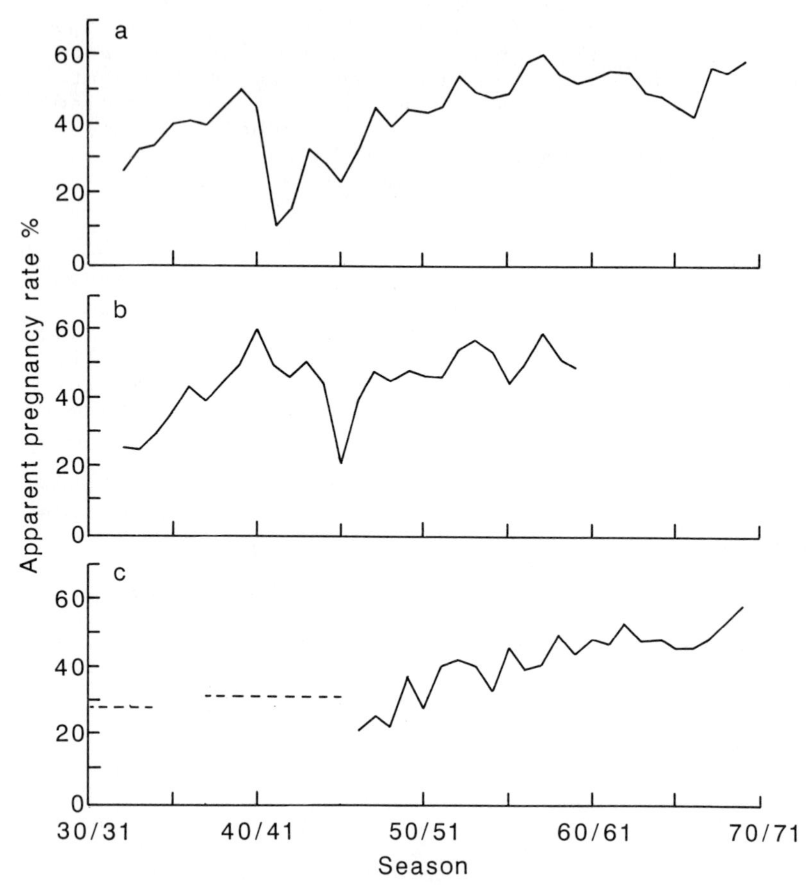

Figure 12.3: Apparent pregnancy rates from the Southern Hemisphere. a: blue whales, b: fin whales, c: sei whales; dashed lines gives the average from limited samples over the periods shown.

Laws (1961) extended that study and concluded that increases in pregnancy rate had occurred, and that fertility was influenced by fishing pressure. Gambell (1973) presented apparent pregnancy rates of blue, fin and sei whales sampled over the years 1930-1970 and these are given in Figure 12.3, along with the estimate of average apparent pregnancy rate of sei whales from 1920/21-1934/35 (0.28) and the average rate over the seasons 1937/38-1945/46 (0.31). Data are mainly from catches in the Antarctic and at South Georgia and show, for sei whales, a near doubling of pregnancy rate and that the pregnancy rate started to increase between 1945 and 1950. The apparent pregnancy rate by Area from 1956 to 1975 was documented by Masaki (1978), and a reduction was found in Area III whereas increases were recorded from Areas IV, V and VI, but compared with the pregnancy rates sampled from the 1920-1940 period those changes recorded after 1956 are all very small.

For the earliest and immediate post-war years Gambell (1973) considered that the trend in pregnancy rates of sei whales were similar to those found in blue and fin whales, but since large scale exploitation of sei whales did not start until the 1960s the observed increases could not be a response to a greater intensity of fishing, as had been postulated for blue and fin whales. It is true that substantial exploitation did not start until the late 1950s, but for all Areas combined Table 12.1 shows that by 1956/57 14% of the total catch ever taken had been removed and that in each of Areas I, II and III the removals were about 20% of the total. Lockyer (1978b) had commented on the relatively small extra amount of food needed for growth changes and that pregnancy rates may respond even more sensitively, and it is then possible to envisage a situation in which the stocks in Areas I-III did not respond to the light pre-1946/47 exploitation but after that their reduction may have been of substantial biological significance. It is thus not necessary to evoke interspecific mechanisms to explain these gross changes found in Areas I-III. Nevertheless, Masaki (1978) has shown that pregnancy rates in Areas IV-VI were high between 1956/57 and 1960/61, and before this period exploitation had been extremely light in these Areas. This may be explained by postulating: (1) a gross mixing of stocks; but this is not substantiated by marking data; Chapters 4, 10; (2) that these are the true "unexploited" values and that Gambell's pre-1956 series is generated by a trend in sampling bias, or (3) that all Areas have been subject to the same stimulus to pregnancy rates, such as interspecific competition. Additional information is given by the consideration of trends in ages at maturity.

CHANGES IN AGE AT MATURITY OF SEI WHALES

The number of laminations, up to the transition layer, in the ear plug of the sei whale was described by year of birth by Lockyer (1974) for catches taken in the South Atlantic, and the data are illustrated in Figure 9.9. A trend can be seen of decreasing age at maturity for male sei whales born after about 1935 and for females born as early as 1925. Gambell (1975) considered this as supportive of the explanation of interspecific competition.

These data were extended in the study by Lockyer (1979), Figure 9.12, and it was possible to investigate the declines in age at maturity by Area. The relative declines in mean ages at maturity of the year classes 1945-50 and of 1955-60 compared with pre-1920 year classes are given in table 9.6 and it can be seen that the earliest and largest declines were found in Areas II and III and the lowest in Areas IV and V. Lockyer (1979) demonstrated that the relative declines were well correlated with the start of whaling for sei whales in each Area, although the start of whaling for fin whales also gives a good correlation. In Area I however the changes are somewhat more than expected on the basis of sei whale exploitation, and it is possible that this is due to mixing of whales between Areas I and II. It would then seem that interspecific effects are not required to explain the existence of the observed trends in the transition layer data, and exploitation of sei whales may be a significant factor.

Table 12.1: Catch of sei whales by Southern Hemisphere IWC Areas, percentage in brackets. Time periods are i: pre 1946/47, ii: 1946/47-56/57, iii: 1900-1976 (numbers from Ohsumi and Yamamura 1978a). © Crown 1980

Area	I	II	III	All
Time	Catch	Catch	Catch	Catch
i	1,452 (15)	4,635 (7)	3,717 (10)	9,832 (6)
ii	536 (6)	9,340 (14)	3,470 (9)	13,737 (8)
iii	9,488 (100)	65,004 (100)	39,059 (100)	168,924 (100)

RESPONSES OF OTHER POPULATIONS

Laws (1977b) calculated that the initial stocks of baleen whales consumed some 190 million tonnes of krill per year whilst, because of depleted populations, they currently consume about 45 million tonnes; such a release of potential food may well have affected many species.

Crabeater seals from the Antarctic peninsula have been subjected to negligible recent exploitation but Laws (1977b) found that ages at maturity in the population had declined during this century. His information came from examination of a transition layer in teeth and, as for the similar material from whales, the study by Bengtson and Laws (1985) has shown that interpretation of the observed trends is not straightforward. Nevertheless, fur seals at South Georgia increased in numbers from less than one hundred during the 1930s to about one million at present (Bonner 1968, M. Payne 1979). Changes have also been seen in bird populations. The chinstrap penguin, which was rare at South Georgia during the nineteenth century has increased in numbers at South Georgia and South Orkney (Stonehouse, 1967; Croxall, Rootes and Price, 1981) and although there have been reductions in some colonies of King penguin there has been an overall increase in numbers (Conroy and White, 1973). Increases in the numbers of gentoo, macaroni and Adelie penguins have also been recorded (Conroy, 1975; Croxall et al 1981; Rankin, 1951).

The ages at maturity, from transition layer material, of minke whales from the Southern Hemisphere have been described by Masaki (1979b), Best (1982) and Kato (1985). Before 1970 the numbers of minke whales caught was insignificant but all these sets of data show substantial declines in mean ages at maturity for whales born from 1930. These data are not without problems of interpretation (Cooke and de la Mare, 1986) but are accepted here because invalidity has not been established and the typical one year breeding cycle of the minke whale, found in the northern and southern hemispheres, leaves only a limited scope for a population response of pregnancy rate. Masaki's data were collected between 1972 and 1977 along with information on pregnancy rates. Using these data Horwood (1980b) ranked mean ages at maturity for pre-1944 year classes for males and females in each Area along with the pregnancy rate. Table 12.2 shows the results, with the lowest ranking representing the lowest age of maturity and highest pregnancy rate.

Table 12.2: Rankings of age of maturity by Area and of pregnancy rates of pre-1944 year-classes of minke whales. © Crown 1980

Area	I	II	III	IV	V	VI
Female	2	1	4	6	3	5
Male	2	1	3	6	5	4
Pregnancy	4	3	2	1	6	5
Sum	8	5	9	13	14	14

The summed ranks show the striking result that the characteristics of Areas I, II and III are much different from those of Areas IV, V and VI. It is likely that these changes have arisen as a response to the earlier and heavier exploitation of baleen whales in Areas II and III.

DISCUSSION

The increases in bird populations vary with species and location and Sladen (1964) considered that the increases were a response to greater food supply brought about by the reduction in the numbers of whales; however Conroy and White (1973) thought that although this hypothesis might be true, there also might be natural increases responding to the early exploitation of penguins during the sealing in the nineteenth century. If a response to early exploitation is a major factor then the time-lags have been very long and Conroy and White considered that the problems of initial colonisation may be responsible for the long time-lags. Stonehouse (1965), Yates (1975) and Taylor (1962) all stressed the importance of climate in breeding success. Stonehouse considered that the existence of the Southern Ocean penguins was a race against time each year and that a late spring or early winter very seriously affected the year's crop of young, but Croxall (pers. comm.) has commented that such a "race against time" might apply to Continental species but is unlikely to exist for species in the Subantarctic areas.

The increase in the numbers of fur seal presents similar problems of explanation to those for the birds, with early, very severe exploitation. Jones (1973) cites Weddell as calculating that by 1822 at least 1.2 million furs had been taken from South Georgia; 100 years later there were still less than 100 seals on South Georgia. A decrease in the age of maturity of the Crabeater seal has been recorded over the year-classes 1940-70 and this population has been only very lightly exploited. Half of this decline in the age at maturity occurred during a period when whaling was prohibited in the Area and would seem to exclude interspecific causes, but mixing in and out of Area I of the whales cannot be excluded. The information from the seals gives equivocable arguments for any one explanation of the changes. The changes in ages at maturity and pregnancy rate of minke whales do however focus attention on factors other than direct exploitation of minke whales.

The pregnancy rates and ages of maturity of sei whales show smaller changes of demographic parameters in Areas where direct exploitation was absent, and the case for interspecific competition is not strong from these data. Laws (1977a, b) considered the evidence for competition as circumstantial and indicated the ecological separation of the different groups in time, space and habit. This is amplified in Chapter 5 and by Figure 5.6 illustrating the different prey species taken by the different whales. However, this may only mean that whales have a specific niche which can be held against others and does not imply that if other zones are left free they could not exploit them. Nevertheless, it was shown in Chapter 5 that from a consideration of the feeding ecology of whales there was likely to be greater competition between sei and right whales than between sei whales and any other group. Right whales are severely depleted from nineteenth century whaling and Kawamura (1978) argued that the sei whales could have benefitted from their depletion, but this fails to explain why rates should have increased in the twentieth century.

There is no evidence that over the feeding season the whales have been short of food for Foxton (1956) found that the standing stocks of zooplankton, less than 20 mm (smaller than sei whale food), were not reduced during the season and, from an analysis of stomach contents collected over the period 1946-71, Kawamura considered that there was no evidence for improved feeding. Mackintosh (1973) found that whales maintained a similar intake of

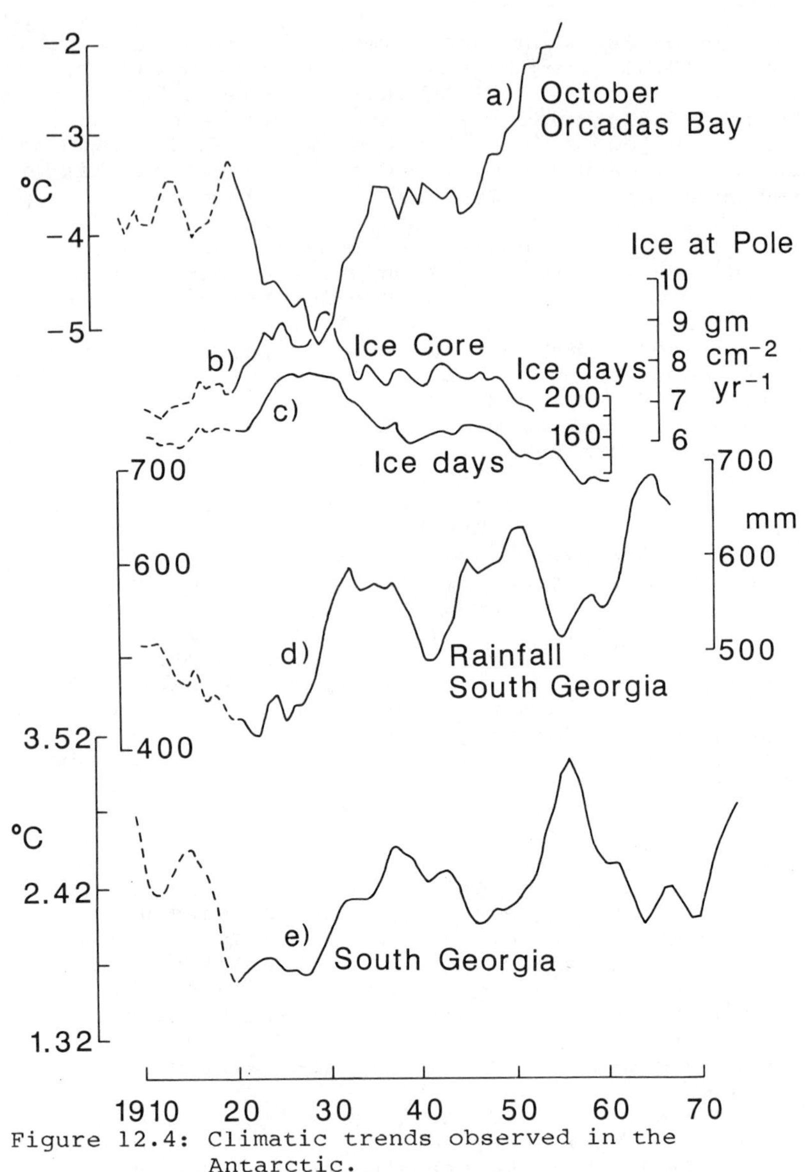

Figure 12.4: Climatic trends observed in the Antarctic.
a. October air temperatures from South Orkney.
b. Accumulation rates of ice cores from the South Pole. c. Days of ice at Orcadas Bay.
d. Summer rainfall at South Georgia.
e. Average air temperatures, March to May, from South Georgia. © Crown 1980

food over the whole season, but Mackintosh (1972) noted that in warmer years there may be an intermediate group of krill and that with a warm season there may be an extended season and a greater area of ice-free sea to feed in.

Climatic changes have been observed in the Antarctic and some of these changes are illustrated in Figure 12.4 from Horwood (1980b). Limbert (1974) presented mean annual air temperature data from Laurie and Signey Is. of South Orkney and a cold period during 1925-35 can be seen followed by a warming, which by 1940 amounted to 2°C. Rubinstein and Polozova (1966) and Fletcher (1969) show similar trends by month from Laurie Island, and Jones (1978) from Grytviken, South Georgia. Piggot (1977) has shown increases in summer rainfall at South Georgia and temperature trends throughout the Antarctic show that the Western Peninsula temperatures have experienced a 2°C increase in mean annual temperature from 1935 to 1975 (British Antarctic Survey, Annual Report 1976-77). The changes are regional. Snow core data show that accumulation rates have varied over this century (Giovetto and Schwerdtfeger, 1966) and so have the days of free ice at Orcada Bay (Schwerdtfeger, 1959; Fletcher, 1969). In general there appears to have been a cold spell over the period 1925-35 followed by an amelioration in climate which may have had a profound effect on the ecology.

Evidence that there have indeed been some demographic responses in sei whales, independent from those inferred from measured reproductive rates, comes from increased sighting and catches per effort from Durban and Donkergat. Best and Gambell (1968), and the IWC assessment meeting (Fukuda 1978) reported that from Durban the catch and aircraft sightings, per unit effort, both increased by 9.3% per annum from 1954 to 1965, and from Donkergat the catch per effort from 1958 to 1965 increased at a rate of 12.5% per annum. Best and Lockyer (1977 ms) reported that sightings data from Donkergat also reflected this trend. It can be seen from the above that these population responses, of the Southern Hemisphere sei whales, cannot be due only to a release of interspecific competition for food. The larger responses found in minke whales, and bird and seal populations can be attributed to their greater dependence on krill, which is the main food of the depleted blue and fin whales. However, in all cases a less severe Antarctic climate could have played a significant role. Consequently, most of the changes in reproductive parameters found in the Southern Hemisphere sei whales can be attributed to intraspecific associations but with some contribution from other sources.

SUMMARY

Changes in reproductive parameters of sei whales have been examined in relation to the exploitation of the baleen whales of the Southern Hemisphere. This is a particularly suitable study since substantial exploitation occurred on blue and fin whales before sei whales were heavily exploited. The results indicate that although changes in reproductive parameters from sei whales caught from some locations occurred early on, most of the changes can be explained through the timing of exploitation of the different stocks of sei whales. Nevertheless it does appear that some additional responses occurred and that these could be associated with depleted stocks of blue and fin whales or a more favourable climate, each leading to richer feeding.

Chapter Thirteen

ESTIMATES OF STOCK SIZE

INTRODUCTION

It is very difficult to count whales. They live in remote areas, are mobile, spend much of their time under water, and consequently indirect techniques of estimation are used. Apart from methods based directly upon the numbers seen, two other general techniques are used to estimate numbers. The first is mark-recapture analysis which relies upon the proportion of marks returned in a catch, in relation to the numbers marked. The second obtains estimates by relating a decline in an index of abundance to the level of catches. Both methods are straightforward in principle, but frequently suffer badly in practice because of statistical variability, or violations of the assumptions inherent in the methods.

The indices of abundance play an important role in determining the status of a stock. They should be proportional to stock size, and if one sees an index continuously decreasing it is evident that the stock size is decreasing and management action can follow without knowing anything of the basic biology or absolute stock size. The most reliable stock estimates are found from situations where a stock has been severely depleted over a short period of time, for in these cases it is only stock size and catches which are of significance, with recruitment playing a minor role. Where exploitation has been continuous and light, then estimates tend to be unreliable because indices will be relatively stable, and essentially the same trend in the index would arise from the true stock size or any larger one. If catches have been taken over a long period then it becomes necessary to model recruitment through the density dependent mechanisms of the stock and, as has been shown, little is known about the form of these functions.

Most stock assessments have been undertaken to provide advice for management, usually through the IWC. It is recognised that little is known about the integrity of populations, and about their response to exploitation, and consequently the assessments are presented as the best given the limited material. The IWC has been advised by a Scientific Committee from its earliest days, but about 1960 the Committee of Three was instigated in order to undertake more quantitative stock assessments; the Committee included K R Allen, D G Chapman and S J Holt, soon to be joined by J G Gulland, and their initial and continuing work has been of great significance in the assessment of whale stocks. This Committee worked with the IWC until 1964 when the assessments were taken over by FAO, and from 1970 an expanded IWC Scientific Committee carried out its own assessments. The following sections draw heavily on the material provided by the IWC, and the main methods and indices of abundance used in the assessments are described, and estimates of stock sizes are given for the three ocean regions.

ASSESSMENT TECHNIQUES AND INDICES OF ABUNDANCE

Mark and recapture analyses

The estimation of population size, from the recovery of marked individuals, has been undertaken for many species and the techniques are reviewed by Seber (1973) and Jones (1976). The underlying idea is simple: if 100 animals are marked and from a catch, of 200, there were 20 recoveries of marked individuals, we assume one tenth of the population is of marked animals and must therefore be of 1000. The actual appliction is not straightforward for there are a large number of assumptions that need to be valid or else need to be compensated for. The essential features are that recaptured individuals are a random subset of the whole population at risk of capture, and that the total number of marked animals at large and at risk or capture can be estimated (Harding 1984).

For most whale populations these assumptions are difficult to justify. Stocks are ill defined, and marking and catching are not random with respect to geography. We have seen that smaller sei whales do not penetrate into the Antarctic and, for example, there is severe sexual segregation of sperm whales, and minke whales marked together have been recovered from the same school. Whales which were marked have been amenable to close contact with vessels and hence may be more easily caught, conversely marking may make them wary. All these invalidate the assumption of randomisation.

The position is little better when it comes to estimation of the number of whales marked at any time. At the time of marking there can be confusion about identification of species, and whether a mark did actually hit a whale. If, as is usually the case, there is selection by the gunners for large whales then the size distribution of the marked whales should be known, but this is difficult to estimate from the marking vessel. It is necessary to know the fate of the marked whales relative to those unmarked; a few are injured at marking, and it is usually assumed that there is a small mortality due to marking. Marks are also shed, but some information can be obtained on this aspect by double marking individuals; Cooke and Rorvik (1985) considered that, for minke whales, half the loss rate of marks could be due to shedding. At the time of recovery of marks the proportion found should be known, and this can be evaluated experimentally, but differences in this proportion between fleets has been identified. It is assumed that the catch is known. Even if all the assumptions are met the estimates of stock size have a high variance for the typically low numbers of marks recovered. The use of mark and recapture techniques for most large whale populations was well summarised by Garrod and Brown (1980) who, in their assessment of sei whales, only claimed a plausible interpretation of the available data.

The most direct application of the marking data is with the Petersen method, wherein the stock size is estimated from one season's experiment. If T is the total marks available for capture at the beginning of a period and m marks are returned from a catch of size c then the Petersen estimate of stock size (N) is given by,

$$N = (T + 1)(c + 1)(m + 1)^{-1} - 1.$$

If marking has been done each season then a sequential analysis can be attempted to provide estimates of survival rate, using the method of Robson (1963); this information can be utilised to obtain improved estimates of the numbers of marks available for capture for use in a Petersen analysis.

DeLury method

The original DeLury method was used to estimate the size of an initial population, or cohort, assuming that there was no natural mortality or recruitment, and a regression was used to relate an index of stock size to cumulative effort. The technique is easily appreciated, since if an index of abundance shows a population to be halved, from a catch of 100, then approximately the population was initially 200. This basic model was modified by Chapman (1974) to include a constant recruitment and a natural mortality, and an index of stock size was then related to cumulated catch. This is a particularly useful model since the age of recruitment can be large, and it will be that number of years before a change in recruitment occurs because of a change in stock size. Consequently if a fishery has operated for only a few years the model is likely to be robust. The model assumes,

$$N(t+1) = [N(t) - C(t)]\ (1 - M) + M.N(o),$$

where $N(t)$ is the size of the recruited component of the stock at time t, C is the catch, M is the proportionate mortality and $N(o)$ the original, unexploited recruited stock size. The constant recruitment can be recognised as $M.N(o)$. An index of stock size during season j, I_j, can be related to the stock size at that time with a constant of proportionality k, such that,

$$I_j = kN(o) - k\ [\sum_{i=1}^{j-1} C(i)\ (1 - M)^{j-1} + C(j)/2].$$

Consequently a linear regression can be used to solve for a and b in,

$$I_i = a + b.X_i\ ,$$

where X_i is the adjusted cumulative catch, the bracketed term above, $a = kN(o)$ and $b = k$; bias and variance of this technique are given by Tillman and Breiwick (1977). A modification to the Chapman-DeLury method was described by Tillman and Grenfell (1980), who considered a situation similar to that above, but for an index of abundance that started some time after exploitation commenced.

It needs to be emphasised that the index of abundance is required to be directly proportional to the recruited stock; consequently an index such as CPUE might be appropriate, but an index from, say, sightings data, may have to be corrected from total whales seen to those from only the recruited component.

General least squares

Several methods of estimating stock size use regression techniques, including those of DeLury and Schaefer (Ricker 1975), but they utilise simple biological models which give rise to linear regressions, and which can then be solved immediately. The increased availability of computers allowed the use of more sophisticated biological models which required relatively large numbers of calculations to solve for initial stock size; K R Allen pioneered these techniques. Most of the models take the form of,

$$N(t+1) = (N(t) - C(t)).S + R(t),$$

where $N(t)$ is the size of the recruited population at the beginning of whaling season t, C is the catch, S the natural survival rate and R the number of recruits to the exploited stock.

Allen (1966b, 1969a) determined the recruitment as, $R(t) = r(t).N(t)$, where $r(t)$, the proportion of new recruits, was obtained from analysis of the relative numbers of catch at age. The catch was assumed to be obtained from the relationship,

$$\text{expected catch} = qE(t)[N(t)-C(t)/2],$$

where E is the effort expended in season t, and q is a catchability coefficient that is assumed constant with time. Joint estimates are obtained of q and $N(o)$, the original stock size, that minimises the difference squared of the expected and actual catch.

This model has subsequently been developed in several respects. The index of abundance is related to the exploited component of the stock, and obviously the catch is taken from this set; however recruitment is more directly related to some more biological component of the population, such as the total stock size, mature stock, or numbers of females. Consequently, the models were expanded to include simulations of numbers at age over a series of years. The initial age composition is assumed to be in an equilibrium state and usually only the two parameters are estimated. As we have seen in the previous chapters the exact nature of any density dependent response of the population is far from known, and a density dependent response is explicitly assumed in the stock assessment model. This assumption is necessary since Allen's original method requires a knowledge of catch by age, and this is often not obtainable, and in addition Horwood (1987) established that Allen's method of estimation of the recruitment rates can be unreliable. The gross recruitment in the baleen whale models is usually related to the female mature population by a general production relationship of the form,

$$R(t) = r_o P(t)[1 - (P(t)/P(o))^k] + M^*.P(t),$$

where P(t) is the total number of mature females at time t, P(o) is the number of mature females in the unexploited stock, and $M^* + r_o$ is the maximum gross recruitment rate. The M* term is introduced to balance the natural losses in the unexploited population. Often an appropriate time lag is introduced.

The parameter k determines the shape of the recruitment function; if k = 1 the model is identical to a Schaefer model with its parabolic yield curve, and as k increases the initial response of recruitment to a decrease in population is more pronounced, and will give a maximum sustainable yield nearer to the unexploited level. The value of k is poorly defined, and for all whale stocks it is thought that k lies between 0.2 and 10.0, that is, at a value which gives maximum net production at reductions in stock size to 40% or 80% respectively. This inference is based upon an examination of density dependent processes in a variety of mammals, but the range is obviously large and reflects our ignorance of the total density dependent responses; nevertheless it is not critical for many stock assessments. It has been critical for subsequent IWC management decisions which are based upon stock levels relative to the maximum sustainable yield level. The stock assessments are more sensitive to the magnitude of the response, and information on its size has come from fitting simple models to data from the fisheries for blue and fin whales, and from observed rates of increase of other whales. An alternative procedure recognises that the particular functional form and parameter values can be difficult to justify, and sampled demographic parameters, such as pregnancy rate or age at maturity, are given as a time series from documented data. The models are described in more detail by Kirkwood and Allen (1978) and Kirkwood and Allen (1979).

Although these models are relatively large, and can accommodate most requirements to provide estimates of stock size, the statistical properties of the model need consideration. For instance the best fit of several parameters can be undertaken but it can be shown that, for instance, estimates of stock size and natural mortality rate are highly correlated and hence independent estimates of mortality should be used. It may be noted that in the above observed and expected catch were compared, and one might assume that similar answers would be obtained if observed and expected catch per effort were used, but this is not the case as the two measures have very different distributional properties. It is important that the assumed statistical properties are recognised for this then allows less biassed estimates to be obtained, and more correct estimates of variance of the parameters. Reviews of these problems in relation to the stock assessment models are given by Kirkwood (1978, 1981).

Sighting estimates

Direct counts of whales can be made from surveys by ships or aeroplanes or, in special cases, from land based viewing platforms; from these counts estimates of population abundance can be obtained. For the ocean living species of whales, surveys by ships are the only surveys practicable, and for these there are several important considerations. Only a limited area can be surveyed and the density in the survey area is usually assumed to be similar to that in the larger area enclosing all the population. In some cases stratification of effort and analysis is necessary. From the viewing platform the ability to see a whale reduces with the distance that a whale is from the ship; consequently the frequency distribution of numbers of whales seen against the distance from the ship tends to be exponential-like. This sightability has to be converted to an effective survey width over which sightability is constant. Not all whales will be seen even if they are close to the ship; this is because of failure of observers, and also because the larger whales can stay submerged for a significantly long time. In addition to these major considerations there are also problems associated with inexactness of recordings of the relative positions of the whale and vessel, the fact that the whales move at almost the same speed as the vessel, problems in identifying the species and counting the numbers in a school. Notwithstanding these significant problems surveys designed to count the numbers of Southern Hemisphere minke whales have been very successful, (Horwood 1981b, Butterworth and Best 1982).

Estimates of populations of large whales from sightings information were made by Mackintosh and Brown (1956), and Nasu and Shimadzu (1970), but a more comprehensive theory of estimating abundance of whales was developed by Doi (1974), who attempted to model the behaviour of observers and whales, and this theory has been used with data from Japanese scouting vessels to give estimates of abundance. The behavioural components are not ideal and recent analyses have tended to use a line-transect methodology (Burnham, Anderson and Laake 1980), in which analysis is based upon the empirical distribution of sightings perpendicular to the track-line of the vessel. More recently techniques based upon radial distributions have been advocated (Hiby 1985). No more details are included here as the analyses are complicated and the more recent techniques have not been used to provide estimates of sei whales.

Indices of abundance

Indices of abundance are used with population models to give estimates of stock size, as earlier described. The indices are most important and, in determining stock size and management options, the indices alone, or with a knowledge of catch, provide a majority of the necessary information; for management a steeply declining index clearly means that reductions in catches are necessary. However, often implicit, is the assumption that the index is proportional to the exploited stock, but this is not generally the case. For instance an index from sightings may include all, including young, whales seen and the index would not decline as rapidly as the exploited component of the stock. Indices used in the earliest assessments were based on the numbers of days that the catcher boats were in the Antarctic (gross catcher days), but this was amended to take out those days when there was no whaling, to give net catcher days; sometimes this was approximated by taking the number of days on which a catcher caught at least one whale. In the analyses it was assumed that this index was proportionately related to exploited stock size, but this is not so. Current operations in the Antarctic Japanese fleet are such that catching is regulated to give a steady catch of minke whales which are processed efficiently, and catching is stopped during the day if there is a build up of whales. This is quite different from the hectic post-war industry where the emphasis was on oil and there was competition amongst fleets. Consequently it can be seen that operations can change with time. Over time the catcher boats became bigger and more powerful, ASDIC assisted in the capture of sperm and baleen whales and the target species changed; catch per catcher day was not a very good index even when attempts were made to compensate for changes in efficiency.

In recent years the indices derived from catching operations have been much improved. It is not necessary that the index be proportional to any one component of the modelled stock, but its precise relationship does need to be known. In a days operation a catcher may have to steam to a locality, search for a whale, chase it, kill and secure it and tow the whale back to the factory or land station; the whale need not be of the target species. Of this operation only the time spent searching is likely to be related to the density of whales. In the Japanese pelagic operations of the 1977/78 season the time spent in these operations for each sei whale were; chasing 1h 23m, handling 30m and towing 18m, in each day 12h 42m were spent working. Beddington (1979) demonstrated that an index of abundance could be developed incorporating the elements of a search proportionate to density, handling times of the different species of whales and the total time worked. Even after such correction the indices of abundance of the whales have been very variable from year to year and regression techniques have been used to attempt to remove environmental influences (Zahl 1985), but without much success. Locally, the whale density may be proportional to the time spent searching but the distribution of whales is contagious. The whaling fleet is likely to spend more time in areas of high local density, but travel through areas of low or average density. This effect distorts the relationship between searching time and density of the total population (Zahl 1983). It should be recognised that the indices of abundance are extremely complicated statistics and their importance in assessments indicates that more effort could be profitably spent on their examination.

NORTH PACIFIC ASSESSMENTS

Historical analyses

The Scientific Committee of the IWC set up a North Pacific working group in 1962, and the first assessment of the stocks of sei whales was by Doi, Nemoto and Ohsumi (1967a, b). They obtained estimates of the natural mortality rate, and then compared the relative declines of age classes from CPUE data from the Kuril islands, Hokkaido and Sanriku, the American coast and five pelagic regions. An estimate based on mark returns was also considered. The results showed an exploited stock size of about 32 thousand for 1964. This analysis was updated by Doi and Ohsumi (1968a) to give an estimate of 30 thousand for 1967. Allen (1969b) obtained an estimate by assuming that the catchability of fin and sei whales was similar, and using an estimate of fin whale abundance, the numbers of sei whales were obtained directly. Doi and Ohsumi (1969b, 1970) also utilised the estimates of fin whales to obtain the number of sei whales from relative sightings and catches. They also obtained a direct estimate of the pelagic population from sightings using the theory proposed by Nasu and Shimadzu (1970). For the coastal Japanese population an estimate of survival rate was obtained from marking studies, and they assumed that the long history of catches of about 600 per year were sustainable, to give a population size of about 8-9 thousand. Ohsumi, Shimadzu and Doi (1971) also considered the catches of sei whales back to 1910, and presented estimates for sei whales on the Asian side (West of 180° longitude), including any coastal component, and on the American side. The combined figures were of an original population of 58-82 thousand, and in 1970 of 34 to 58 thousand. This assessment was updated by Ohsumi and Masaki (1972) and Wada (1973).

The general switch of whaling from fin to sei whales, the confusion with Bryde's whales, and the IWC desire to set quotas by species led to an IWC special assessment meeting in 1974; these studies and that of Oshumi and Fukuda (1975b) confirmed that the stocks had been relatively heavily depleted and led to the North Pacific sei whales being protected from whaling in 1976. The following assessment is based largely on the review by Tillman (1977).

Indices of abundance

Sightings. Since 1965 sightings surveys have been conducted regularly in the North Pacific, by Japanese whalers, as described by Ohsumi and Yamamura (1982), and data from these surveys have been published in a series of reports by Wada (1975-1981). The indices of abundance from Wada (1981) are given in Table 13.1 by rectangles of ten degrees of latitude and twenty degrees of longitude. The index is based on whales seen per mile travelled and corrected for the different sizes of the rectangles. Figure 13.1 gives the plot of the index with year for the North Pacific and the decline from 1965 to 1975 is pronounced, after 1975 there is an increase. Sightings East and West of 180° longitude were given by Tillman (1977), as perhaps representative of the American and Asian stocks and these data are given in Table 13.2. However the geographic distribution of effort has varied somewhat over the years and the intensity has declined to about a tenth of that at its peak.

Catch per effort. The catch of sei, fin and Bryde's whales from the North Pacific is given in Table 13.3, for the years 1963 to 1974 and by latitudinal zones. Prior to 1967 the emphasis was on fin whales and they were caught mainly north of 50°N, from 1967 there was a change to sei whales being of greater importance, and with the catch being taken in the latitudes 40°-50°N (Zone N). In the later years Bryde's whales were also important.

In the North Pacific, Japanese pelagic effort has been mainly directed towards taking baleen whales rather than sperm, and consequently has tended to be used in preference to Soviet data in the construction of indices of abundance. Tillman (1977) extracted the data on Japanese pelagic catch and effort expended in Zone D for the seasons 1967 to 1974 and corrected it for increases in the tonnage of the average catcher boat, and the CPUE series is given in Table 13.4. Tillman (1977) considered the effort too small in the first two years to be able to give reliable indices, but the remainder of the series shows a large and consistent decline as illustrated in Figure 13.1. Although the use of data from only Zone N means that sei whales were the most important species taken, significant numbers of others were caught, and in addition handling times are not incorporated; consequently the stock was likely to have been declining at a faster rate than the series indicates.

Least squares and De Lury estimates

An estimate of the size of the exploited component of the North Pacific population was obtained by Tillman (1977), from a minimisation of observed and predicted catches from a model with recruitment obtained from Allen's (1966b) formulation. Horwood (1987) has shown that these recruitment rates can be unreliable because of changes in the age specific selection, but an examination of the age composition

Table 13.1: Whales seen per 10,000 miles in the North Pacific, 1965-1978. P: 50-60°N, N: 40-50°N, 23-29: 160-170°E, ..., 140-130°W. Numbers in brackets are extrapolations.

Year	P23	N23	N24	N25	N26	P27	P28	P29	N29	Total
1965	0	96	25	134	237	138	82	108	(473)	1293
1966	41	362	209	24	43	44	111	43	66	943
1967	193	187	328	217	78	30	0	0	(0)	1033
1968	32	195	230	64	35	166	146	(16)	72	956
1969	3	95	89	217	138	101	79	27	54	803
1970	15	48	107	34	83	13	78	31	290	699
1971	8	103	96	24	35	16	24	28	191	525
1972	5	37	41	23	43	5	0	0	22	176
1973	0	62	9	9	98	0	0	5	26	209
1974	0	4	19	34	22	0	0	0	7	86
1975	(2)	19	7	0	14	0	0	0	16	58
1976	0	13	58	0	11	0	14	0	29	125
1977	0	25	9	19	26	0	21	89	31	220
1978	0	52	39	0	63	0	0	(6)	25	185

Table 13.2: Sightings indices of abundance for the Asian and American components of the North Pacific sei whales.

YEAR	INDEX	
	Asian	American
1967	2.07	-
1968	1.94	-
1969	1.02	2.96
1970	1.00	1.97
1971	1.28	1.06
1972	0.80	0.96
1973	0.83	1.10

Table 13.3: Pelagic catch of North Pacific fin, sei and Bryde's whales by zone, and total pelagic and coastal catch. L: 20-30°N, ..., Q: 60-70°N.

Year		L	M	N	P	Q	Total Catch
1963	Fin						2503
	Sei	←	data not available		→		2590
	Bryde						0
1964	Fin						3991
	Sei	←	data not available		→		3642
	Bryde						0
1965	Fin			464	2414	20	3165
	Sei			382	1696	10	3172
	Bryde						8
1966	Fin			167	2095	343	2885
	Sei			573	2906	263	4406
	Bryde						63
1967	Fin			791	1205	35	2272
	Sei			3501	1966		6053
	Bryde						63
1968	Fin			526	1260	65	1942
	Sei		59	4089	772		5740
	Bryde						171
1969	Fin			890	279		1276
	Sei			4126	304	251	5157
	Bryde						89
1970	Fin			698	232		1012
	Sei			3852	163		4503
	Bryde			66			139
1971	Fin			525	207		802
	Sei		15	2568	132		2993
	Bryde		638	109			919
1972	Fin		10	455	211		758
	Sei		828	1265	19		2327
	Bryde		76				201
1973	Fin		5	350	62		455
	Sei		635	1170	8		1856
	Bryde	526	127				724
1974	Fin		32	257	100		413
	Sei		359	862	10		1280
	Bryde	351	825				1363

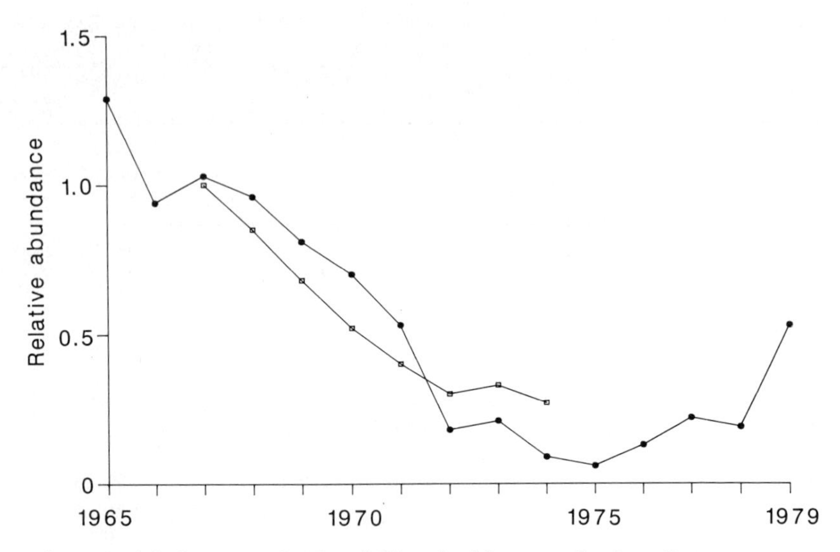

Figure 13.1: North Pacific indices of abundance. Dots: sightings. Squares: CPUE.

Table 13.4: Catch, effort (catcher-ton days in thousands), and CPUE from Japanese pelagic operations in Zone N (40°-50°N).

Year	Catch	Effort	CPUE
1965	65	72	0.91
1966	455	119	3.81
1967	2085	557	3.74
1968	3110	979	3.18
1969	3381	1323	2.56
1970	3071	1564	1.96
1971	2287	1541	1.48
1972	1221	1076	1.13
1973	1089	874	1.25
1974	837	824	1.02

data given by Tillman (1977 Table 2) revealed a relatively consistent age specific selectivity; consequently the least squares estimates are included. The effort used in the analysis is that given in Table 13.4 The results gave estimates of the recruited component of the population, for the beginning of 1974, of 7 thousand, and for 1963 of 40 thousand. If the CPUE series is used in a De Lury analysis, assuming a constant recruitment prior to 1967, then similar estimates are obtained of 45 thousand in 1963 and of 8.5 thousand in 1974.

In both the above analyses an estimate of the catchability is obtained and this was similar from both techniques. If these catchabilities are used with the given CPUE then estimates of stock size can be obtained. Tillman (1977) gives these as 26-27 thousand for the 1967 population and 8 thousand for the 1974 population size. Clearly they are not independent of the above estimates.

Tillman (1977) also presented a De Lury estimate using the sightings data of Table 13.2 for the Asian and American groups, and a combined series for the years 1968 to 1973. Estimates for each group were not presented separately but adding the estimates from the 2 groups, analysed independently, gave an estimate of the exploited population in 1968 of 27 thousand, compared with 26 thousand if the North Pacific data were treated as a unit; these values are similar to those obtained using the CPUE series. The sightings data indicated that in the late 1960s the stock on the Asian side was about 2.5 times larger than on the American side.

Combined estimates

All the methods employed by Tillman (1977) gave rather similar results which were summarised, by him, as an exploited stock size in 1963 of 42 thousand, in 1967 of 30.5 thousand and in 1974 of 8.6 thousand. These results are similar to those given by Ohsumi and Fukuda (1975b) who used a De Lury analysis with a slightly different index of abundance. Over this period a substantial decline in the population is evident. Prior to 1967 there was a long history of coastal catches from Japan and declines relative to an earlier unexploited level will be greater. Since whaling ceased the indices from sightings have increased, but the rate of increase is larger than might be reasonably expected, and although an increase might be anticipated as yet information is insufficient to confirm any increase. In the above analysis interspecific associations, if any, have not been incorporated, but the steep declines and lack of variability in the indices of abundance indicate that these estimates are likely to be robust to many reinterpretations.

NORTH ATLANTIC ASSESSMENTS

There are no estimates of the numbers of sei whales that migrate along the European coast.

From the area of the Iceland-Denmark Strait catches of about 60 per year have been taken since 1949. The catches are very erratic, the sei whales seemingly at the northern extent of their feeding range, and catches are also affected by the availability of fin whales. Consequently no indices of abundance are available for this stock. Chapter 3 shows only eleven recoveries from 59 marked sei whales and any mark-recapture estimate would be unreliable. A coarse estimate would indicate a stock of a few thousand but the status of this population is essentially unknown.

A tentative assessment is available of the sei whales off Nova Scotia and Labrador from Mitchell and Chapman (1977). From Nova Scotia 825 sei whales were caught from 1966 to the closure of the whaling station in 1972, and three marked whales were recovered from a tagged group of between nine and fourteen; estimates of stock size ranged between 1400 and 2200. An alternative estimate was based on the number of sei whales seen during surveys. Mitchell (1974a) reported a similar analysis for fin whales in which he assumed that fin whales were seen to a perpendicular distance of 4 miles; no distance is given for sei whales but it can be assumed that the same distance was used. Based on 39 sei whales seen an estimate of 2078 was given for the whales in the Northwest Atlantic. Those off Nova Scotia comprised 870 and off Labrador 965.

It is astonishing that with over a century of catches of sei whales no estimates are available for most of the North Atlantic and that those for the Northwest Atlantic are so unsatisfactory.

SOUTHERN HEMISPHERE ASSESSMENTS

Historical analyses

The first assessment of the numbers of Southern

Hemisphere sei whales was by the Committee of Four, which reported its finding to the IWC in 1964 (Chapman 1965). They relied upon estimates of the size of the stocks of fin whales, and relative indices of abundance of fin and sei whales. Based upon catches, the numbers of sei whales was estimated as about 20 thousand, excluding those found in regions not exploited. Relative sightings gave an estimate of 50 thousand over the period 1933-1939. From then until 1974 similar assessments were undertaken by the Committee of Four, FAO and Japanese scientists; the results are given in Table 13.5. However it was recognised that the main sei whale grounds were further north than those of fin whales, and that comparisons of the indices at high latitudes were invalid. As effort switched onto sei whales more reliance was placed on the use of sei whale CPUE, and estimates could be obtained of sei whales from Areas where a decline in CPUE was found, and then only intraspecific comparisons were needed for the other Areas.

Following the decision of the IWC in 1974 to manage whales on a sustainable basis, under the revised management procedure, a special meeting of the Scientific Committee was held in late 1974 to assess the numbers of sei whales. This meeting took particular advantage of the modified De Lury technique of Chapman (1974) and improved indices of abundance, and the results from various analyses are given in Table 13.6. These assessments resulted in protection being given to sei whales in Area III from 1975 onwards. Subsequent special assessment meetings were held in 1977 and 1979 and in which more sophisticated models were employed (Anon 1978, Breiwick 1978); these assessments are described in the next sections.

Indices of abundance

Sightings. Data from Japanese whale sightings were given by Horwood (1980a) by management Area and latitudinal Series for those strata where there was adequate information. In Areas II-V data from Series D (40-50°S) and E (30-40°S) were combined to increase sample size, and because densities were about the same in each Series; the density was about half in Series A (50-60°S) and half again in Series B (60-70°S). The sightings are given in Table 13.7 and illustrated in Figure 13.2. Information is scarce from Area I and variable generally, but

Table 13.5: Estimates of exploited stock size of Southern Hemisphere sei whales obtained from 1965 to 1972.

Source	Reference Year	Exploited Stock thousands	Method
Chapman (1965)	1964	20	catch rel. to fin
	1936	50	sightings rel. to fin
FAO (1965)	1964	60	CPUE rel. to fin
	1957	85	
FAO (1967a)	1964	180	as above but including a northern component
	1966	60-130	CPUE rel. to Area II
FAO (1967b)	1967	66+Area I	various
Doi et al (1967)	1966	100-130	various
Doi and Ohsumi (1968b)	1967	107	various
FAO (1969)	1967	59+Area I	various
Doi and Ohsumi (1969a)	1967	100-110	various
FAO (1970)	initial	131	various
	1969	74	
Chapman (1971)	initial	150	De Lury
	1969	75	
Doi et al (1971)	initial	150	De Lury
	1970	83	
Ohsumi and Masaki (1972)	1965-71	321+Area I	absolute sightings
	1971	82	De Lury

severe decreases can be seen in Areas III-IV from 1965 to 1978.

Sightings from spotter planes operating from Donkergat and Durban are given by Best and Gambell (1968) for the period 1954 to 1968. From Brazil, sightings from whaling vessels were reported by Da Rocha (1980) for the years 1966 to 1978.

Table 13.6: Estimates of exploited stock size of Southern Hemisphere sei whales presented to the IWC 1974 special assessment meeting. Numbers in thousands estimated in October of the year given.

Area	1961	1973	Source
II	43 - 47	20 - 24	1
	40	17	2
III	20 - 22	5 - 7	1
	25	6	2
IV	20 - 36	6 - 22	1
	27	13	2
V	22	12	1,2
I-VI combined	121 - 135	52 - 66	1
	125	53	2
	98	57	3
	132 - 135	52 - 54	4

1: Tillman and Breiwick (1977)
2: Chapman (1974)
3: Borodin (1974)
4: Ohsumi and Fukuda (1975a)

Table 13.7: Whales seen and miles steamed in the Southern Hemisphere by Area and Series.

Area	Series		65/66	66/67	67/68	68/69	69/70	70/71	71/72	72/73	73/74	74/75	75/76	76/77	77/78
I	A	Seen	3	18	—	—	0	—	—	—	—	0	20	27	0
		Miles	2,851	209	0	0	654	0	0	0	0	934	5,688	2,856	494
II	D+E	Seen	3,888	63	—	—	468	—	—	—	—	0	232	0	0
		Miles	76,813	4,004	0	0	35,572	0	0	0	0	268	13,226	495	147
III	D+E	Seen	1,104	4,485	1,540	1,002	613	643	428	642	368	56	—	59	84
		Miles	32,972	78,704	29,894	50,637	59,295	60,539	53,786	47,894	57,350	24,088	0	15,146	7,261
IV	D+E	Seen	—	148	1,455	1,292	2,453	4,705	2,302	1,635	770	700	441	253	103
		Miles	0	4,017	44,909	63,097	39,872	111,131	110,000	79,931	59,071	69,722	28,333	29,124	21,029
V	D+E	Seen	—	977	716	1,491	—	494	2,310	1,796	745	495	405	321	52
		Miles	0	7,463	21,351	41,489	0	21,009	33,011	48,894	32,934	47,536	41,909	25,071	3,103
VI	D	Seen	—	7	823	256	—	10	—	12	37	148	5	34	44
		Miles	0	1,256	7,456	14,456	0	2,115	0	2,155	832	10,207	1,731	3,026	2,155
VI	A	Seen	—	125	653	281	—	1	—	212	281	198	96	22	13
		Miles	0	2,470	15,810	12,124	0	1,452	0	8,543	17,198	15,213	8,455	4,758	1,800
VI	B	Seen	—	—	1	68	—	—	—	326	100	20	10	9	0
		Miles	0	0	1,340	4,418	0	0	0	9,847	12,093	4,049	4,427	912	659

© Crown 1980

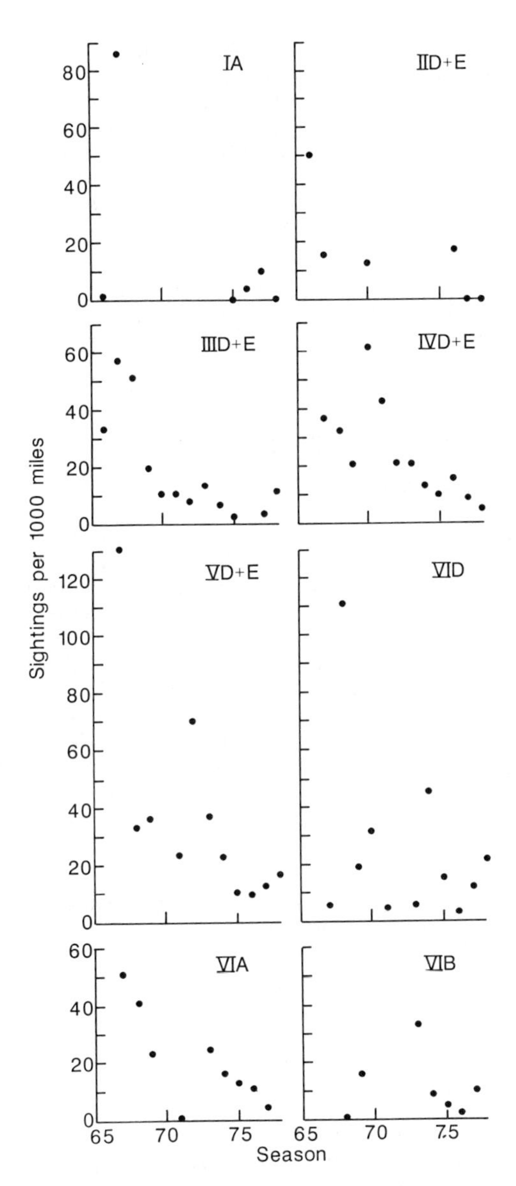

Figure 13.2: Sightings index. (whales seen per 1000 miles), against Antarctic season, (65 = 1965/66 season), by Area and series. E: 30°-40°S, D: 40°-50°S, A: 50°-60°S, B: 60°-70°S. Areas I: 60°-120°W, II: 0°-60°W, III: 0°-70°E, IV: 70°-130°E, V: 130°E-170°W, VI: 120°W- 170°W. © Crown 1980

Catch per effort. An index of abundance can be obtained from estimating the catch of sei whales per time spent searching for whales. The searching time can be estimated from the number of hours spent operating in a season, less the time spent chasing, handling and towing whales that were caught. These latter parameters can be different with species. The searching time of a fleet in season i, E_i, can then be calculated from,

$$E_i = \Sigma_j(CDW_{ij} \times d_j) - \Sigma_k h_{ik} C_{ik},$$

where CDW_j is the catcher-days worked in month j, d is the operating hours per day, h_k is the total handling time per whale of species k and C_k the catch of that species. These data were available from Japanese operations from 1961 and the parameter values are given in Table 13.8 with the catch per effort given in Table 13.9. From Table 13.9 the first value is negative, emphasising that these are estimated values. More significantly the increase and subsequent decrease of the index in Areas II and III shows that the index is imperfect and that the earlier catches of the other species have not been adequately incorporated, and the latter part of the series is likely to be better because of the concentration on sei whales. In interpreting the index it can be seen that it is of crucial significance that the start of the series be correctly identified. If the start is at the high values of Areas II and III then substantial decreases in the index are found. The data from Area IV are very variable, but if the 1973 and 1974 values of series D are inspected it

Table 13.8: Time budget information from Japanese Antarctic operations, times in hours.

Species	Seasons	Chasing	Handling	Towing	Resting
Sei	1961/62-62/63	0.68	0.22	3.1	7.3
	1976/77	0.77	0.34	1.0)	10.5
	1977/78	1.38	0.29	0.3)	
Fin	-	1.02	0.24	3.1	
Blue	-	1.11	0.24	3.1	
Sperm	-	0.52	0.20	3.1	

can be seen that catch and index are highly correlated. Area V shows a decrease.

Although the data presented above are the best available it is clear that they are not perfect and their use must be questionable, and more inspection of the original log-books will be necessary to provide a better measure. Additional but much less refined data are available from land stations in Brazil and South Africa.

Mark and recapture estimates

The marking of Southern Hemisphere sei whales and subsequent recovery of marks is described in Chapter 3, and 71 marked whales were recovered from a total of 848 sei whales marked since the season of 1955/56. An estimate of the size of the exploited population was attempted by Garrod and Brown (1980) using these data. Because of the availability of the data they used information from only the International Scheme of marking which consisted of 395 whales marked and 51 recovered, 43 of these by the Japanese fleet. As has been shown in Chapter 10, whilst there is some degree of independence of the six Antarctic management Areas, there is mixing between Areas. This, and the fact that the total number of recoveries is small and could not be used to give reliable estimates for each of the Areas, led Garrod and Brown to assess the total numbers in the Antarctic in one exercise.

The mark-recapture methodology assumes a random mixing of marked and unmarked whales. In order to allow some dispersal of marked whales, whilst at the same time not eliminating too much of the data, they only considered recoveries that were made over 10 days after marking. Over these years the efficiency of detection of marks from the carcass was high in Japanese operations and so only those recoveries were considered. These two considerations reduced the number of recoveries to 32, but the nominal number of marks recovered each year wás increased to give the same ratio of total marks in the total catch to Japanese recoveries in the Japanese catch. The total of nominal recoveries after 10 days after marking was 53. An estimate of the total numbers of marked whales caught each year was thus obtained, but for the analysis it is necessary to allocate each estimated recovery to a year of marking. The numbers marked (not corrected for those caught within the 10 days) and the nominal pattern of recoveries is given in Table 13.10 along with the

Table 13.9: Southern Hemisphere Japanese catch, hours searched and CPUE, by Area and latitudinal zone. 1965 is the season 1964/65.

Year	Catch	Hours	CPUE
Area I:	40-50°S		
1965	5	-224	-22
1976	54	850	64
	50-60°S		
1964	1	213	5
1965	35	678	52
1976	64	526	122
Area II:	40-50°S		
1963	8	841	10
1964	377	3748	101
1965	4430	8683	510
1966	9068	15995	567
1970	1109	6594	168
1975	4	189	21
1976	147	2634	56
	50-60°S		
1962	2	972	2
1963	86	7476	12
1964	1108	24998	44
1965	5662	30247	187
1966	7	647	11
Area III:	40-50°S		
1962	165	9551	17
1963	666	21075	32
1964	33	4647	7
1965	201	3353	60
1966	2230	8218	271
1967	6114	19804	309
1968	1092	3566	306
1969	539	7757	69
1970	785	10151	77
1971	441	9006	49
1972	602	11920	51
1973	429	6050	71
1974	391	9919	39
1975	93	5168	18

Year	Catch	Hours	CPUE
Area III:	50-60°S		
1962	170	28140	6
1963	209	24724	8
1964	104	6910	15
1965	69	411	168
1966	4	819	5
1968	35	570	61
1971	1	186	5
1972	18	809	22
Area IV:	40-50°S		
1962	76	1034	73
1963	68	238	286
1967	440	2292	192
1968	1210	6442	188
1969	753	8069	93
1970	1535	3561	431
1971	3672	13869	265
1972	2557	11637	220
1973	1439	8345	172
1974	3112	4148	750
1975	1045	7942	132
1976	381	2109	181
1977	383	1323	290
1978	253	4657	54
	50-60°S		
1962	520	12871	40
1963	55	7505	7
1967	2103	3841	548
1968	987	8169	121
1969	231	5907	39
1970	66	3755	18
1971	23	1471	16
1972	21	2084	10
1973	77	2811	27
1974	4	854	5
1975	26	2254	12
1976	20	79	252

Table 13.9 (cont'd)

Year	Catch	Hours	CPUE	Year	Catch	Hours	CPUE
Area V:	40-50°S			Area VI:	40-50°S		
1964	259	4379	59	1968	713	1691	422
1968	1183	4400	269	1969	145	786	185
1969	1379	4545	303	1975	344	2346	147
1972	1122	3379	332	1976	10	618	16
1973	893	3800	235	1977	86	507	170
1974	440	2735	161		50-60°S		
1975	478	3537	135	1968	1758	5492	320
1976	351	3800	92	1969	266	1917	139
1977	548	2587	212	1973	33	1752	19
	50-60°S			1974	409	3304	124
1964	52	534	92	1975	254	4933	51
1968	40	272	147	1976	135	2697	50
1977	4	369	11	1977	180	1123	160

Robson estimate of proportion of the population surviving between periods. The periods considered are pairs of seasons to increase the sample size, hence the first survival rate of 0.91 indicates that catching and natural mortality removed 9% of the population from the start of the 1961/62 season to the beginning of the 1963/64 season.

The results from Table 13.10 enabled better estimates to be made of the numbers of marks surviving to each period and hence Petersen estimates of stock size could be obtained. These are shown in Table 13.11 and it can be seen that the stock size was estimated at some 400 thousand in 1963 to 1966 but by 1974 had fallen to 100 thousand. The coefficients of variation are large, but even so they do not account for the various procedures that led to the estimated number of marked whales. The effect of assuming that the entire Antarctic population can be treated as an entity cannot be judged; in fact Garrod and Brown (1980) regard this assessment as the most plausible under the necessity of having to provide an estimate. Garrod (1980) extended this analysis to the marks and recoveries from the Soviet Scheme and a similar range of estimates with time was found. They both illustrated a substantial decrease of the population over the period from 1962 to 1977 and from 1966 there is a good correlation between the estimates from marking and from indices of abundance from sightings data.

Table 13.10: Total of sei whales marked in the Southern Hemisphere and nominal recaptures of those at liberty over 10 days. S is the Robson survival rate between period t and t+1. © Crown 1980

Marking season	Nos. marked	Recapture Period 1	2	3	4	5	6	7	8	9	10	11	Total	S
55/56-56/57	6	1											1	-
57/58-58/59	13												0	-
59/60-60/61	13												0	-
61/62-62/63	19					1	1	2					4	0.91
63/64-64/65	69					7	5	3			1		16	0.97
65/66-66/67	35						2	1	1			1	5	0.55
67/68-68/69	66							2	2	4		1	9	0.56
69/70-70/71	44								3	2	1		6	0.58
71/72-72/73	54									4	3	1	8	0.59
73/74-74/75	45										4	1	5	-
75/76-76/77	31												0	-
Estimated total		1	0	0	0	8	8	8	6	10	9	4	53	

Table 13.11: Petersen estimates of exploited stock size in the Southern Hemisphere.
T: total surviving marked whales, m: nominal recoveries, c: catch, N: Petersen estimate in thousands, cv: coefficient of variation.
© Crown 1980

Season	T	m	c	N	cv
1963/64-64/65	109	8	32151	393	0.38
1965/66-66/67	122	8	31340	428	0.38
1967/68-68/69	129	8	16437	237	0.38
1969/70-70/71	117	6	12101	204	0.45
1971/72-72/73	122	10	9385	105	0.34
1973/74-74/75	117	9	8435	96	0.36

Least squares analysis

An estimate of abundance can be obtained by fitting a model of the stock dynamics to indices of abundance. This was undertaken by Horwood (1980a) and modified at the last assessment of sei whales by the Scientific Committee of the IWC. The population model of Kirkwood and Allen (1979) was used with the pregnancy rates given as a known series with time for each of the six management Areas. Provided that all responses have been monitored this has the advantage of including any possible interspecific effects.

Inspection of catch data revealed that mean ages at recruitment varied between 5 and 9 years depending on Area and sex, and in the subsequent analysis a common age at recruitment of 8 years was used. Hence the estimates of stock size are of the population of age 8 and older; the total population will be about 1.5 times greater than this estimate. The age at maturity was obtained using data from the

Table 13.12: Female age at maturity by season and Area. 1940 = 1939/40 season. © Crown copyright

	I	II	III	IV	V	VI
1940	12.2	11.1	11.1	11.2	11.3	11.5
1941	12.2	11.1	11.1	11.2	10.4	11.5
1942	12.2	11.1	10.8	10.6	10.3	11.0
1943	12.2	11.1	10.7	10.3	9.9	11.0
1944	12.2	11.0	10.6	10.0	9.3	10.5
1945	12.2	11.0	10.6	10.0	9.3	10.5
1946	12.2	10.5	10.6	9.8	10.0	10.5
1947	12.2	10.4	10.9	9.7	11.0	11.0
1048	12.2	9.7	10.9	9.7	11.6	11.0
1949	12.2	9.7	10.9	9.7	11.6	11.5
1950	12.2	9.7	10.9	9.7	11.6	11.5
1951	12.2	9.5	10.5	10.0	11.6	11.5
1952	11.4	9.4	10.3	10.5	11.5	11.1
1953	11.4	9.4	9.7	11.0	11.3	11.1
1954	11.0	9.4	9.7	11.0	10.9	10.6
1955	10.6	8.8	9.2	11.0	10.9	10.6
1956	10.6	8.8	9.2	11.0	10.9	10.6
1957	10.6	8.3	8.6	11.0	10.7	10.4
1958	10.6	8.3	8.6	11.0	10.6	10.1
1959	10.6	8.3	8.6	11.2	10.6	10.1
1960	10.6	8.4	8.4	11.2	10.6	10.1
1961	9.8	8.6	8.1	11.2	9.9	9.9
1962	9.8	8.6	8.1	10.7	9.9	9.8
1963	9.1	8.6	8.1	10.6	9.3	9.7
1964	9.1	8.6	8.1	10.3	9.3	9.7
1965	9.1	9.0	8.3	10.3	9.3	9.7
1966	9.2	9.0	8.7	10.2	9.3	9.4
1967	9.6	9.0	8.7	10.2	9.2	9.1
1968	9.6	9.0	8.7	9.9	9.2	9.1
1969+	9.6	9.0	8.7	9.9	9.2	9.1

Table 13.13: Estimated "true" pregnancy rates by season and Area. 1930=1929/30

	I	II	III	IV	V	VI
1930	.27	.27	.27	.27	.27	.27
1931	.27	.28	.27	.28	.27	.27
1932	.27	.29	.28	.28	.28	.27
1933	.27	.31	.28	.29	.28	.27
1934	.27	.32	.29	.29	.28	.27
1935	.27	.34	.29	.30	.29	.27
1936	.27	.35	.30	.30	.29	.27
1937	.27	.37	.30	.31	.29	.27
1938	.27	.38	.30	.32	.30	.27
1939	.28	.40	.31	.33	.30	.27
1940	.28	.41	.31	.33	.30	.27
1941	.28	.31	.32	.34	.31	.27
1942	.28	.31	.32	.35	.30	.27
1943	.28	.31	.32	.35	.31	.27
1944	.28	.31	.32	.35	.31	.27
1945	.28	.31	.32	.35	.31	.27
1946	.28	.31	.32	.35	.30	.27
1947	.28	.28	.32	.35	.30	.27
1948	.28	.28	.32	.35	.31	.27
1049	.28	.29	.32	.35	.32	.27
1950	.28	.30	.32	.36	.34	.27
1951	.28	.31	.33	.36	.34	.27
1952	.28	.31	.34	.37	.34	.27
1953	.28	.33	.33	.37	.35	.27
1954	.28	.33	.33	.38	.36	.27
1955	.28	.34	.34	.38	.36	.27
1956	.28	.35	.34	.38	.37	.31
1957	.29	.36	.34	.38	.37	.32
1958	.31	.37	.34	.39	.37	.33
1959	.36	.37	.34	.39	.37	.34
1960	.37	.38	.35	.39	.38	.36
1961+	.37	.39	.35	.39	.38	.39

analysis of ear plug material of Lockyer (1979), and this was converted approximately to mean age at maturity by year (Horwood 1980a); they are given in Table 13.12. An adult mortality rate of 0.06 y^{-1} was accepted and it was assumed that the mortality rate of immature whales would be higher. An arbitrary value of 0.066 y^{-1} was used which, in equilibrium, with an age of first parturition (one year older than the age of maturity) gave an initial, unexploited true pregnancy rate of 0.27 y^{-1}. Apparent pregnancy rates had been sampled with time but it was noted that they were highly correlated with the estimated biomass of fin and blue whales. The sampled data were first corrected to account for variation in the mean date of capture and for any missed fetuses and then converted to an absolute pregnancy rate as described in Chapter 9. The resulting time series of estimated true pregnancy rates is given in Table 13.13.

Estimates of stock size were obtained by calculating best fits to the sightings and CPUE data. Only CPUE series showing a decline could be used and consequently subsets of the CPUE series were selected; this is obviously an unsatisfactory and biassed procedure but was the only alternative to providing no estimates. The results are given in Table 13.14. Using both Japanese and Brazilian data failed to give an estimate for Area II, (negative stock sizes being predicted), but the original stock size must have been in excess of 38 thousand to allow any catches in the later years. A comparison of relative sightings between Areas indicated that by 1980 the Area II population was at most 25% of its size in 1910. In Area IV the estimates were similar except for that based on the sightings in Series B. The final, accepted estimates of the stock size in 1979 are given in Table 13.14.

Combined estimates

The estimates of exploited stock size in the late 1970s, from the least squares analyses, indicate a Southern Hemisphere stock of about 15 thousand; the estimates from the Japanese and Soviet marking schemes are in the region of 50-100 thousands. Clearly these figures are very different and reflect the fact that substantial problems exist in the use of both techniques; nevertheless they do provide some perspective on current numbers. The analyses do however allow us to make a stronger statement about the relative numbers of sei whales over the

Table 13.14: Estimates of exploited stock size in thousands, from a least squares analysis. Series of index in brackets.

Area and data used	stock size 1910	1930	1960	1979	accepted 1979
Area I					1.60
sightings (A)	8.0	6.9	6.2	1.6	
Area II		no results obtained			0.6-9.2
Area III					1.15
sightings (D+E)	24.6	22.4	21.4	1.2	
CPUE(D)	24.5	22.4	21.3	1.1	
Area IV					5.70
sightings (D+E)	-	14.7	17.4	5.7	
Area V					2.00
sightings (D+E)	-	11.2	11.8	1.1	
CPUE(D)	-	12.3	13.0	2.9	
Area VI					0.40
sightings(D)	-	7.9	6.5	0.3	
sightings(A)	-	8.1	6.7	0.6	
sightings(B)	-	57.1	55.9	65.9	
CPUE(A)	-	7.9	6.5	0.3	

seasons. Both techniques show the population to have substantially declined since 1960. Figure 13.3 gives a plot of relative population size from the marking data (excluding one point) and of the mature population from the Southern Hemisphere excluding Area II. The correlation is good and provides evidence to support a belief in the relative trends. The difference in absolute numbers can be explained by a variety of factors, such as a lower efficiency of finding marks in carcasses, which would still give approximately correct trends. Although the least squares model appears biologically very complicated its essential underlying assumptions, in the light of catches large relative to stock size, are simple and robust and provided the indices of abundance are valid the estimates from this technique

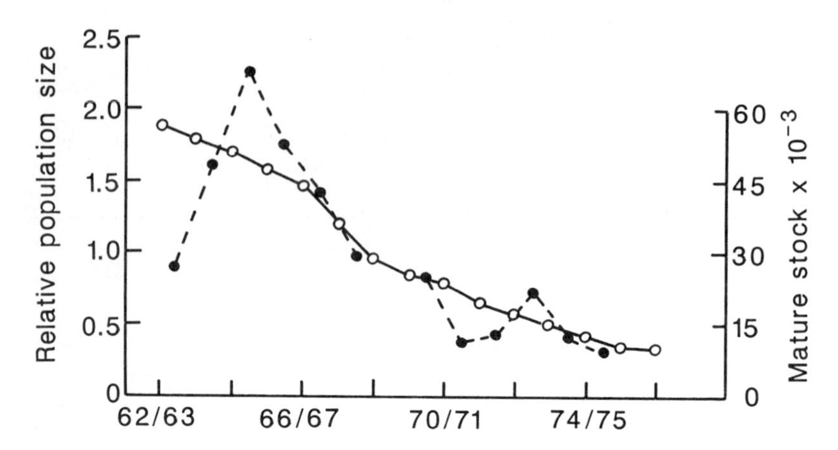

Figure 13.3: Relative size of the exploited sei whale population in the Southern Hemisphere from the mark-recapture analysis (dots and left axis) and the estimate of the mature stock excluding that in Area II (circles and right axis).

will be the more correct. However it should not be forgotten that only indices that did show declines were used and that several assessments predicted negative numbers. It can tentatively be concluded that the original exploited size of the populations in the Southern Hemisphere was about 67 thousand and that the stock in 1979 was about 16 thousand.

SUMMARY

No large scale, systematic sightings surveys have been undertaken to count sei whales. Mark and recapture analyses have been attempted for several areas but the number of returned marks is small, giving large variances even when the assumptions of the analysis hold; in general we know many of the assumptions are violated. The most reliable estimates come from a comparison of catches with declines in an index of abundance when a series of catches have rapidly depleted the stock. This occurred in both the North Pacific and Southern Hemisphere following restrictions on the catching of fin whales. The correct identification of the relationship between the index of abundance and a particular component of the stock is crucial, and for CPUE data this relationship is imperfectly known. In some instances the assessments have required the simulation of population trajectories over many years and in these cases a knowledge of the demographic parameters with time, or of the inter- or intra specific responses is necessary, and the correct identification of relatively isolated populations becomes of some importance; most of these latter aspects are known very imperfectly. It is mainly because of the rapid reductions of the North Pacific and Southern Hemisphere populations that the estimation of stock sizes has been possible.

In the Southern Hemisphere the exploited stock size prior to 1930 was about 67 thousand and the total stock would have been some 100 thousand; by 1979 it had been reduced to about 16 thousand exploited or 24 thousand total. In the North Pacific the exploited stock size in 1963 was 42 thousand exploited, 63 thousand total, and was reduced to 8.6 thousand exploited, about 13 thousand total by 1974. If the stock increased at 4% per year after whaling ceased the total population in the two oceans would have been about 50 thousand in 1985 compared with a pre-exploitation size of about 170 thousand, still relatively depleted. The status of the North Atlantic stocks is unknown.

There is scarcely an area which could not profit by more research, and an obvious target is the North Atlantic where extensive surveys are required. Too little has been gleaned from the extensive marking of fin whales and an investigation of the entire set of marking data may improve the interpretation of sei whale marking data. A refining of the indices of abundance is necessary. To be able to explain and predict the longer term population trajectories a greater knowledge of all aspects of population biology is needed, and the lack of knowledge about the magnitude of population homoeostatic mechanisms means that predictions of stock size with time, especially in response to exploitation, are necessarily imprecise.

Chapter Fourteen

CONCLUSION

INTRODUCTION

This study has been prepared to describe the state of knowledge of the population biology of the sei whale, and to a lesser degree its significance for the management of sei and other large whales. Why focus on the sei whale? By the turn of the century pre-modern whaling had reduced the stocks of right and bowhead whales to low numbers and collections of biological material from these species are rare. Humpback whales were taken by pre-modern whaling, but modern whaling quickly reduced their numbers; similarly Figure 12.1 shows the rapid depletion of Southern Hemisphere blue whales over 15 years of intensive exploitation. Early studies by biologists of the southern blue and humpback whales, especially those of the Discovery Investigations, detailed much of the basic reproductive biology of the Balaenopteridae, but the limited availability of material meant that population responses could not be assessed. As shown in Figure 12.1 the exploitation of southern fin whales started along with that of blue whales but continued for longer, and over the years more biologists worked with the whaling operations, and the quantity and quality of material increased, and much more information was obtained about fin whales. The interest in sei whales occurred later, Figures 13.1-13.3, at a time of increased biological sampling at land stations and with the pelagic fleets. Consequently biologists sampled over the periods of decline of the sei whale, and a more complete picture of the biological responses with time might be expected than that recorded for the other species. However, I do not argue that this is necessarily the case for there is more extensive material on fin whales. However this has not been subjected to some of the later and better analytical techniques; further studies may show that the fin whale data has more to offer. In addition it is probable that interspecific effects disturbed the populations of sei whales, and these changes occurred over the earlier times of imperfect sampling. It is likely that a study combining the information from sei whales with new analyses of the fin whale material would be rewarding. Nevertheless I considered that, as a species, the sei whale was likely to illustrate a large proportion of what is known about population processes in the great whales.

This study has concentrated on presenting, examining and interpreting information largely independent of that known from other similar species. Of course there are aspects where interspecific considerations are essential, as in the discrimination of species and where changes in observed vital rates can only be explained through changes in the ecosystem. However it is necessary to be aware of exactly what can be concluded from information on the species per se. It is only too easy to accept as general truths information gleaned from examinations of similar species. Within the Balaenopteridae the large annual migrations are general and well known, resulting in the typical minimum two year breeding cycle, but the Bryde's whale undergoes a very limited seasonal movement and the minke whale has a shortened breeding cycle. It is important to be aware of the subtle difference between these close relatives. This may, or even should, seem obvious, but a problem arises when advice is sought for management and, as is most often the case in all natural resource management, necessary elements of information are lacking. In order that any advice may be given, information is taken from related species with related life stvles. This is obviously sensible, but over the years there is a tendency to forget that these are not established facts. For instance the age of minke whales has been estimated from laminae in the ear plug but independent studies have not as yet confirmed that one dark and one light layer are formed each year. Consequently the review below, of the more important aspects of the state of knowledge of the sei whale's population biology, is made in this light and minimises inferences from the biology of other species.

RÉSUMÉ

Taxonomy and stock identity

The anatomy of the sei whale has been described sufficiently to distinguish it from other balaenopterids, and large remains can be identified. The structure of a range of proteins has been described using electrophoretic techniques from the Antarctic and North Atlantic, and to a lesser degree from the North Pacific, and basic differences with other species have been recognised, however no comparative exercise has yet been fully presented, although a study is in progress (Wada, pers. comm.). The existence of polymorphisms indicates a possibility of intraspecific differentiation of separate populations. The morphological characteristics have been well documented and the sei whale can be easily identified through several features, and in particular it can be distinguished from the somewhat similar Bryde's whale by the fineness of the baleen and extent of the ventral grooves. It is possible that the sei whale exists in a northern and southern form as sub-species, with the nothern sei whale being smaller, but this is inadequately described from anatomical and morphological features, and has not been established from the limited electrophoretic studies.

As described in Chapter 10 the concept of a stock is complex, but the first task in defining stocks is to examine aspects of intraspecific variability. Limited comparative morphometrics have generally been inconclusive, undertaken opportunistically rather than systematically and with the studies not usually repeated, and at best, possible differences between catches from different areas have been described. Examination of the distribution of white scars on sei whales revealed that there is limited mixing throughout the Antarctic in summer, and indicates the existence of at least partially isolated groups. The distribution of recoveries of marks shows that there is a tendency to return to the same feeding areas each year, and examination of the timing of changes in reproductive rates reveals differences between geographic areas in the Southern Hemisphere. Differences in reproductive parameters have been found between Iceland and Canada, and between California and Japan. The evidence suggests that, within oceans, different populations exist and are relatively isolated.

The above techniques do have the ability to detect differences where they exist but suffer from two major failings. The first is that much of the information is obtained on the feeding grounds and the relationship between breeding and feeding aggregations is unknown. The second is that differences may only be manifest at a more subtle genetic level and it is evident that our knowledge of the range of genotypes of the sei whale is poor. Even if information were available from the breeding areas, it has been demonstrated for fin whales that matings are not equally successful amongst all genotypes; such details are lacking for sei whales.

Distribution and migration

The sei whale is an oceanic species which is found from the equator to the poles but it avoids the coldest waters and the marginal seas. There is a pronounced seasonality in distribution with the majority of the population spending summer feeding in the poleward temperate waters and mating and calving in winter in the tropics and subtropics. Coastal whaling operations have taken sei whales from along, or just offshore of, the shelves of the continents, and it is possible that there are narrow migration paths along the edge of the shelves. During migration there exists a segregation by age and sexual classes. Within this broad description there are local feeding aggregations associated with permanent and transitory oceanographic features that provide an enriched and suitable prey fauna. Aggregations of sei whales have been reported to disperse following changes in weather and this may be due to a dislike of colder conditions or the dispersion of transitory enriched areas. However their seasonal occurrence off some whaling stations has been erratic, and the sei whale has not been seen for many years off North Norway; changes in the migratory path have been suggested.

Although much is known of the distribution and migration, there are some significant omissions. Firstly, the winter distributions are not well known and at this time the sei and Bryde's whales are found in the same latitudes. Consequently early reports of catches and sightings from these waters are unreliable and only recent sightings by experts can be accepted. Secondly, the migration paths are not well defined. Catches from the coastal stations tend to be along the shelf edge, but except from the North Pacific there is little information from further offshore at this time. Not only might such distributional information help in understanding the mechanisms of navigation, it is also very significant for management. It is frequently argued that coastal whaling, with a necessarily limited range of operations, can do only little damage to an oceanic species, but if a large proportion of the population pass along a narrow migratory strip then clearly substantial depletion can be achieved. Similar points are raised about "preferred" feeding areas, where whale density may remain locally high, and catch rates may remain high until the total population has been drastically reduced. Thirdly the marking data have provided only limited data on movement between the breeding and feeding areas, and too little is known of the movement of known individuals throughout a year; recent success with tracking by satellite will be of immense help.

Natural mortality rates

Chapter 8 reviews the information on mortality and it was concluded that instantaneous rates of natural mortality of adults are about 0.06-0.07 per year. The estimates are very imprecise with uncertain bias due to non-random catches of sei whales and incompletely examined methodologies. In addition there is insufficient information to determine whether adult mortality rates vary with age or whether there have been changes with time. A more detailed examination of material currently held may improve the situation to some degree. There are no quantified estimates of the contribution of different factors, such as predation or disease, to natural mortality.

Reproduction and interspecific effects

Biologists have worked extensively with the whaling operations and as Chapter 9 shows a great wealth of information has been gained. However, because material has come from the whaling operations samples have seldom been taken systematically from the population, and it is often difficult to relate information from the sample to the entire population. It is not the intention here to review all those aspects described in Chapter 9, and only those which are particularly important to understanding population responses will be focussed upon. Two aspects of the reproductive activity of the whales are particularly helpful in understanding the population processes, and these are the persistence of corpora in the ovary and the presence of a layer indicating age at maturity in the ear plug. These give valuable information on the reproductive history of the whale, a feature uncommon in most mammals.

The most important documented change of the reproductive processes is that of pregnancy rate. The data do not reveal any reduction of pregnancy rate with age, although from the North Pacific it was found that pregnancy rate was much lower in those whales with large numbers of corpora. However large changes of pregnancy rates with time have been recorded, and from the Southern Hemisphere rates increased from 0.27 y^{-1} to 0.38 y^{-1} over the period of exploitation of sei and other whales. The evidence for this change is not complete, and it relies heavily on the validity of early reporting of fetuses by the whaling industry. Nevertheless low pregnancy rates were also found in sei whales caught off Japan in the late 1940s, low ovulation rates occurred in sei whales from eastern Canada, and possibly from older sei whales caught from the pelagic North Pacific, and these support the interpretation of a substantial change in pregnancy rate.

No changes have been observed in length at maturity over the years, and ages at maturity, have been difficult to interpret from direct observations because of selection for the larger whales by the industry. However, trends with time, of ages at maturity, have been described from the transition phase of the ear plug, and declines are recorded from the three main ocean areas. The greatest reductions are found from the South Atlantic where mean ages at maturity have declined by one to three years. Although substantial problems with the interpretation of these data remain unresolved, it is quite evident that, from the Southern Hemisphere, the data show negligible proportions of the sei whales maturing at young ages, from early year classes, and significantly greater proportions later on. From the early year classes, which were only sampled some 20 years after birth, the early maturing whales apparently disappeared: perhaps being faster growers they were caught before sampling began, or perhaps the transition phase has been identified differently in younger and older whales, or else there has been a real reduction in the age at maturity. For sei whales the latter explanation seems more plausible or at least is likely to have some substantial contribution.

It is the magnitude and timing of these changes in pregnancy rates and ages at maturity, particularly in sei whales, which led to Gambell's hypothesis that, in the Southern Hemisphere, the changes were caused by interspecific effects. However, in Chapters 5 and 12, it is shown that the ecological niche of the sei whale is largely separate from those of the other rorquals, and that the timing of the changes also reflects the pattern of catches of sei whales. Most of the changes can then be attributed to intraspecific effects, with a lesser contribution from interspecific or environmental effects. However the existence of some element of interspecific effect, combined with the fact that the assessments of numbers have used the observed reproductive rates, has meant that the construction of a relationship between reproductive rates and sei whale numbers has not been attempted. Nevertheless the largest changes occurred early on in time, with substantial exploitation following later, and one can conclude that the sei whale responded to a relatively small depletion with a substantial increase in reproductive effort. In this respect it is significant that Lockyer et al (1985) have shown large year to year variations in pregnancy rates in rorquals to be associated with annual variability in body condition and food supply; clearly implying that pregnancy rate can be sensitive to change.

The magnitudes of the changes in pregnancy rates and ages at maturity appear large, with the whales maturing up to three years earlier and the calving interval reducing from four years to two and a half years, and this might give the impression of relatively large rates of population increase. Observed rates of increase of large whales are documented from South Africa where indices of abundance, from the late 1950s, showed sei whales to be increasing at 9-13% per annum, and sightings of right whales to be increasing by 7%, sightings of gray whales off California show increases of 3.7% or over per annum. Can these observed rates of increase be explained by increases in pregnancy rates and reductions in age at maturity? Let us consider that natural mortality rates have remained unchanged with time and that the adult annual proportional survival is 0.94. If the population was originally in equilibrium with an age at first parturition of 11 years and a pregnancy rate (P) of 0.27 y^{-1} then the balance equation of survivals to births determines the unobserved juvenile survival rate as 0.825 y^{-1} over the first year of life. If these constant parameters are put into a Leslie population matrix the dominant root of the matrix is unity as we have arranged the parameters such that births equal deaths. If we now alter P and the age at first parturition the dominant root gives us the equilibrium rate of population increase. If the pregnancy rate is increased to 0.38 the dominant root is 1.013; so this large reduction in calving interval only allows the population, in equilibirum, to increase at a rate of 1.3% per annum. Retaining the original fecundity but reducing the age at first parturition to 8 years gives a root of 1.008, an increase of less than 1% per annum. A combination of the two gives a population increase of 2.4% per annum. If the lower age at maturity is retained but an extreme two year calving interval is assumed, then this would result in an increase of 3.9% per annum.

Clearly the previous observed rates of population increase cannot be explained solely by the changes in reproductive rates. The differences can be explained through two mechanisms. The first is that undetected, and probably undetectable, reductions in natural mortality occurred. The second is that the observed high, population growth rates are a transitory phenomenon associated with a non- equilibrium age structure and, or else, a sub-set of the total populations were counted, and only observations over a longer period of time are likely to reconcile these alternatives.

It must be noted that the previous rates refer to equilibrium population growth rates, and if harvesting takes place on only a component of the population then sustainable harvest rates will differ from the population rate. For example, in the case of the population of sei whales increasing at 1.3% per year, if harvesting was on only those of age 10 and older, then a harvest rate of about 2.4% could be sustained. As an aid to management, harvesting at a rate of 2% to 3% is likely to be sustainable and the higher observed rates should not be used as a basis for setting quotas until the observed population increase are validated.

Status of stocks

The stocks referred to here are of whales contained within different management regions. In the North Pacific and Southern Hemisphere exploitation of sei whales continued at modest levels until the early 1960s, at which time the decline of fin whales resulted in a switch to catches of sei whales. Over the next decade catches which were large relative to the stock size were taken, and caused the rapid depletion of these stocks to some 20-25% of their original levels, and protection afforded through the IWC revised management scheme stopped further declines. It was the rapid decline of these stocks that has allowed relatively robust estimates of stock size, as given in Chapter 13, for errors in predicted recruitment rates have little effect in comparison with the high catch rates. However, the estimates are obviously sensitive to the correct interpretation of indices of abundance.

Over the longer periods of time, from the turn of the century, our knowledge is much less sure, for we do not know how natural mortality rates have responded with time, and there is some argument over the magnitude of changes in reproductive parameters. Simulations over 50 years, common in assessments of large whales, are sensitive to errors in these terms. That our knowledge is imperfect is exemplified by the last assessment of sei whales in Area II of the Southern Hemisphere, where any adequate reconciliation of the history of catches, biological information and indices of abundance resulted in a prediction that the stock should have been exterminated; subsequent catches clearly show this to be untrue. Much time has been expended in examination of the biological material but relatively little in examination of indices of abundance, and it is likely that improvements in assessments could be made after further examinations. Long term predictions could also be improved with a better understanding of the nature of the biological stock boundaries.

The status of sei whale stocks in the North Atlantic is very uncertain and there are no adequate assessments of absolute or relative numbers. There has been a long history of erratic catches, but catches have been smaller than from the North Pacific. At the time of writing Iceland has recorded the intention to continue to catch sei whales under the system of scientific permits of the IWC.

The stocks are affected by factors other than direct exploitation. It is possible that the Southern Hemisphere sei whales have prospered from the decline of blue and fin whales, but, as described in Chapter 12, this is probably to a limited extent; nevertheless the supposed increase of the other species, now that they are protected, may adversely affect the projected recovery of the sei whales. This can be regarded as a reduction in the "carrying capacity" of the stocks, the size of the stock that the environment will allow, and it is then easy to imagine that exploitation of the food species of the sei whale, by man, may reduce this carrying capacity. Here one may consider the recent fishery for krill in the Southern Ocean, and the fisheries in the seas of the Northern Hemisphere, but unfortuantely our knowledge is too limited to quantitatively predict the effects on sei whale numbers of these fisheries. The habitat of other whales has been adversely affected by industry and tourism, but these are particularly the species that breed or feed in coastal areas, such as the gray and bowhead whales, but the more oceanic sei whale will be less affected by coastal disturbances. Nevertheless the ocean environment is subject to the low levels of chemical and radioactive pollution found throughout the surface waters of the oceans and whose effects are scarcely known.

IMPLICATIONS FOR MANAGEMENT

The present management of large whales is under the auspices of the IWC, particularly those aspects relating to exploitation, although many nations have their own legislation of the coastal seas, and there are also related international agreements such as the Convention on Trade in Endangered Species, and the Migratory Species Convention. The management scheme adopted by the IWC, over the last decade, followed international pressure to avoid risks of extinction of the great whales, and the procedure devised was one in which overexploitation should not occur and in which catches were to be taken only on a sustainable basis; the details are given in Chapter 11. The apparent great strength of the management scheme is that, given the details arising from stock assessments, recommendations concerning the level of catches were virtually automatic, using a simple algorithm. Subsequent management decisions might involve sensitive negotiation but the principles are relatively explicit and room for consideration of other aspects are deliberately restricted. The scheme could be presented to the public as an example of concerned and rational management; however the application falls short of this ideal. As will be appreciated from the previous chapters, the Scientific Committee of the IWC is unable to provide some of the details required in the management procedure, and others are known only approximately. We do not know the maximum sustainable yield level of sei or any other whale species or stock and, although catch limits are reduced from 100% to zero over a 10% range of stock size, most stock assessments are not accurate to within 10%. The proper application of the management scheme is therefore obviously impracticable, and in particular the present management scheme fails to explicitly deal with the problem of imprecise assessments.

In order to overcome the obvious problems, the IWC has urged the Scientific Committee to present more unified and precise opinions, for, after all, the procedure emerged through members of this Committee, and to some degree the scientists have felt that, given the spirit of the management procedure, they were in the best position to provide the final estimates which would ultimately lead to a specific quota. For instance they may have two estimates of a stock and for a mangement decision one is needed; they can accept one or the other or some weighted average, or even feel both totally inadequate. The evidence for any one interpretation of the status of a stock is seldom definitive, and, lacking crucial evidence, the final choice often owes more to professional judgement than logical and objective argument. Judgements of that kind are in their turn vulnerable to the charge of being politically motivated rather than scientific: whether that accusation is true or false is immaterial. This situation opens up the IWC to the charge that the scientists are taking crucial decisions which are more properly the task of the Commission itself; indeed some of the scientists themselves believe this to be true. It also admits scope for argument within the Commission on the quality of scientific recommendations. There are therefore very real difficulties in the IWC decision process which call in question the relative roles of scientist, administrators and politicians, and the quality of information on which each sector can act.

Management of whale stocks in the long term is a particular problem as we do not understand the nature and extent of interspecific effects or how the "carrying capacity" is likely to vary with climate or exploitation of food species. Our present knowledge can only produce a range of plausible alternatives and the concept of a managed ecosystem, as indicated in the Antarctic Treaty, has, at present, little meaning. The acceptance of any one model of the ecosystem is likely to lead to extremes of management responses and the single most important factor is to recognise our lack of knowledge and hence to allow for a flexibility in management. The problems of habitat can be considered within longer term considerations, and comments on this aspect have been made in the previous section. Management in the short term can be presented in two aspects. Firstly the previous chapters have shown that it is possible to obtain approximate estimates of stock size and sustainable rates of harvest. From these we may predict a sustainable yield that will be correct within a factor of two or three. Temporary quotas can be set safely on this basis, but only maintained providing that there is continued monitoring of stock trends and other biological elements that will ultimately reveal the status of the stocks; without this monitoring management will remain too imprecise. Secondly, management schemes can be devised which, in the short term, are largely independent of the population biology, and these will take the form of feed-back control policies. Two elements are necessary, the first an approximate value for a sustainable catch, and the second, an index that can be related to stock size; it is of course crucial that the index be interpreted correctly. This is an area suitable for research for the natural delays are likely to introduce instabilities into the control system but, in essence, this is how management has operated over the previous decade; irrespective of what we think we know of stock sizes or harvesting rates, if indices of abundance started to decline more than expected then the scientists usually advised a reduction in catches.

Finally, over the last twenty years, nations have withdrawn from whaling and, due to overexploitation and to political philosophies, catches of large whales have been reduced to insiginificant levels. Associated with this has been a reduction in national support for research in Cetacean population biology, and the consequent loss of individual experts in this field. Under a moratorium and with no commercial whaling it must be recognised that financial support will be less with no commercial interest, but the scientific expertise is still needed for broader management issues and it is negligent to squander this expertise.

FUTURE RESEARCH

In the résumé I have considered the strengths and weaknesses of our knowledge of the population dynamics of the sei whale. In this section particular projects are identified as being both important to the resolution of current problems and capable of being implemented. In respect of the latter it is appreciated that future whaling on all large whales may be of a limited extent or even absent. Consequently the projects tend to be of either the reanalysis of existing materials, or else using material gained by non-lethal techniques. Throughout the previous texts it will be obvious that many aspects can profit by more research and so the following is not exhaustive, but represents projects of a combination of personally perceived importance and interest.

Stock identity and genetics

Present knowledge of genetical variability in sei whales is slight, and basic studies are needed to describe the extent of this variability. Such information is useful in interpreting the phylogeny of the species and will aid in interspecific identification; however once a range of polymorphisms have been identified then more intensive studies can be undertaken to delineate populations, and comparisons can be made from different breeding units and social groupings. Material for analyses can be obtained from catches or with the use of non-lethal techniques. The present evidence indicates that such studies will be profitable, since phenotypic differences have been indicated, and genotypic differences have been found in several other species of whales. Any new information will provide not only basic descriptions but will lead to better divisions for management, may provide insights into how management should operate, and will help in the interpretation of past records. The validity of the postulated existence two sub-species in the two hemispheres should be tested. It is possible that such data will indicate the magnitude of mixing rates between groups, but of particular importance is the identification of mechanisms of isolation. The studies on maternal-fetal incompatibility undertaken with fin whales should be extended to sei whales if catches are available; if not then non-lethal techniques can be used on cow and calves and, in cases when small and identifiable breeding groups can be recognised, by also sampling from bulls.

Distribution and migration

Two aspects need further studies. As explained earlier the pelagic whaling fleets have not operated in the warmer waters of the Atlantic and baleen catches have been small in the warmer waters of the Pacific and Indian Oceans; consequently the distribution of sei whales during the winter breeding season is imprecisely described. Bryde's whales are also found in these regions and many older records suffer from unreliable identification. If voyages are designed to describe the winter distribution of sei whales then it can be anticipated that much time will be spent in ensuring correct identification. This may have its advantages in that aspects of breeding behaviour might be observed simultaneously; a subject about which there is minimal information. This is contrasted with systematic sightings surveys where it is wished to maximise searching effort in order to obtain precise estimates of stock size, and here time spent in identification is at a cost of a higher variance. If breeding and distributional studies are undertaken as a prime object then there would be little "wasted" time in such operations. Nevertheless one must recognise that with finite resources this area of study is not the most important. The second aspect is that of migration and movement. Most marks have been placed and recovered from the feeding grounds and only limited returns exist to show the connection between summer and winter aggregations. With modern marks sending frequent returns to satellites a great deal will be revealed about the migration paths, extent of mixing of various groups and localised movements. This is clearly a rich area for future work and information will be of direct relevance to identifying better management and genetic stocks. Correlation of movement and local environmental conditions may also explain the erratic movements of herds of sei whales.

Reproduction and mortality

The single most important area for investigation of the reproductive responses and possibly of any aspect of whale biology, is that associated with the interpretation of the pattern of ages at maturity, as determined from the transition phase of the ear plug. Not only is this important for the study of the sei whales, it is of immense significance for the understanding of the population biology of the other large whales, seals and the interactions amongst all the large animals of the Southern Ocean. At present declines in the ages at maturity have been found in many populations of whales and seals, with the greatest declines found in those regions subject to the greatest exploitation. However empirical evidence, with some analytical support, has indicated that the "year of capture" of the samples is more significant than the "year of birth" in determining these trends. The observed pattern can be generated by a true decline in age at maturity, other aspects of the life history which mean that young maturing animals are not available for capture at old ages, biases introduced through sampling, reading, analysis, etc, or combinations of these. To date critical investigations of methodology have been undertaken only with material from minke whales and that of a cursory nature. The utilisation of large amounts of invaluable material, held at laboratories throughout the world, is held in limbo until these issues are resolved.

Although the balance of evidence suggests a substantial increase in the pregnancy rates of Southern Hemisphere sei whales, the increase largely rests upon placing a reliance on the greater body of information from the whaling fleets compared with the few samples taken by biologists at South Georgia, both sets of information obtained in the early years of Antarctic exploitation. It is possible that some of the whaling records are affected by a variable quality in the finding and reporting of fetuses, and it would therefore be instructive to examine the original whaling records, where they exist, for evidence of differential reporting by individuals, factory ships and fleets. Of particular interest would be comparisons of records from trained and less expert inspectors. Associated with this problem is the recognition of different ovulation rates, from sei whales stocks, at different locations and times. A more systematic study is warrented to compare these records and particularly to confirm the existence of low ovulation and pregnancy rates.

The range of reproductive rates gives some insight into possible rates of population growth, although our lack of knowledge of actual or potential changes in mortality rates means that this information is of uncertain value. Nevertheless, many observed rates of population increases could not be generated by changes in reproduction rates solely, and consequently, adequate monitoring, over a long enough period to overcome transient phenomena, may be able to shed some light on responses of mortality rates; however because of the difficulties in monitoring this is unlikely to be achieved with sei whales, although practicable for other species. Direct observations of calf mortality where this is possible may be particularly useful; however most information on mortality rates has come from the analyses of catch data, and because of problems of selection and segregation, only approximate estimates of adult mortalities have been obtained. Although further examinations may be worthwhile it seems unlikely that these adult mortality estimates can be much refined.

Other topics

The vast majority of catches of sei whales have been recorded by date, location, sex and length but details from the earlier years are incomplete and, in addition, records of catches from the warmer waters have frequently failed to distinguish between catches of sei and Bryde's whales. With perseverance and recourse to original log-books it should be possible to produce more definitive sets of national records, associated with which could be analyses of the likely proportions of sei and Bryde's whales.

One of the most profitable areas of study for sei and other species is the examination of indices of abundance for these play a vital role in the estimation of stock sizes, and are clearly important in any management scheme. Over recent years refinements have been made of indices of CPUE, to account for time spent on operations other than searching, and for time spent dealing with other than the target species, and for sei whales this is particularly important as sei and fin whales were frequently caught in the same years; nevertheless the indices of Chapter 13 show that more analysis is necessary. Less attention has been given to critical examination of time series of sightings records. Finally, and related to the previous problems associated with interpretation of indices of abundance, is the need for examination and design of schemes for the management of exploitation. It is hoped that this study has given some insight into the extent and quality of the knowledge of whale population numbers and dynamics, and it has been shown that present management schemes have failed to adequately recognise the strengths and weaknesses in our present understanding of whale population biology. Management schemes more fully incorporating these features are needed, but in such schemes indices of abundance will play an important part.

REFERENCES AND FURTHER READING

Aguayo A L. 1974 Baleen whales off continental Chile. p 209-217. In (ed W E Schevill) The Whale Problem. Harvard University Press, Mass. pp420.

Aguilar A & C Sanpera. 1982 Reanalysis of Spanish sperm, fin and sei whale catch data. (1957-1980). Rep. int. Whal. Commn. 32:465-470.

Aguilar A & J Pelegri. 1980 Notas sobre cetacos de las aguas Ibericas.3. Sobre un ejemplar de Balaenoptera borealis Lesson 1828 capturado frente a las costas gallegas. Inm. Cien 13.

Aguilar A & S Lens. 1981 Preliminary reports on Spanish whaling activities. Rep. int. Whal. Commn. 31:639-643.

Aguilar A, Grau E, Sanpera C, Jover L & G Donovan. 1983. Report of the 'Balaena' whale marking and sighting cruise in the waters off western Spain. Rep int. Whal. Commn. 33: 649-655.

Allen G M. 1916 The whalebone whales of New England. Mem Boston Soc. Nat. Hist. 8(2):107-322

Allen K R. 1966a A method of fitting growth curves of the von Bertalanffy type to observed data. J. Fish. Res. Bd. Canada, 23(2):163-179.

Allen K R. 1966b Some methods for the estimation of exploited populations. J. Fish. Res. Bd. Canada. 23:1153-1174.

Allen K R. 1969a An application of computers to the estimation of exploited populations. J. Fish. Res. Bd. Canada. 26:179-189.

Allen K R. 1969b Further estimates of whale populations and sustainable yields in North Pacific areas. Rep. int. Whal. Commn. 19:120-122.

Allen K R. 1980 Conservation and Management of Whales. University of Washington Press. pp 107.

Allen K R & D G Chapman. 1977 Whales. p 335-358, In (ed. J Gulland) Fish Population Dynamics. Wiley and Sons, Lond.

Anderson J. 1878 Anatomical and zoological researches : comprising an account of the zoological results of the expeditions to western Yunnan in 1868 and 1875. London, B Quaritch, 551-564.

Andrews R C. 1916a Monographs of the Pacific Cetacea, II. The sei whale (Balaenoptera borealis Lesson). 1. History, habits, external anatomy, osteology and relationships. Mem. Amer. Mus. Nat. Hist., New Series Vol.1, Part V :289-388.

Andrews R C. 1916b Whale Hunting with Gun and Camera. Appleton and Co. London. pp333.

Anon. 1920 Report on the Interdepartmental Committee on Research and Development in the Dependencies of the Falkland Islands. HMSO, 1920 :pp 164. Command paper 657.
Anon. 1953 Fin whale (Balaenoptera physalus) with six foetuses. Norsk Hvalfangst Tid. 42(12):685-686
Anon. 1960 Proceedings of joint scientific meeting of ICNAF, ICES and FAO on fishing efforts, the effect of fishing on resources and the selectivity of fishing gear. Spec. Publ. Int. Commn. NW Atlantic Fish. 1(2):1-45.
Anon. 1969 Report of the meeting on age determination in whales. Rep. int. Whal. Commn. 19:131-137.
Anon. 1977 Report of the Working Group on North Atlantic Whales. Rep. int. Whal. Commn. 27:369-387.
Anon. 1980 Report of the Special Meeting on Southern Hemisphere sei whales. Rep. int. Whal. Commn. 30:493-505.
Anon. 1984 Report of the minke whale ageing workshop. Rep. int. Whal. Commn. 34:675-699.
Anon. 1951 Whaling operations from Gabon. Norsk. Hvalfangst Tid. 40(10):516.
Anon. 1978 Report of the Scientific Committee. Sei whales, Southern Hemisphere. Rep. int. Whal. Commn. 28:47-54.
Arnason A & J H Sigurosson. 1983 An electrophonetic study of protein and enzyme marks of the blood in three species of whale, B. physalus, B. borealis and P. macrocephalus. Int. Whal. Comm. SC/34/08 (unpubl.) pp 11.
Arnason U. 1970 Karyotype of a male sperm whale (Physeter catadon L.) and a female sei whale (Balaenoptera borealis Lesson). Hereditas 64:291-293.
Arnason U. 1972 The role of chromosomal rearrangement in mammalian speciation, with special reference to Cetacea and Pinnipedia. Hereditas 70(1):113-118.
Arnason U. 1974 Comparative chromosome studies in Cetacea. Hereditas 77:1-36.
Arriaga L G. 1981 Actividad ballenera en el Pacifico sur-oriental. FAO Fisheries Series,5, Vol 3:311-319.
Ash C E. 1953 Weight of Antarctic humpback whales. Norsk. Hvalfangst Tid. 42(7):387-391.
Atwood R P & L M Razavi. 1965 Chromosomes of the sperm whale. Nature 207 (4994):328.
Australia 1985 Australia progress report on cetacean research, June 1983 to April 1984. Rep.

int. Whal. Commn. 35 :158-161.
Bada J L, Brown S & P M Masters. 1980 Age determination of marine mammals based on aspartic acid racemization in the teeth and lens nucleus. Rep. int. Whal. Commn. (Sp. Is. 3) :113-118.
Baldridge A. 1972 Killer whales attack and eat a gray whale. J. Mammal. 53:898-900.
Bannister J L & A C Baker. 1967 Observations on food and feeding of baleen whales at Durban. Norsk Hvalfangst Tid. 56:78-82.
Bannister J L & R Gambell. 1965 The succession and abundance of fin, sei and other whales off Durban. Norsk. Hvalfganst Tid. 54:45-60.
Barlow J. 1984 Mortality estimation: biassed results from unbiassed ages. Can. J. Fish. Aquat. Sci. 41(12):1843-1847.
Beddington J R. 1979 On some problems of estimating population abundance from catch data. Rep. int. Whal. Commn. 29:149-154.
Bengtson J L & R M Laws. 1985 Trends in crabeater seal age at maturity: an insight into Antarctic marine interactions. p 669-675. In (ed W R Siegfried, P R Condy & R M Laws) Antarctic Nutrient Cycles and Food Webs. Springer, Berlin.
Bennett A G. 1920 On the occurrence of diatoms on the skins of whales. Proc. Roy. Soc. B, 91:352-357
Bertalanffy L. 1938 A quantative theory of organic growth. Hum. Biol. 10(2):181-213.
Berzin A A. 1978 Whale distribution in tropical eastern Pacific waters. Rep. int. Whal. Commn. 28:173-178.
Berzin A A & N V Doroschenko. 1982 Distribution and abundance of right whales in the North Pacific. Rep. int. Whal. Commn. 32:381-383.
Berzin A A & V L Vladimirov. 1981 Changes in abundance of whalebone whales in the Pacific and Antarctic since the cessation of exploitation. Rep. int. Whal. Commn. 31:495-499.
Berzin A A, Vladimirov V L & N V Doroshenko. 1985 The distribution and numbers of Cetacea in the Okhotsk Sea: results from aerial surveys. IWC doc SC/37/05 pp7 (mimeo).
Best P B. 1967a Distribution and feeding habits of baleen whales off the Cape Province. Investl. Rep. Div. Sea Fish. S Africa 57:1-44.
Best P B. 1967b The sperm whale (Physeter catodon) off the west coast of South Africa. 1. Ovarian changes and their significance. Investl. Rep. Div. Sea Fish. S. Afr. 61:1-27.
Best P B. 1969 The sperm whale (Physeter catodon) off the west coast of South Africa. 3. Reproduc-

tion in the male. Investl. Rep. Div. Sea Fish. S. Afr. 72:1-20.
Best P B. 1970 Exploitation and recovery of right whales, Eubalaena australis, off the Cape Province. Investl. Rep. Div. Sea Fish., Un. S. Afr. 80:1-20.
Best P B. 1974a The status of the whale populations off the west coast of South Africa and current research. p53-81. In (ed W E Schevill) The Whale Problem. Harvard University Press, Mass. pp420.
Best P B. 1974b Status of whale stocks off Natal, 1972. Rep. int. Whal. Commn. 24:127-141.
Best P B. 1975 Status of whale stocks off South Africa. Rep. int. Whal. Commn. 25:198-207.
Best P B. 1976 Status of whale stocks off South Africa, 1974. Rep. int. Whal. Commn. 26:264-286.
Best P B. 1977a. Two allopatric forms of Bryde's whale off South Africa. Rep. int. Whal. Commn. (Spec. Issue 1):10-38.
Best P B. 1977b. Status of whale stocks off South Africa. Rep. int. Whal. Commn. 27:116-121.
Best P B. 1979 Social organization in sperm whales. In (ed H E Winn and B L Olla) Behaviour of Marine Animals Vol 3: 227-289. Plenum Press.
Best P B. 1982 Seasonal abundance, feeding, reproduction, age and growth in minke whales off Durban. Rep. int. Whal. Commn. 32:759-786.
Best P B. 1985 The use of natural markings to determine calving intervals in right whales off South Africa. IWC doc SC/37/PS9, pp26. (mimeo).
Best P B & C H Lockyer. 1977ms The biology of sei whales (B. borealis) off the west coast of South Africa. (unpubl.)
Best P B & R Gambell. 1968 The abundance of sei whales off South Africa. Norsk Hvalfangst Tid. 57:168-174.
Betesheva E I. 1955 The baleen whales feeding in the Kurile Isles area (by materials of 1953). Trudy Inst. Okeanol., Acad. Sci. USSR 18:78-85.
Betesheva E I. 1955 Food of Whalebone Whales in the Kurile Island Region. Trans. Inst. Oceanogr. Acad. Sci. USSR 18: 78-85.(in Russian).
Beverton R J H & S J Holt. 1959 A review of the lifespans and mortality rates of fish in nature. p 142-177. In (ed G Holme & M O'Connor) The Lifespan of Animals, Vol 5. Churchill London.
Beverton R H J & S J Holt. 1957 On the dynamics of exploited fish populations. MAFF Fish. Invest. Ser. 2,19:1-533.
Birnie P. 1985 International Regulation of Whaling. Vols 1 & 2. Oceana Publications Inc. Lond. pp 1033

Bjarnason I & P Lingaas. 1954 Some weight measurements of whales. Norsk Hvalfangst Tid. 43:8-11.
Bonner W N. 1968 The fur seal of South Georgia. Br. Antarctic Surv. Sci. Rep. 56:1-81.
Bonner W N. 1980 Whales. Blondford Press, Dorset. pp 278.
Borisov V I. 1980 Haptoglobin polymorphism of minke whales and sei whales. Int. Whal. Comm. SC/32/013 :pp 11. (mimeo).
Borisov V I. 1981 A comparative analysis of the electrophoretic spectra of Antarctic whale proteins. Zool. Zhur. LX(3):438-441. (in Russian).
Borodin R G. 1974 Application of the method of virtual population for assessment of sei whale stocks in the Antarctic. IWC doc. SC/SP74/4. (mimeo).
Bree P J H van & A M Husson. 1974 Strandingen van Cetacea op de Nederlandse kust in 1972 en 1973. Lutra,16:1-10.
Bree P J H van. 1975 Preliminary list of Cetaceans from the Southern Caribbean. Studies on the fauna of Curacao and other Caribbean Islands. 160:79-87.
Bree P J H van. 1977 On former and recent strandings of Cetacea on the coast of the Netherlands. Zeit. fur Sougertierkunde, 42:101-107.
Breiwick J. 1978 Southern Hemisphere sei whale stock sizes prior to 1960. Rep. int. Whal. Commn. 28:179-186.
Brinkmann A. 1967 The identification and names of our fin whale species. Norsk. Hvalfangst Tid. 56(3):49-56.
Brown S G. 1960 Swordfish and whales. Norsk Hvalfangst Tid. 49(8):345-351.
Brown S G. 1965 The colour of the baleen plates in Southern Hemisphere sei whales. Norsk. Hvalfangst Tid. 54(6):131-135.
Brown S G. 1968a The results of sei whale marking in the Southern Ocean to 1967. Norsk Hvalfangst Tid. 57:77-83.
Brown S G. 1968b Feeding of sei whales at South Georgia. Norsk. Hvalfangst Tid. 57(6):118-125.
Brown S G. 1974 Notes on the coordination of the International Whale Marking Scheme by the Institute of Oceanographic Sciences. Rep. int. Whal. Commn. 24:127-135.
Brown S G. 1976a Whale marking progress report 1975. Rep. int. Whal. Commn. 26:31-38.
Brown S G. 1976b Modern whaling in Britain and the North Atlantic Ocean. Mammal Review 6:25-36.
Brown S G. 1977a Some results of the sei whale

marking in the Southern Hemisphere. Rep. int. Whal. Commn. (Sp. Is. 1):39-43.
Brown S G. 1977b Whale marking: a short review. p 569-581. In (ed M Angel) A Voyage of Discovery. Pergamon Press. Oxford.
Brown S G. 1977c Whale marking in the North Atlantic. Rep. int. Whal. Commn. 27:451-455.
Brown S G. 1978 Sei whale marking data. Rep. int. Whal. Commn. 28:369-372.
Brown S G. 1979 Whale marking in the North Atlantic 1950- 1978. ICES CM 1979/N 11 pp6. (mimeo).
Brown S G & C H Lockyer. 1984. Whales. p 717-781. In (ed R M Laws) Antarctic Ecology, 2. Academic Press, London. pp 850.
Brownell R L. 1981 Review of coastal whaling by the Republic of Korea. Rep. int. Whal. Commn. 31:395-402.
Budker P. 1958 Whales and Whaling. G. Harrup, London. pp182.
Budylenko G A. 1970 On some problems of the sei whale biology in the South Atlantic. 'The Whales of the Southern Hemisphere'. Trudy Atlant NIRO, 29:17-33.
Budylenko G A. 1973 The observations on the sei whale behaviour in the Southern Ocean. 'Fishery Investigations in the Atlantic Ocean'. Trudy Atlant NIRO, 51:142-149.
Budylenko G A. 1977 Distribution and composition of sei whale schools in the Southern Hemisphere. Rep. int. Whal. Commn. (Sp. Is. 1):121-123.
Budylenko G A. 1978a Distribution and migration of sei whales in the Southern Hemisphere. Rep. int. Whal. Commn. 28:373-377.
Budylenko G A. 1978b On sei whale feeding in the Southern Ocean. Rep. int. Whal. Commn. 28:379-385
Burfield S T. 1912 Report of the Committee appointed to investigate the biological problems incidental to the Belmullet Whaling Station. Rep. of the British Assoc. for the Adv. of Sci. 81:121-125.
Burfield S T. 1913 Report of the Committee Appointed to investigate the biological problems incidental to the Belmullet Whaling Station. Rep. of the British Assoc. for the Adv. of Sci., Sect. D (Dundee):1-42.
Burfield S T. 1915 Report of the Committee appointed to investigate the biological problems incidental to the Belmullet Whaling Station. Rep. of the British Assoc. for the Adv. of Sci., 84:125-161.
Burnham K P, Anderson D R & J L Laake. 1980

Estimation of density from line transect sampling of biological populations. Wildlife Monographs. 71:1-202.
Butterworth D S & P B Best. 1982 Report of the Southern Hemisphere minke whale assessment cruise 1980/81. Rep. int. Whal. Commn. 32:835-874.
Cabrera A. 1925 Los grandes cetaceos del Estrecho de Gibralter:su pesca y explotacion. Trab. Mus. Cien. Nat. Madrid, Serie Zoologica 52:1-48.
Casinos A & J R Vericad. 1976 The Cetaceans of the Spanish coasts: a survey. Mammalia 40(2):267-289
Caughley G. 1966 Mortality patterns in mammals. Ecology 47: 906-918.
Cawthorn M W. 1984 New Zealand progress report on cetacean research. Rep. int. Whal. Commn. 34:213-215.
Cawthorn M W. 1985 New Zealand progress report on cetacean research 1984-1985. IWC doc SC/37/Prog. Rep. pp9. (mimeo).
Chapman D G. 1964 Reports of the committee of three scientists on the special scientific investigations of the Antarctic whales. Rep. int. Whal. Commn. 14:32-106.
Chapman D G. 1965 Report of the committee of four scientists. Rep. int. Whal. Commn. 15:47-60.
Chapman D G. 1970 Reanalysis of Antarctic fin whale population data. Rep. int. Whal. Commn. 20:54-59
Chapman D G. 1971 Analysis of 1969/70 catch and effort data for Antarctic baleen whale stocks. Rep. int. Whal. Commn. 21:67-75.
Chapman D G. 1974 Estimation of population size and sustainable yield of sei whales in the Antarctic. Rep. int. Whal. Commn. 24:82-90.
Chapman D G. 1983 Some considerations on the status of stocks of Southern Hemisphere minke whales. Rep. int. Whal. Commn. 33:311-314.
Chapman D G. & D S Robson. 1960 The analysis of a catch curve. Biometrics 16(3):354-368.
Chittleborough R G. 1954 Studies on the ovaries of the humpback whale, M. nodosa (Bonnaterre), on the Western Australian coast. Aust. J. mar. Freshw. Res. 5(1):35-63.
Chittleborough R G. 1955 Aspects of reproduction in the male humpback whale Megaptera nodosa (Bonnaterre). Aust. J. Mar. Freshw. Res. 6(1):1-29
Chittleborough R G. 1959 Determination of age in the humpback whale, Megaptera nodosa Bonnaterre. Aust. J. mar. Freshwat. Res. 10(2):125-143.
Chittleborough R G. 1960 Marked humpback whale of known age. Nature, Lond. 187:164.
Chittleborough R G. 1965 Dynamics of two popula-

tions of the humpback whale Megaptera novaeangliae (Borowski). Aust. J. Mar. Freshwat. Res. 16(1):33-128.

Christensen I. 1968 Studier over den sydlige seihvals (B. borealis) biologi. Hovedfagsoppgave varsemestret, pp65.

Christensen I. 1977 Observations of whales in the North Atlantic. Rep. int Whal. Commn. 27:388-399.

Christensen I. 1980 Observations of large whales (minke not included) in the North Atlantic 1976-78 and markings of fin, sperm,and humpback whales in 1978. Rep. int. Whal. Commn. 30:205-208.

Christensen I. 1981 Age determination of minke whales from laminated structures in the tympanic bullae. Rep. int. Whal. Commn. 31:245-254.

Christensen I & C J Rorvik. 1981 Availability of minke whales in the Barents Sea and adjacent waters. Rep. int. Whal. Commn. 31:259-262.

Clark W G. 1982 Historical rates of recruitment to Southern Hemisphere fin whales. Rep. int. Whal. Commn. 32:305-324.

Clark W G. 1983 Apparent inconsistencies among countries in measurments of fin whale lengths. Rep. int. Whal. Commn. 33:431-434.

Clarke R. 1962 Whale observation and whale marking off the coast of Chile in 1958 and from Ecuador towards and beyond the Galapagos Islands in 1959. Norsk Hvalfangst Tid. 51(7) :268-287.

Clarke R. 1980 Catches of sperm whales and whale-bone whales in the Southwest Pacific between 1908 and 1975. Rep. int. Whal. Commn. 30:285-288

Clarke R. 1981 Whales and dolphins of the Azores and their exploitation. Rep. int. Whal. Comm. 31:607-615.

Clarke R & A Aguayo. 1965 Bryde's whales in the south-east Pacific. Norsk Hvalfangst Tid. 54(7):141-148.

Clarke R, Aguayo A L & S B del Campo. 1978 Whale observations and whale marking off the coast of Chile in 1964. Sci. Rep. Whales Res. Inst. 30:117-177.

Cockrill W R. 1960 Pathology of the Cetacea : a veterinary study on whales. Br. Vet. J. 116: 133-144 & 175-190.

Collett R. 1886 On the external characteristics of Rudolphi's rorqual (Balaenoptera borealis). Proc. Zool. Soc. London, p 243-265 + plates 25-26

Conroy J W H. 1975 Recent increases in penguin populations in the Antarctic and sub-Antarctic. In (ed B Stonehouse). The Biology of Penguins. Macmillan, London. pp555.

Conroy J W H & M G White. 1973 The breeding status of the King penguin. Br. Antarctic Surv. Bull. 32:31-40.
Cooke J G. 1984 The effects of depletion on the length distribution of baleen whale catches. Rep. int. Whal. Commn. 34:399-402.
Cooke J G & J R Beddington. 1981 Biasses in estimates of total mortality rates from age data and how to avoid them. IWC doc SC/33/05. pp9. (mimeo).
Cooke J G & W K de la Mare. 1984 A note on the estimation of time-trends in the age at sexual maturity in baleen whales from transition layer data, with reference to North Atlantic fin whale. Rep. int. Whal. Commn. 34:701-709.
Cooke J G & C J Rorvik. 1985 Estimate of mark shedding in the Northeast Atlantic stock. Rep. int. Whal. Commn. 35:98.
Croxall J P, Rootes D M & R A Price. 1981 Increases in penguin populations at Signey Island, South Orkney Islands. Br. Antarc. Surv. Bull. 54:47-56.
Cushing D H. 1982 Climate and Fisheries. Academic Press, London, pp373.
Cushing J E. 1964 The blood groups of marine animals. Advances in Marine Biology (2):85-132.
Cuvier G. 1823 Rechershes sur les Ossements Fossiles. Nouv. ed V. Paris. pp 405.
Dagerbol M. 1940 Mammalia. Zoology of the Faroes, Copenhagen.
Dailey M D. 1985 Diseases of mammalia: Cetacea. p805-847. In (ed O Kinne) Diseases of Marine Animals, IV(2). Biologische Anstalt Helgoland, Hamburg, pp 884.
Dawbin W H. 1959 Evidence on growth rates obtained from two marked humpback whales. Nature, Lond. 183:1749-1750.
Dawbin W H. 1966 The seasonal migratory cycle of humpback whales. p 145-170. In (ed K S Norris) Whales, Dolphins and Porpoises. pp 789. Univ. of California Press, Los Angeles.
DeMaster D P. 1978 Calculation of the average age of sexual maturity in marine mammals. J. Fish. Res. Bd. Canada 35(6):912-915.
DeMaster D P. 1984 Review of techniques used to estimate the average age at attainment of sexual maturity in marine mammals. Rep. int. Whal. Commn. (Sp. Is. 6):175-179.
Doi T. 1974 Further development of whale sighting theory. p 359-368. In (ed W E Schevill) The Whale Problem. Harvard University Press. pp 419.
Doi T & S Ohsumi. 1968a Fourth memorandum on

results of Japanese stock assessment of whales in the North Pacific. Rep. int. Whal. Commn. 18:62-66.

Doi T & S Ohsumi. 1968b Memorandum of further study on population assessment of sei whales in the Antarctic. Rep. int. Whal. Commn. 18:67-72.

Doi T & S Ohsumi. 1969a The present status of sei whale population in the Antarctic. Rep. int. Whal. Commn. 19:118-120.

Doi T & S Ohsumi. 1969b Fifth memorandum of results of Japanese stock assessment of whales in the North Pacific. Rep. int. Whal. Commn. 19:123-129

Doi T & S Ohsumi. 1970 Sixth memorandum on results of Japanese stock assessment of whales in the North Pacific. Rep. int. Whal. Commn. 20:97-111.

Doi T, Ohsumi S & T Nemoto. 1967 Population assessment of sei whales in the Antarctic. Norsk Hvalfangst Tid. 56(2):26-29.

Doi T, Nemoto T & S Ohsumi. 1967a Third memorandum on results of Japanese stock assessment of whales in the North Pacific. Rep. int. Whal. Commn. 17:89-92.

Doi T, Nemoto T & S Ohsumi. 1967b Memoradum on results of Japanese stock assessment of whales in the North Pacific. Rep. int. Whal. Commn. 17:111-115.

Doi T, Ohsumi S & Y Shimadzu. 1971 Status of stocks of baleen whales in the Antarctic 1970/71. Rep. int. Whal. Commn. 21:90-99.

Donovan G P. 1983 Identification of Bryde's whales if ridges are not seen. IWC doc SC/35/Ba 9. Appendix 2. pp1.

Donovan G P. 1984 Blue whales off Peru, December 1982, with special reference to pygmy blue whales. Rep. int. Whal. Commn. 34:473-476.

Dorsey E M. 1981 Exclusive home ranges in individually identified minke whales, B. acutorostrata. IWC SC/33/Mi4 (mimeo)

Duguy R. 1975 Rapport annuel sur les cetaces et pinnipedes trouves sur les cotes de France.III annee 1974. Mammalia 39:132-142.

Duguy R. 1978 Rapport annuel sur les cetaces et pinnipedes trouves sur les cotes de France.VII annee 1977. Ann. Soc. Sc. Nat. Charente-Marit. 6:308-317.

Duguy R & D Robineau. 1973 Ceteces et phoques des cotes de France. Ann. Soc. Sc. Nat. Charente-Marit., Suppl. pp 93.

Duguy R & H Aloncle. 1974 Note preliminaire a l'etude des cetaces du nord-est Atlantique. Reun. Cons. Perm. Int. Explor. Mer. 62:1-9.

Eberhardt L L. 1985 Assessing the dynamics of wild populations. J. Wildl. Manage. 49:997-1012.
Erdman D S. 1970 Marine mammals from Puerto Rico to Antigua. J. Mammal. 51(3):636-639.
Erdman D S, Harmes J & M M Flores. 1973 Cetacean records from the northeastern Caribbean region. Cetology. 17:1-14.
Evans P G H. 1980 Cetaceans in the British waters. Mammal Review 10(1):1-52.
Evans W E & A V Yablokov. 1978 Intraspecific variation of the colour pattern of the killer whale (Orcinus orca). In (ed V E Sokolov & A V Yablokov), pp 102-115, Advances in Studies of Cetaceans and Pinnipedes. Novka, Moscow.
Evans W E, Yablokov A V & A E Bowles. 1982 Geographic variation in the colour pattern of killer whales (Orcinus orca). Rep. int. Whal. Commn. 32:687-694.
FAO 1967a Report on the effects on whale stocks of pelagic operations in the Antarctic during the 1965/66 season and on the present status of those stocks. Rep. int. Whal. Commn. 17:47-69.
FAO 1967b Report on the effects on whale stocks of pelagic operations in the Antarctic during the 1966/67 season and on the present status of those stocks. FAO Fisheries Circular 113:1-19.
FAO 1969 Report on the effects on the baleen whale stocks of pelagic operations in the Antarctic during the 1967/68 season, and on the present status of those stocks. Rep. int. Whal. Commn. 19:29-38.
FAO 1970 Report on the effects on the baleen whale stocks of pelagic operations in the Antarctic during the 1968/69 season, and on the present status of those stocks. Rep. int. Whal. Commn. 20:21-32.
FAO 1965 Report on the effects on the whale stocks of pelagic operations in the Antarctic during the 1964/65 season, and on the present status of those stocks. Norsk. Hvalfangst Tid. 54(5):101-110.
Fairley J. 1981 Irish Whales and Whaling. Blackstaff Press Belfast pp218.
Fevolden S E. 1980 A note on Norwegian observations of whales in the Antarctic. Rep. int. Whal. Commn. 30:385-387.
Filella S. 1974 Esquema comparativo para la identificacion de las 4 especies de balaenopteridos citadas en las costas de la Peninsula Iberica. Misc. Zool.,3(4):171-176.
Fischer P. 1881 Ceteces du Sud-Ouest de la France. Act. Soc. Linn. Bordeaux 35:5-220.

Fletcher J O. 1969 Ice Extent on the Southern Ocean and its Relations to World Climate. RM-5993-NSF Rand Co. USA pp106.
Flower W H. 1883 On a specimen of Rudolphi's rorqual (Balaenoptera borealis, Lesson), lately taken off the Essex coast. Proc. Zool. Soc. London, p 513-517.
Ford J K B & H D Fisher. 1982 Killer whale (Orcinus orca) dialects as an indicator of stocks in British Columbia. Rep. int. Whal. Commn. 32:671-679.
Fowler C W. 1980 Comparitive population dynamics of large mammals. U.S. Dept. Commerce NTIS pp 330.
Foxton P. 1956 The distribution of the standing crop of zoo-plankton in the Southern Ocean. Discovery Rep. 28:191-236.
Fraser F C. 1974 Report on Cetacea stranded on the British coasts from 1948 to 1966. Br. Mus. (Nat. Hist.). Lond. 14.
Frazer J F D & A St G Huggett. 1974 Species variation in the foetal growth rates of eutherian mammals. J. Zool., Lond. 174:481-509.
Fujino K. 1953 On the seriological constitution of the sei, fin, blue and humpback whales. Sci. Rep. Whales Res. Inst. 8:103-126.
Fujino K. 1954 On the seriological constitution of the sperm and Bairds beaked whales. Sci. Rep. Whales Res. Inst. 9:105 -120.
Fujino K. 1955 On the body weight of the sei whale located in the adjacent waters of Japan. Sci. Rep. Whales Res. Inst. 10:133-142.
Fujino K. 1956 On the seriological constitution of the fin whale. Sci. Rep. Whales Res. Inst. 11:85-98.
Fujino K. 1958 On the seriological constitution of the fin whale. Sci. Rep. Whales Res. Inst. 13:171-184.
Fujino K. 1960 Immunogenetic and marking approaches to identifying subpopulations of the North Pacific whales. Sci. Rep. Whales Res. Inst. 15:85-142.
Fujino K. 1962 Blood types of some species of the Antarctic whales. Amer. Nat. 96(889):205-210.
Fujino K. 1963a Population genetics of whales with reference to blood types. Bul. Jap. Soc. Sci. Fisheries 29(12):1133-38, 1149-50.
Fujino K. 1963b Intra-uterine selection due to maternal-fetal incompatability of blood types in the whales. Sci. Rep. Whales Res. Inst. 17:53-65.
Fujino K. 1964a Fin whale subpopulations in the Antarctic whaling areas II,III and IV. Sci. Rep. Whales Res. Inst. 18 :1-28.
Fujino K. 1964b The report of biological study of

whales in the North Pacific in 1963 season. Whales Res. Inst. Tokyo. p 1-95. (In Japanese, Masaki 1976a).

Fujino K. 1964c Survey reports on whale stocks in the North Pacific, 1963. The identification of subpopulations. Whales Res. Inst. Tokyo, p 79-83. (in Japanese).

Fukuda Y. 1978 Report of the special meeting on Southern Hemisphere sei whales. Rep. int. Whal. Commn. 28: 335-343.

Gallardo V A & L Pasterne. 1983 A short note on the catch of "sei" whales in the eastern South Pacific,1946-1979. IWC doc SC/35/Ba3, ppl. mimeo.

Gallardo V A, Arcos D, Salamanca M & L Pastene. 1983 On the occurrence of Bryde's whales (Balaenoptera edeni Anderson 1878) in an upwelling area off central Chile. Rep. int. Whal. Commn. 33:481-488.

Gambell R. 1966 The dorsal fin of sei and fin whales. Norsk. Hvalfangst Tid. 9:177-180.

Gambell R. 1965 The sei whale stock exploited at South Georgia. IWC/17/ mimeo doc. pp9.

Gambell R. 1968 Seasonal cycles and reproduction in sei whales of the southern hemisphere. Discovery Rep. 35:31-134.

Gambell R. 1973 Some effects of exploitation on reproduction in whales. J. Reprod. Fert., Suppl. 19:531-551.

Gambell R. 1974 The fin and sei whale stocks off Durban. p82-86. In (ed W E Scheville) The Whale Problem. Harvard University Press, Mass. pp420.

Gambell R. 1975 Variations in reproductive parameters associated with whale stock sizes. Rep. int. Whal. Commn. 25:182-189.

Gambell R. 1976 Population biology and the management of whales. Appl. Biol. 1:247-343.

Gambell R. 1977 A revision of population assessments of Antarctic sei whales. Rep. int. Whal. Commn. (Sp. Is. 1): 44-52.

Gambell R. 1985 Sei whale. Balaenoptera borealis Lesson 1828. p155-170. In (ed S Ridgway & R Harrison) Handbook of Marine Mammals. Vol. 3. Sirenians and Baleen Whales. Academic Press Lond. pp362.

Gambell R, Best P B & D W Rice. 1975 Report on the International Indian Ocean whale marking cruise 24 November 1973 -3 February 1974. Rep. int. Whal. Commn. 25:240-252.

Garrod D J. 1980 Sei whale: analysis of marking data. Rep. int. Whal. Commn. 30:504-505.

Garrod D J & S G Brown. 1980 Southern hemisphere sei whales: a population estimate from marking

data. Rep. int. Whal. Commn. 30:507-511.
Gaskin D E. 1968 The New Zealand Cetacea. Fish. Res. Bull. N.Z.(ns) 1:1-92.
Gaskin D E. 1976 The evolution, zoogeography and ecology of Cetacea. Oceanogr. Mar. Biol. Ann. Review 14:247-346.
Gaskin D E. 1977 Sei and Bryde's whales in waters around New Zealand. Rep. int. Whal. Commn. (Sp. Is. 1):50-52.
Gaskin D E. 1982 The Ecology of Whales and Dolphins. Heinemann, Lond. pp 459.
Gill C D & S E Hughes. 1971 A sei whale, Balaenoptera borealis, feeding on Pacific saury, Cololabis saira. Cal. Fish. and Game 57:218-219.
Gillooly J F & N D Walker. 1984 Spatial and temporal behaviour of sea-surface temperatures in the South Atlantic. S. Afr. J. Sci. 80(2):97-100.
Giovetto M B & W Schwerdtfeger. 1966 An analysis of a 200 year snow accumulation series from the South Pole. Archiv. Met. Geophys. Bioklim. 15(2):227-250
Golubovsky Yu P, Yuktiov V L, Shevchenko V I, Flerov A I & V I Neizhko. 1972 On sei whales from the subtropical waters of the Southern Ocean. (Atlantic Ocean). Report of the 5 All-Union meeting on the study of marine mammals,p 151-153.
Gordon A L & E J Molinelli. 1982 Southern Ocean Atlas. Thermohaline and chemical distributions and the atlas data set. Columbia University Press, New York.
Gray J E. 1866 Catalogue of seals and whales in the British Museum. Br. Mus. (Nat. Hist.) pp 402 (2nd edition).
Gray J E. 1871 Supplement to the catalogue of seals and whales in the British Museum. Br. Mus. (Nat. Hist.) pp 103.
Gray R W. 1932 Peterhead sealers and whalers, a contribution to the history of the whaling industry. The Scottish Naturalist. p 197-200.
Guldberg G A. 1884 Sur l'existence d'une quartieme espece du genre Balaenoptera dans les mers septentrionales de l' Europe. Bull. Acad. Roy. de Belgique, 3-Series 7:360-374.
Gulland J A. 1974 Distribution and abundance of whales in relation to basic productivity. pp27-52. In (ed W E Schevill) The Whale Problem. Harvard Univ. Press, Mass. pp420.
Gulland J A. 1983 Fish Stock Assessment. John Wiley and Sons. Chichester. pp223.
Gunaratna R, Obeysekera N & R Hahn. 1985 Sightings of cetaceans off the Sri Lankan southwest coast, May 1985. IWC doc SC/37/O7, pp8. (mimeo).

Gunter G & J Christmas. 1973 Stranding records of a finback whale, Balaenoptera physalus, from Mississippi and the goose beaked whale, Ziphius cavirostris, from Louisiana. Gulf Res. Rep. 4(2):169-173.

Gunter G & R Overstreet. 1974 Cetacean notes. 1. Sei and rorqual whales on the Mississippi coast, a correction. 2. A dwarf sperm whale in the Mississippi Sound and its helminth parasites. Gulf Res. Rep. 4(3):479-481.

Haldane R C. 1904 Whaling in Shetland. Ann. Scott. Nat. Hist. p 74-77.

Haldane R C. 1905 Notes on Whaling in Shetland, 1904. Ann. Scott. Nat. Hist. 54:65-72.

Haldane R C. 1907 Whaling in Scotland. Ann. Scott. Nat. Hist. p 10-15.

Haldane R C. 1908 Whaling in Scotland for 1907. Ann. Scott. Nat. Hist. 66:65-72.

Haldane R C. 1909 Whaling in Scotland for 1908. Ann. Scott. Nat. Hist. 70:65-69.

Haldane R C. 1910 Whaling in Scotland for 1909. Ann. Scott. Nat. Hist. 73:1-2.

Handcock D. 1965 Killer whales kill and eat a minke whale. J. Mammal. 46:341-342.

Harding E F. 1984 Population estimates by mark-recapture for minke whales. Rep. int. Whal. Commn. 34:97-98.

Harmer S F. 1928 The History of Whaling. Proc. Linn. Soc. Lond. p 51-95.

Harmer S F. 1931 Southern Whaling. Proc. Linnean Soc. Lond. Session 142 1929-30:85-163.

Harrison R J. 1979 Whales and Whaling. p 391-431 In Whales by E J Slijper, Cornell Univ. Press, New York. pp 511.

Harrison R J & B J Weir. 1977 Structure of the mammalian ovary. p 113-217. In (ed. S Zuckerman and B J Weir) The Ovary, Vol.1, Academic Press, Lond. pp 517.

Hart J. 1935 On the diatoms of the skin film of whales and their possible bearing on problems of whale movements. Discovery Rep. 10:247-282.

Hay K. 1982 Aerial line-transect estimates of abundance of humpback,fin and long finned pilot whales in the Newfoundland Labrador area. Rep. int. Whal. Commn. 32:475-486.

Heincke F. 1913 Investigations of the plaice. General report 1. The plaice fishery and protective measures, part 1. Rapp. P.-v. Reun. Cons. Perm. int. Explor. Mer 17A:1-153.

Henderson J R & L J Hansen. 1983 Stranded marine mammals recovered by the Southwest Fisheries

Center, 1966-80. Southwest Fisheries Center Admin. Rep. LJ-83-07. pp31.
Hershkovitz P. 1966 Catalogue of living whales. Smithsonian Inst., U.S. Nat. Mus., Bull.246. Washington.
Hiby A R. 1985 An approach to estimating population densities of great whales from sighting surveys. IMA J. Math. Appl. Med. Biol. 2:201-220.
Hiby A R, Martin A R & F Fairfield. 1984 IDCR cruise aerial survey in the North Atlantic 1982: aerial survey report. Rep. int. Whal. Commn. 34:633-644.
Hinton M A C. 1925 Report on the papers left by the late Major Barret-Hamilton, relating to the whales of South Georgia. Crown Agents for the Colonies, London, p 57-209. (cited in Matthews, 1938).
Hjort J. 1933 Whales and whaling. Hvalr. Skr. 7:7-29
Hjort J & J T Rudd. 1929 Whaling and fishing in the North Atlantic. Rapp. P-verb. Cons. Int. Explor. Mer 56(1):1-123.
Hjort J, Lie J & J T Ruud. 1932 Norwegian pelagic whaling in the Antarctic I. Whaling grounds in 1929-1930 and 1930-1931. Hvaldrad. Skr. 3:1-37.
Holt S J. 1980 Procedure to eliminate some biases from estimates of pregnancy rates derived from BIWS data. Rep. int. Whal. Commn. 30:575-580.
Holt S J. 1981 Maximum sustainable yield and its application to whaling. Mammals in the Sea. FAO Fish. Ser. 5, 3:21-56.
Horwood J W. 1978 Sei whale catch statistics and estimated replacement rates. Rep. int. Whal. Commn. 28:391-399.
Horwood J W. 1980a Population biology and stock assessment of southern hemisphere sei whales. Rep. int. Whal. Commn. 30:519-530.
Horwood J W. 1980b Competition in the Antarctic? Rep. int. Whal. Commn. 30:513-517.
Horwood J W. 1981a Management and models of marine multi- species complexes. p339-360. In (ed C W Fowler & T D Smith) Dynamics of Large Mammal Populations. John Wiley NY. pp477.
Horwood J W. 1981b Results from the IWC/IDCR minke whale marking and sighting cruise 1979/80. Rep. int. Whal. Commn. 31:287-314.
Horwood J W. 1986 The lengths and distribution of Antarctic sei whales (Balaenoptera borealis Lesson). Sci. Rep. Whales Res. Inst. 37:47-60.
Horwood J W. 1987. Bias and variance in Allen's recruitment rate method. Fish. Bull. US. 85(1): 117-125.
Horwood J W, Donovan G P & R Gambell. 1980 Pregnancy

rates of the southern hemisphere sei whale (Balaenoptera borealis). Rep. int. Whal. Commn. 30:531-535.

Horwood J W, Best P B & S Ohsumi. 1981 International Whaling Commission, International Decade of Cetacean Research, Southern Hemisphere minke whale assessment cruise 1979-1980. Polar Record 20(129):565-569.

Hosokawa H. 1950 On the cetacean larynx, with special remarks on the laryngeal sac of the sei whale and the aryteno epiglottis of the sperm whale. Sci. Rep. Whales Res. Inst. 3:23-62.

Hosokawa H. 1951 On the pelvic cartilages of the Balaenoptera foetuses with remarks on the specificial and sexual differences. Sci. Rep. Whales Res. Inst. 5:5-15.

Hosokawa H & T Kamiya. 1971 Some observations on the cetacean stomachs, with special considerations on the feeding habits of whales. Sci. Rep. Whales Res. Inst. 23:91-101.

Huggett A St G & W F Widas. 1951 Foetal growth. J. Physiol. 114:306-317.

IUCN. 1979 Pirate whaling. IUCN Bulletin, June 1979:45-48.

Ichihara T. 1959 Formation mechanisms of ear plug in baleen whales in relation to glove finger. Sci. Rep. Whale Res. Inst. 14:107-135.

Ichihara T. 1962 Prenatal dead foeteus of baleen whales. Sci. Rep. Whales Res. Inst. 16:47-60.

Ichihara T. 1964 Prenatal development in ear plug of baleen whale. Sci. Rep. Whales Res. Inst. 18:29-48.

Ichihara T. 1966a The pygmy blue whale, Balaenoptera musculus brevicandata, a new subspecies from the Antarctic. p79-113. In (ed K Norris) Whales, Dolphins and Porpoises. Univ. of California Press, Los Angeles. pp 78.

Ichihara T. 1966b Criterion for determining age of fin whale with reference to ear plug and baleen plate. Sci. Rep. Whales Res. Inst. 20:17-81.

Ingebrigtsen A. 1929 Whales caught in the North Atlantic and other seas. Rapp. Cons. Explor. Mer 56:1-26.

Ivashin M V. 1958 Method of identification of the traces of yellow bodies of pregnancy and of ovulation in the humpback whale. Tr. Vse. Nauchno-Issled. Inst. Morsk. Ryb. Khoz. i Okean (VNIRO) 33:161-172.

Ivashin M V. 1967 Kit-Puteshestvennik. Piroda 8:105-107.

Ivashin M V. 1971 Some results of the marking of

the whales conducted from board Soviet ships in the Southern Hemisphere. Zool. Zh. 50(7):1063-1078

Ivashin M V. 1973 Marking of whales in the Southern Hemisphere(Soviet materials). Rep. int. Whal. Commn. 23:174-191.

Ivashin M V. 1980 Marking of sei whales (Balaenoptera borealis) in the southern hemisphere. (The soviet marking programme). Rep. int. Whal. Commn. 30:549-556.

Ivashin M V. 1984 Characteristics of ovarian corpora in dolphins and whales as described by Soviet scientists. Rep. int. Whal. Commn. (Sp. Is. 6):433-444.

Ivashin M V. 1985 USSR progress report on cetacean research, June 1983 to May 1984. Rep. int. Whal. Commn. 35: 186-188.

Ivashin M V & A A Rovnin. 1967 Some results of Soviet marking in the waters of the North Pacific. Norsk Hvalfangst Tid. 56:123-135.

Ivashin M V & Yu P Golubovsky. 1978 On the cause of appearance of white scars on the body of whales. Rep. int. Whal. Commn. 28:199.

Ivashin M V & G M Veinger. 1979 Distribution of whales in the North Pacific on expedition data of 1965/66, 1969/70 and 1975. Rep. int. Whal. Commn. 29:341-342.

Ivashin M V & L M Votrogov. 1982 Occurrence of baleen and killer whales off Chukotka. Rep. int. Whal. Commn. 32:499-501.

Jamieson A. 1973 Genetic 'tags' for marine fish stocks. p91- 99 In (ed F R Harden Jones) Sea Fisheries Research. Elek Science, London.

Japan. 1984 Japan Progress Report on Cetacean Research. Rep. int. Whal. Commn. 34:203-209.

Japan. 1985 Japan Progress Report on Cetacean Research, June 1983 to April 1984. Rep. int. Whal. Commn. 35:168-172.

Jenkins J T. 1921 A History of the Whale Fisheries. Witherby. Lond. pp 336.

Johnson C O. 1915 Evidence to the interdepartmental Committee on Whaling and the Protection of Whales. Minutes of evidence etc. Misc. 298. Colonial Office pp 210.

Jones A G E. 1973 Voyages to South Georgia 1795-1820. Br. Antarctic Surv. Bull. 32:15-22.

Jones E. 1971 Isistius brasiliensis, a squaloid shark, the probable cause of crater wounds on fishes and small cetaceans. Fish. Bull. U.S. 69(4):791-798.

Jones R. 1976 The use of marking data in fish population analysis. FAO Fish. Tech. Pap. 153. pp42.

Jones R. 1978 Estimating growth rates in sei whales. Rep. int. Whal. Commn. 28:405-410.

Jong C de 1983 The hunt for the Greenland whale: a short history and statistical sources. Rep. int. Whal. Commn. (Sp. Is. 5):83-106.

Jonsgard A. 1966 The distribution of Balaenopteridae in the North Atlantic Ocean. p114-124 In (ed K S Norris) Whales, Dolphins and Porpoises. Univ. Calif. Press. LA. pp 789.

Jonsgard A. 1953 Fin whales with six foetuses. Norsk Hvalfangst Tid 12:685-686.

Jonsgard A. 1966b Biology of the North Atlantic fin whale. Hvalrad. Skr. 49:1-62.

Jonsgard A. 1974 On whale exploitation in the eastern part of the North Atlantic Ocean. p97-107. In (ed W E Schevill) The Whale Problem. Harvard Univ. Press. Camb. Mass. pp 420.

Jonsgard A. 1977 Tables showing the catch of small whales (including minke whales) caught by Norwegians in the period 1938-1975 and large whales caught in different North Atlantic waters in the period 1868-1975. Rep. int. Whal. Commn. 27:413-426.

Jonsgard A & K Darling. 1977 On the biology of the eastern North Atlantic sei whale, Balaenoptera borealis Lesson. Rep. int. Whal. Commn. (Sp. Is. 1):124-129.

Jonsson J. 1965 Whales and whaling in Icelandic waters. Norsk Hvalfangst Tid 54(11):245-253.

Jurasz C M & V P Jurasz. 1979 Feeding modes of the humpback whale, Megaptera novaeangliae, in Southwest Alaska. Sci. Rep. Whale Res. Inst. 31:69-83.

Kapel F O. 1979 Exploitation of large whales off West Greenland in the twentieth century. Rep. int. Whal. Comm. 29:197-214.

Kapel F O. 1984 Whale observations off West Greenland in June-September 1982. Rep. int. Whal. Commn. 34:621-628.

Kapel F O. 1985 On the occurance of sei whale (Balaenoptera borealis) in West Greenland waters. Rep. int. Whal. Commn. 35:349-352.

Kapel F O & F Larsen. 1982 Whale sightings from a small-type whaling vessel off west Greenland, June -August 1980. Rep. int. Whal. Commn. 32:521-530.

Kapel F O & F Larsen. 1983 Whale sightings off West Greenland in June-September 1981. Rep int. Whal. Commn. 33:657-666.

Kasahara H. 1950 Whaling in the adjacent waters to Japan and its resourses. Rep. Nippon Fish. Co. Ltd. 4. (in Japanese). pp103+figs.

Kasuya T. 1966 Karyotype of a sei whale. Sci. Rep.

Whales Res. Inst. 20:83-88.
Kasuya T & H Marsh. 1984 Life history and reproductive biology of the short-finned pilot whale, Globicephala macrorhynchus, off the Pacific coast of Japan. Rep. int. Whal. Commn. (Sp. Is. 6):259-310
Kato H. 1984 Readability of Antarctic minke whale earplugs. Rep. int. Whal. Commn. 34:719-722.
Kato H. 1985 Further examination of the age at sexual maturity of the antarctic minke whale as determined from earplug studies. Rep. int. Whal. Commn. 35:273-277.
Kato H & Y Shimadzu. 1983 The foetal sex ratio of the Antarctic minke whale. Rep. int. Whal. Commn. 33:357-359.
Katona S K & S Kraus. 1979 Photographic identification of individual humpback whales (Megaptera novaeangliae): evaluation and analysis of the technique. US Marine Mammal Commission, MMC-77/17. NTIS publ. PB-298 740.
Katona S K, Baxter B, Brazier O, Kraus S, Perkins J & H Whitehead. 1979 Identification of humpback whales by fluke photograps. p33-44 In (ed H E Winn & B L Olla) Behaviour of Marine Animals. Vol 3. Plenum Press. New York.
Kawamura A. 1970 Food of sei whales taken by Japanese whaling expeditions in the Antarctic season 1967/68. Sci. Rep. Whales Res. Inst. 22:127-152.
Kawamura A. 1973 Food and feeding of sei whales caught in the waters south of 40N in the North Pacific. Sci. Rep. Whales Res. Inst. 25:219-236.
Kawamura A. 1974 Food and feeding ecology in the southern sei whale. Sci. Rep. Whales Res. Inst. 26:25-144.
Kawamura A. 1980 A review of food of Balaenopterid whales. Sci. Rep. Whales Res. Inst. 32:155-198.
Keller R N, Leatherwood S & S J Holt. 1982 Indian Ocean cetacean surveys, Seychelle Islands, April through June 1980. Rep. int. Whal. Commn. 32:503-513.
Kellogg R. 1928 What is known of the migrations of some of the whalebone whales. Ann. Rep. Smithsonian Inst. 467-494.
Kellogg R. 1931 Whaling statistics for the Pacific coast of North America. J. Mammal. 12(1): 73-77.
Kendrew J C, Parrish R G, Marrack J R & E S Orlans. 1954 The species specificity of myoglobin. Nature, Lond. 174:946-949.
Kimura S. (Ohsumi). 1957 The twinning in southern fin whales. Sci. Rep. Whales Res. Inst. 12:103-

125.
Kirby VL & S H Ridgway. 1984 Hormonal evidence of spontaneous ovulation in captive dolphins, Tursiops truncatus and Delphinus delphis. Rep. int. Whal. Commn. (Sp. Is.d):459-464.
Kirkwood G P. 1978 Maximum likelihood estimation of population sizes using catch and effort data. IWC doc. SC/30/12.
Kirkwood G P. 1981 Estimation of stock size using relative abundance data - a simulation study. Rep. int. Whal. Commn. 31:729-735.
Kirkwood G P & K R Allen. 1978 Program to calculate time series of sei whale population components for given catches (SEI). Rep. int. Whal. Commn. 28: 227-231.
Kirkwood G P & K R Allen. 1979 Program to estimate baleen whale population sizes (BALEEN). Rep. int. Whal. Commn. 29:367-368.
Kirpichnikov A. 1950 The observations on the distribution of the large Cetacea in the Atlantic Ocean. Priroda, 10:63-64.
Klevezal' G A. 1980 Layers in the hard tissues of mammals as a record of growth rhythms of individuals. Rep. int. Whal. Commn. (Sp. Is. 3):89-94.
Klevezal' G A & E D Mitchell. 1971 Annual layers in baleen whale bone. (In Russian). Zool. Zhurnal 50(7):1114-1117.
Klumov S K. 1963 Food and helmintho fauna of the baleen whales in the world oceans. Trudy Inst. Okeanol. Acad. Sci. USSR. 71:94-194.
Kozicki V M & E D Mitchell. 1974 Permanent and selective chemical marking of Mysticete ear plug lamination with quinacrine. Rep. int. Whal. Commn. 24:142-149.
Lambertsen R H. 1983 Internal mechanisms of rorqual feeding. J. Mamm. 64(1):76-88.
Landa A, Ramirez P & H Tovar. 1983 Result of partial results of the 1982 IWC/IDCR cruise marking and population assessment of whales, Peruvian area, Nov-Dec 1982. IWC SC/35/Ba13.
Larsen F. 1981 Observations of large whales off west Greenland. Rep. int. Whal. Commn. 31:617-623
Larsen F. 1985 Preliminary results of an aerial survey off West Greenland,1984. IWC doc SC/37/O19 pp11. (mimeo).
Laws R M. 1959 Foetal growth rates of whales with special reference to the fin whale, Balaenoptera physalus Linn. Discovery Rep. 29:281-308.
Laws R M. 1960 Laminated structure of bones from some marine mammals. Nature, Lond. 187:338-339.
Laws R M. 1961 Reproduction, growth and age of

southern fin whales. Discovery Rep. 31:327-486.
Laws R M. 1962 Some effects of whaling on the southern stocks of baleen whales. p137-158 In (ed E D LeCren and M W Holgate) The Exploitation of Natural Animal Populations. Blackwell, Oxford.
Laws R M. 1977a The significance of vertebrates in the Antarctic marine ecosystem. p411-438 In (ed G A Llano) Adaptions within Antarctic Ecosystems. Proc. 3rd SCAR Symp. on Ant. Biol. Smithsonian Inst. Washington.
Laws R M. 1977b Seals and whales of the Southern Ocean. Phil. Trans. R. Soc. B. 279:81-96.
Layne J N. 1965 Observations on marine mammals in Florida waters. Bull. Florida State Mus. 9(4): 131-181.
Leatherwood S, Caldwell D K & H E Winn. 1976 Whales, dolphins and porpoises of the western North Atlantic. A guide to their identification. NOAA Tech. Rep. NMFS Circ.396. pp176.
Lesson R P. 1828 Histoire naturelle generale et particuliere des mammiferes et des oiseaux. Cetaces. Paris.
Lichter A & A Hooper. 1983 Guia para el reconocimiento de cetaceos del Mar Argentino. Fund. Vida Silvestre Argen. p96.
Lillie D G. 1910 Observations on the anatomy and general biology of some members of the larger cetacea. Proc. Zool. Soc. London, p 779-792.
Lillie D G. 1915 Cetacea. Nat. Hist. Rep. Br. Antarct. Terra Nova Exped. Zool. 1(3):85-124.
Limbert D W S. 1974 Variations in the mean annual temperature for the Antarctic peninsula 1904-1972. Polar Rec. 17(108): 303-306.
Liouville J. 1913 Cetaces de l'Antarctique. Deuxieme Expedition Antarctique Francais, p 100-110.
Lockyer C. 1972a The age at sexual maturity of the southern fin whale (Balaenoptera physalus) using annual layer counts in the ear plug. J. Cons. int. Explor. Mer, 34(2):276-294.
Lockyer C. 1972b A review of the weights of Cetaceans with estimates of the growth and energy budgets of the large whales. M. Phil. Thesis. Uni. of London. pp196.
Lockyer C. 1974 Investigation of the ear plug of the southern sei whale, Balaenoptera borealis, as a valid means of determining age. J. Cons. int. Explor. Mer, 36(1):71-81.
Lockyer C. 1976 Body weights of some species of large whales. J. Cons. int. Explor. Mer, 36(3):259-273.
Lockyer C. 1977a Some possible factors affecting

the age distribution of the catch of sei whales in the Antarctic. Rep. int. Whal. Commn. (Sp. Is. 1):63-70.

Lockyer C. 1977b Some estimates of the growth in the sei whale, Balaenoptera borealis. Rep. int. Whal. Commn. (Sp. Is. 1):58-62.

Lockyer C. 1977c Mortality rates for Southern Hemisphere sei whales. Rep. int. Whal. Commn. (Sp. Is. 1):53-57.

Lockyer C. 1977d A preliminary study of variations in age at sexual maturity of the fin whale with year class in six areas of the Southern Hemisphere. Rep. int. Whal. Commn. 29:141-147.

Lockyer C. 1978a A preliminary investigation of age,growth and reproduction of sei whales off Iceland. Rep. int. Whal. Commn. 28:237-241.

Lockyer C. 1978b A theoretical approach to the balance between growth and food consumption in fin and sei whales with special reference to the female reproductive cycle. Rep. int. Whal. Commn. 28:243-250.

Lockyer C. 1979 Changes in a growth parameter associated with exploitation of southern fin and sei whales. Rep. int. Whal. Commn. 29:191-196.

Lockyer C. 1981a Growth and energy budgets of large baleen whales from the Southern Ocean. p379-487. In Mammals in the Sea. Vol.3, FAO, Rome. pp 504.

Lockyer C. 1981b The age at sexual maturity of fin whales off Iceland. Rep. int. Whal. Commn. 31:389-393.

Lockyer C. 1984a Age determination by means of the ear plug in baleen whales. Rep. int. Whal. Commn. 34:692-696.

Lockyer C. 1984b Review of baleen whale (Mysticeti) reproduction and implications for management. Rep. int. Whal. Commn. (Sp. Is. 6):27-50.

Lockyer C. 1986 Body fat condition in northeast Atlantic fin whales Balaenoptera physalus and its relationship with reproduction and food resource. Can. J. Fish. Aq. Sci. 43(1):142-147.

Lockyer C & S G Brown. 1979 A review of recent biological data for the fin whale population off Iceland. Rep. int. Whal. Commn. 29:185-189.

Lockyer C & S G Brown. 1981 The migration of whales. p105- 137 In (ed D J Aidley) Animal Migration. Soc. for Expt. Biol. Seminar Series 13. Cambridge University Press. pp264.

Lockyer C & A R Martin. 1983 The sei whale off western Iceland. 2. Age,growth and reproduction. Rep. int. Whal. Commn. 33:465-476.

Lockyer C & D Butterworth. 1984 A note on age at

maturity time-trends in Icelandic fin whales. Rep. int. Whal. Commn. 34:121.
Lockyer C & C G Smellie. 1985 Assessment of reproductive status of female fin and sei whales taken off Iceland, from an histological examination of the uterine mucosa. Rep. int. Whal. Commn. 35:343-348.
Lockyer C, Gambell R & S G Brown. 1977 Notes on age data of fin whales taken off Iceland, 1967-74. Rep. int. Whal. Commn. 27:427-450.
Lockyer C, McConnell L C & T D Waters. 1984 The biochemical composition of fin whale blubber. Can. J Zool. 62:2553-2562.
Lockyer C, McConnell L C & T D Waters. 1985 Body condition in terms of anatomical and biochemical assessment of body fat in North Atlantic fin and sei whales. Can. J. Zool. 63:2328-2338.
Lockyer C & T D Waters. 1986 Weights and anatomical measurments of northeastern Atlantic fin and sei whales. Marine Mammal Sci. 2(3):169-185.
Lowery G H. 1974 The Mammals of Louisiana and its Adjacent Waters. Louisiana State University Press. pp565.
Lund J. 1938 Whale tribes in the Antarctic. Iodine value determination. Norsk Hvalfangst Tid 6:257-261.
Lund J. 1951 Charting of whale stocks in the Antarctic 1950/51 on the basis of iodine values. Norsk Hvalfangst Tid 40(8):384-6.
Lydekker R. 1895 Mammals. J.F.Shaw & Co., London. pp 340.
Machida S. 1970 A sword-fish sword found from a North Pacific sei whale. Sci. Rep. Whales Res. Inst.,22:163-164.
Machida S. 1974 Surface temperature field in the Crozet and Kerguelen whaling grounds. Sci. Rep. Whales Res. Inst.,26 :271-287.
Mackintosh N A. 1942 The southern stocks of whalebone whales. Discovery Rep. 31:327-486.
Mackintosh N A. 1946 The natural history of whalebone whales. Biol. Review 21:60-74.
Mackintosh N A. 1965 The Stocks of Whales. Fishing News (Books) Ltd. Lond. pp232.
Mackintosh N A. 1966 The distribution of southern blue and fin whales. p125-144 In (ed K S Norris) Whales, Dolphins and Porpoises. Univ. California Press, Los Angeles. pp789.
Mackintosh N A. 1972 Life cycle of Antarctic krill in relation to ice and water circulation. Discovery Rep. 36: 1-94.
Mackintosh N A. 1973 Distribution of post-larval

krill in the Antarctic. Discovery Rep. 36:95-156.
Mackintosh N A. 1974 Sizes of krill eaten by whales in the Antarctic. Discov. Rep. 36:157-178.
Mackintosh N A & J F G Wheeler. 1929 Southern blue and fin whales. Discovery Rep. 1:257-540.
Mackintosh N A & S G Brown. 1956 Preliminary estimates of the southern populations of the larger baleen whales. Norsk. Hvalfangst Tid. 9:469-480.
Major P F. 1979 An aggressive encounter between a pod of whales and billfish. Sci. Rep. Whales Res. Inst. 31:95-96.
Makino S. 1948 The chromosome of Dall's porpoise Phocoenoides dallii (True), with remarks on the phylogenetic relation of the Cetacea. Chromosoma, 3:220-231.
Mare W K de la. 1985a On the estimation of mortality rates from whale age data with particular reference to minke whales (Balaenoptera acutorostrata) in the Southern Hemisphere. Rep. int. Whal. Commn. 35:239-250.
Mare W K de la. 1985b Some evidence for mark shedding with Discovery whale marks. Rep. int. Whal. Commn. 35:477-486.
Martin A R. 1982 A link between the sperm whales occurring off Iceland and the Azores. Mammalia 46(2):259-260.
Martin A R. 1983 The sei whale off western Iceland. 1. Size, distribution & abundance. Rep. int. Whal. Commn. 33:457-463
Martin A R, Hembree D, Waters T D & J Sigurjonsson. 1984 IDCR cruise/aerial survey in the North Eastern Atlantic 1982: cruise report. Rep. int. Whal. Commn. 34:645-653.
Masaki Y. 1968 A trial for reading of ear plug lamination of the sei whale by means of a soft X-ray. IWC doc. to Age Determination Meeting, Oslo. (mimeo).
Masaki Y. 1976a Biological studies on the North Pacific sei whale. Bull. Far Seas Fish. Res. Lab. 14:1-104.
Masaki Y. 1976b Japanese pelagic whaling and sighting in the Antarctic. Rep. int. Whal. Commn. 26:333-349.
Masaki Y. 1977a The separation of stock units of sei whales in the North Pacific. Rep. int. Whal. Commn. (Sp. Is. 1): 71-79.
Masaki Y. 1977b Japanese pelagic whaling and whale sighting in the Antarctic 1975/76. Rep. int. Whal. Commn. 27:148-155
Masaki Y. 1978 Yearly changes in the biological parameters of the Antarctic sei whale. Rep. int.

Whal. Commn. 28:421-429.
Masaki Y. 1979a Japanese pelagic whaling and whale sighting in the 1977/78 Antarctic season. Rep. int. Whal. Commn. 29:225-251.
Masaki Y. 1979b Yearly changes in the biological parameters for the Antarctic minke whale. Rep. int. Whal. Commn. 28:421-430.
Masaki Y. 1980 Additional comments on the result of whale sighting by scouting boats in the Antarctic whaling season from 1965/66-1977/78. Rep. int. Whal. Commn. 30:339-357.
Masaki Y & Y Fukuda. 1975 Japanese pelagic whaling and sighting in the Antarctic 1973/74. Rep. int. Whal. Commn. 25:106-128.
Masaki Y & K Yamamura. 1978 Japanese pelagic whaling and whale sighting in the 1976/77 Antarctic season. Rep. int. Whal. Commn. 28:251-262.
Masters P B, Bada J L & J S Zigler. 1977 Aspartic acid racemization in the human lens during ageing and cataract formation. Nature, Lond. 268:71.
Matsuura Y. 1940 On the multiple pregnancies and multiple ovulations of the baleen whales. Jap. J. Zool. 52(11):407-414. (in Japanese).
Matsuura Y & K Maeda. 1942 Biological investigations of the Northern Pacific whales. Reports for Whaling, 9(1). Japan Whal. Fish. Assoc., Tokyo.
Matthews L H. 1932 Lobster krill. Discovery Rep. 5:467-484.
Matthews L H. 1938 The sei whale, Balaenoptera borealis. Discovery Rep. 17:183-290 + 2 plates.
Matthews L H. 1948 Cyclic changes in the uterine mucosa of balaenopterid whales. J. Anat. 82(4): 207-232.
Maturana R. 1981 Study of the living pattern of the sei whale. IWC Doc. SC/33/Ba10.
Maturana R. 1981 Chile progress report on cetacean research 1966-May 1980. Rep. int. Whal. Commn. 31:181-183.
Mead J G. 1977 Records of sei and Bryde's whales from the Atlantic coast of the United States, the Gulf of Mexico, and the Caribbean. Rep. int. Whal. Commn. (Sp. Is. 1): 113-116.
Mikhaliev Yu A. 1978 Occurrence and distribution of cetaceans in the Pacific sector of the Antarctic regions according to the results of observations, 1973/74 and 1974/75 seasons. Rep. int. Whal. Commn. 28:263-268.
Mikhaliev Yu A. 1979 Revealing the differences of baleen whale stocks on the basis of the 'Jacobson's organ' position. Rep. int. Whal. Commn. 29:343-346.

Mikhaliev Yu A. 1980 General regularities in pre-natal growth in whales and some aspects of their reproductive biology. Rep. int. Whal. Commn 30:249-254.
Mikheev B I. 1965 On the biology and fishing of some fishes from the Patagonian shelf (Falkland area) and the Scotia sea. In The Antarctic Krill. Trudy AtlantNIRO, p 84-89.
Millais J G. 1906 The Mammals of Great Britain and Ireland. Vol 3. London.
Miller G S. 1924 A pollack whale from Florida presented to the National Museum by the Miami Aquarium Association. Proc. U.S. Natn. Mus. 66(9):1-15.
Miller G S. 1927 A pollack whale on the east coast of Virginia. Proc. Biol. Soc. Wash. 40:111-112.
Miller G S. 1928 The pollack whale in the Gulf of Compeche. Proc. Biol. Soc. Washington, 41:171.
Miscellaneous 298. 1915 Interdepartmental Committee on Whaling and the Protection of Whales, Minutes of evidence etc. Colonial office, October 1915: pp 210.
Mitchell E D. 1974a Present status of northwest Atlantic fin and other whale stocks. p108-169 In (ed W E Schevill) The Whale Problem. Harvard University Press, Mass.pp420.
Mitchell E D. 1974b Trophic relationships and competition for food in Northwest Atlantic whales. Proc. Can. Soc. Zool. Ann. Meeting :123-133.
Mitchell E D. 1974c Canadian progress report on whale research 1972 to 1973. Rep. int. Whal. Commn. 24: 196-213.
Mitchell E D. 1975 Preliminary report of Nova Scotia fishery for sei whales (Balaenoptera borealis). Rep. int. Whal. Commn. 25:218-225.
Mitchell E D. 1976 Scientific Committee IDCR research proposal for the North Atlantic. (Convenor). Rep. int. Whal. Commn. 26:142-179.
Mitchell E D. 1983 Potential of whaling logbook data for studying aspects of social structure in the sperm whale, Physeter macrocephalus, with an example - the ship MARINER to the Pacific, 1836-1840. Rep. int. Whal. Commn. (Sp. Is. 5):63-80.
Mitchell E D & V M Kozicki. 1974 The sei whale (Balaenoptera borealis) in the northwest Atlantic Ocean. (m.s.) IWC doc SC/SP74/32. pp48 + 48 figs.
Mitchell E D & D G Chapman. 1977 Preliminary assessment of stocks of Northwest Atlantic sei whales (Balaenoptera borealis). Rep. int. Whal. Commn. (Sp. Is. 1):117-120.
Mitchell E D & R R Reeves. 1983 Catch history,

abundance and present status of Northwest Atlantic humpback whales. Rep. int. Whal. Commn. (Sp. Is. 5):153-209.

Miyashita T. 1985 Bryde's whales stocks in the western North Pacific estimated using sightings data. IWC doc SC/37/Bal, pp10. (mimeo).

Miyazaki I. 1952 Study on maturity and body length of the whales caught in the waters adjacent to Japan. Fisheries Agency of Japan, May 1952 pp27.

Miyazaki N & S Wada. 1978 Observations of Cetacea during whale marking cruise in the western tropical Pacific. Sci. Rep. Whales Res. Inst. 30: 179-195.

Mizroch S A. 1980 Some notes on Southern Hemisphere baleen whale pregnancy rate trends. Rep. int. Whal. Commn. 30:561-574.

Mizroch S A. 1985 On the relationship between mortality rate and length in baleen whales. Rep. int. Whal. Commn. 35:505-510.

Mizroch S A & J M Breiwick. 1984 Variability in age at length and length at age in Antarctic fin, sei and minke whales. Rep. int. Whal. Commn. 34:723-732.

Mizue K. 1950 Factory ship whaling around Bonin Island in 1948. Sci. Rep. Whales Res. Inst., 3: 106-118.

Mizue K. 1951 Food of Whales. Sci. Rep. Whales Res. Inst.3: 81-90.

Mizue K & H Jimbo. 1950. Statistical study of the foetuses of whales. Sci. Rep. Whales Res. Inst. 3:119-131.

Mizue K & T Murata. 1951 Biological investigations on the whales caught by the Japanese Antarctic whaling fleets, season 1949/50. Sci. Rep. Whales Res. Inst. 6:73-132.

Moerzer-Bruyns W F J 1971 Field Guide of Whales and Dolphins. Uitgeverij tor, Amsterdam. p 258.

Morch J A. 1911 On the natural history of whalebone whales. Proc. Zool. Soc. Lond. 661-670.

Nakai J & T Shida. 1948 Sinus hairs of the sei whale (Balaenoptera borealis). Sci. Rep. Whales Res. Inst. 1:41-47.

Nasu K. 1957 Oceanographic conditions of the whaling grounds adjacent to Aleutian Islands and Bering Sea in the summer of 1955. Sci. Rep. Whales Res. Inst. 12:91-102.

Nasu K. 1966 Fishery oceanographic study on the baleen whaling grounds. Sci. Rep. Whales Res. Inst. 20:157-210.

Nasu K & Y Masaki. 1970 Some biological parameters for stock assessment of the Antarctic sei whale.

Sci. Rep. Whales Res. Inst. 22:63-74.
Nasu K & Y Shimadzu. 1970 A method of estimating whale population by sighting observation. Rep. int. Whal. Commn. 20:114-129.
Negus N C & R Chipman. 1956 A record of the piked whale, Balaenoptera acutorostrata, off the Louisiana coast. Proc. La. Acad. Sci. 19:41-42.
Nemoto T. 1955 White scars on whales, I. Lamprey marks. Sci. Rep. Whales Res. Inst. 10:69-78.
Nemoto T. 1957 Foods of baleen whales in the Northern Pacific. Sci. Rep. Whales Res. Inst. 12:33-89
Nemoto T. 1959 Food of baleen whales with reference to whale movements. Sci. Rep. Whales Res. Inst. 14:149-290.
Nemoto T. 1962 Food of baleen whales collected in recent Japanese Antarctic whaling expeditions. Sci. Rep. Whales Res. Inst. 16:89-103.
Nemoto T. 1963 Some aspects of the distribution of Calanus cristatus and Calanus plumcrus in the Bering Sea and its related waters,etc. Sci. Rep. Whales Res. Inst. 17:157-170.
Nemoto T. 1964 School of baleen whales in the feeding areas. Sci. Rep. Whales Res. Inst. 18:89-110.
Nemoto T. 1970 Feeding pattern of baleen whales in the ocean. p241-252 In (ed J H Steele) Marine Food Chains. University of California Press, Berkeley and Los Angeles. pp552.
Nemoto T & K Nasu. 1958 Thysanoessa macrura as a food of baleen whales in the Antarctic. Sci. Rep. Whales Res. Inst. 13:193-199.
Nemoto T & A Kawamura. 1977 Characteristics and food habits and distribution of baleen whales with special reference to the abundance of North Pacific sei and Bryde's whales. Rep. int. Whal. Commn. (Sp. Is. 1):80-87.
Nerini M K. 1983 Age determination of fin whales (Balaenoptera physalus) based upon aspartic acid racemization in lens nucleus. Rep. int. Whal. Commn. 33:447-448.
Nishimoto S, Tozawa M & T Kawakami. 1952 Food of sei whales (Balaenoptera borealis) caught in the Bonin Island waters. Sci. Rep. Whales Res. Inst. 5:79-85.
Nishiwaki M. 1950a Determination of the age of Antarctic blue and fin whales by colour changes in crystalline lens. Sci. Rep. Whales Res. Inst. 4: 115-161.
Nishiwaki M. 1950b Age characteristics in baleen plates. Sci. Rep. Whales Res. Inst. 4:162-183.
Nishiwaki M. 1950c On the body weight of whales. Sci. Rep. Whales Res. Inst. 4:184-209.

Nishiwaki M. 1951 On the periodic mark on the baleen plates as the sign of annual growth. Sci. Rep. Whales Res. Inst. 6:133-152.
Nishiwaki M. 1952 On the age determination of Mystacoceti, chiefly blue and fin whales. Sci. Rep. Whales Res. Inst. 7 :87-119.
Nishiwaki M. 1957 Age characteristics of ear plugs of whales. Sci. Rep. Whales Res. Inst. 12:23-32.
Nishiwaki M. 1966 Distribution and migration of the larger cetaceans in the North Pacific as shown by Japanese whaling results. p171-191 In (ed K S Norris) Whales, Dolphins and Porpoises. Univ. of California Press, Los Angeles . pp 78.
Nishiwaki M. 1974 Systematic relationship between sei (Balaenoptera borealis) and Bryde's whales (Balaenoptera edeni). IWC doc. SC/SP 74/22 (mimeo).
Nishiwaki M & T Kasuya. 1971 Osteological note of an Antarctic sei whale. Sci. Rep. Whales Res. Inst. 23:83-90.
Nishiwaki M, Ichihara T & S Ohsumi. 1958 Age studies of fin whales based on ear plug. Sci. Rep. Whales Res. Inst. 13: 155-169.
Nishiwaki M, Hibiya T, Kimura S & S. Ohsumi. 1954 On the sexual maturity of the sei whale of the Bonin Waters. Sci. Rep. Whales Res. Inst. 9:165-77
Norman J R & F C Fraser. 1948 Giant Fishes, Whales and Dolphins. Putnam, London. pp 376.
Nowosielski-Slepowron B J A & A D Peacock. 1955 Chromosome number in the blue, fin and sperm whales. Proc. Roy. Soc. Edinb. B 65(III):358-368.
Ogawa T & S Arifuka. 1948 On the acoustic system in the cetacean brains. Sci. Rep. Whales Res. Inst. 2:1-20.
Ogawa T & T Shida. 1950 On the sensory tubucles of lips and of oral cavity in the sei and fin whale. Sci. Rep. Whales Res. Inst. 3:1-16.
Ohsumi S. 1962 Laminae Formation. Norsk. Hvalfangst Tid. 51(5):192-198.
Ohsumi S. 1964a Examination on age determination of the fin whale. Sci. Rep. Whales Res. Inst. 18:49-88.
Ohsumi S. 1964b Comparison of maturity and accumulation rate of corpora albicantia between left and right ovaries in Cetacea. Sci. Rep. Whales Res. Inst. 18:123-148.
Ohsumi S. 1969 Occurrence and rupture of vaginal band in the fin, sei and blue whales. Sci. Rep. Whales Res. Inst. 21: 85-94.
Ohsumi S. 1977 Bryde's whales in the pelagic whaling grounds of the North Pacific. Rep. int.

Whal. Commn. (Sp. Is. 1): 140-149.
Ohsumi S. 1978 Estimation of natural mortality rate, recruitment rate and age at recruitment of Southern Hemisphere sei whales. Rep. int. Whal. Commn. 28:437-448.
Ohsumi S. 1979a Provisional report of the Bryde's whales caught under special permit in the southern hemisphere in 1977/78 and a research programme for 1978/79. Rep. int. Whal. Commn. 29:267-273.
Ohsumi S. 1979b Interspecific relationships among some biological parameters in cetaceans and estimation of natural mortality coefficient of the Southern Hemisphere minke whale. Rep. int. Whal. Commn. 29:397-406.
Ohsumi S. 1980a Catches of sperm whales by modern whaling in the North Pacific. Rep int. Whal. Commn. (Sp. Is. 2):11-18.
Ohsumi S. 1980b Population study of the Bryde's whale in the southern hemisphere under scientific permit in the three seasons 1976/77-1978/79. Rep. int. Whal. Commn. 30:319-331.
Ohsumi S. 1986 Yearly changes in age and body length at sexual maturity of a fin whale stock in the eastern North Pacific. Sci. Rep. Whales Res. Inst. 37:17-30.
Ohsumi S & Y Masaki. 1972 The eighth memorandum on the results of Japanese stock assessments of whales in the North Pacific. Rep. int. Whal. Commn. 22:91-95.
Ohsumi S & Y Masaki. 1974 Status of whale stocks in the Antarctic,1972/73. Rep. int. Whal. Commn. 24: 102-113.
Ohsumi S & S Wada. 1974 Status of whale stocks in the North Pacific,1972. Rep. int. Whal. Commn. 24: 114-126.
Ohsumi S & Y Fukuda. 1975a On the estimates of exploitable population size and replacement yields for the Antarctic sei whale by the use of catch and effort data. Rep. int. Whal. Commn. 25:102-105
Ohsumi S & Y Fukuda. 1975b A review on population estimates for the North Pacific sei whales. Rep. int. Whal. Commn. 25:95-101.
Ohsumi S & Y Masaki. 1975 Japanese whale marking in the North Pacific,1963-1972. Bull. Far Seas Fish. Res. Lab. 12 :171-219.
Ohsumi S & Y Masaki. 1978 Age length keys and growth curves of the Southern Hemisphere sei whale. Rep. int. Whal. Commn. 28:431-436.
Ohsumi S & K Yamamura. 1978a A review on catch of sei whales in the southern hemisphere. Rep. int. Whal. Commn. 28:449-458.

Ohsumi S & K Yamamura. 1978b Catcher's hours work and its correction as a measure of fishing effort for sei whales in the Antarctic. Rep. int. Whal. Commn. 28:459-467.
Ohsumi S & K Yamamura. 1982 A review of the Japanese whale sightings system. Rep. int. Whal. Commn. 32:581-586.
Ohsumi S, Nishiwaki M & T Hibiya. 1958 Growth of fin whales in the North Pacific. Sci. Rep. Whales Res. Inst. 13:97-134.
Ohsumi S, Shimadzu Y & T Doi. 1971 The seventh memorandom on the results of Japanese stock assessments of whales in the North Pacific. Rep. int. Whal. Commn. 21:76-89.
Olsen O. 1913 On the external characters and biology of Bryde's whale (B. brydei), a new rorqual from the coast of South Africa. Proc. Zool. Soc. London, p 1073-1090.
Omura H. 1950a Whales in the adjacent waters of Japan. Sci. Rep. Whales Res. Inst. 4:27-114.
Omura H. 1950b On the body weight of sperm and sei whales located in the adjacent waters of Japan. Sci. Rep. Whales Res. Inst. 4:1-13.
Omura H. 1958 North Pacific right whale. Sci. Rep. Whales Res. Inst. 13:1-52.
Omura H. 1959 Bryde's whales from the coast of Japan. Sci. Rep. Whales Res. Inst. 14:1-33.
Omura H. 1962 Bryde's whale occurs on the coast of Brazil. Sci. Rep. Whales Res. Inst. 16:1-5.
Omura H. 1964 A systematic study of the hyoid bones in the baleen whales. Sci. Rep. Whales Res. Inst. 18:149-170.
Omura H. 1966 Bryde's whales in the northwest Pacific. p 70-78 In (ed K S Norris) Whales Dolphins and Porpoises. University of California Press, Los Angeles. pp 789.
Omura H. 1971 A comparison of the size of vertebrae among some species of the baleen whales with special reference to whale movements. Sci. Rep. Whales Res. Inst. 23:61-70.
Omura H. 1973 A review of pelagic whaling operations in the Antarctic based on the effort and catch data in 10 degree squares of latitude and longitude. Sci. Rep. Whales Res. Inst. 25:105-204.
Omura H. 1978 Preliminary report on morphological study of pelvic bones of the minke whale from the Antarctic. Sci. Rep. Whales Res. Inst. 30:271-280
Omura H & K Fujino. 1954 Sei whales in the adjacent waters of Japan, II. Further studies on external characters. Rep. Whales Res. Inst. 9:89-103.
Omura H & S Ohsumi. 1964 A review of Japanese whale

marking in the North Pacific to the end of 1962, with some information on marking in the Antarctic. Norsk Hvalfangst Tid. 53(4):90-112.
Omura H, Ichihara T & T Kasuya. 1970 Osteology of pygmy blue whale with additional information on external and other characters. Sci. Rep. Whales Res. Inst. 22:1-28.
Omura H, Nishimoto S & K Fujino. 1952 Sei whales (Balaenoptera borealis) in the adjacent waters of Japan. Japan Whaling Assoc.,pp 79.
Omura H, Ohsumi S, Nemoto T, Nasu K & T Kasuya. 1969 Black right whales in the North Pacific. Sci. Rep. Whales Res. Inst. 21:1-78.
Paiva M P. 1961 Recursos basicos da pesca maritima no nord-este Brasiliero. Bol. Est. Biol. Mar. Univ. Ceara 3: pp10.
Paiva M P & B F Grangeiro. 1965 Biological investigations on the whaling seasons 1960-1963 off northeastern coast of Brazil. Arq. Est. Biol. Mar. Uni. Fed. Ceara. 5(1):29-64.
Paiva M P & B F Grangeiro. 1970 Investigations on the whaling seasons 1964-67, off northeastern coast of Brazil. Arq. Est. Biol. Mar. Uni. Fed. Ceara, Fortaleza. 10(2):111-126.
Paloheimo J E. 1961 Studies on estimation of mortalities,1: comparison of a method described by Beverton and Holt and a new linear formula. J. Fish. Res. Bd. Canada. 18:645-662.
Pasterne L, Acevedo M & V A Gallardo. 1983 A note on Chilean Bryde's whales. IWC doc SC/35/Ba4,pp8. (mimeo).
Paulsen H B. 1939 Fetus measurements and occurrences of twins and multiple foetuses. Norsk Hvalfangst Tid 12:464-471.
Payne M R. 1979 Growth in the Antarctic fur seal, Arctocephalus gazella. J. Zool. 187:1-20.
Payne R. 1979 Humpback whale songs as an indicator of 'stock'. 3rd Bienn. Conf. Biol. Mar. Mammals, 46 (abstract).
Payne R, Brazier O, Dorsey E M, Perkins J S, Rowntree V J & A Titus. 1983 External features in southern right whales (Eubalena australis) and their use in identifying individuals. p371-445 In (ed R Payne) Communication and Behaviour of Whales. AAAS Selected Symposia Series, 76. Westward Press, Colorado.
Perkins I S, Balcomb K C, Nichols G, & M DeAvilla. 1982b Ecological study of whales, seabirds and the marine environment off West Greenland. Ocean Research & Education Soc.,Cruise Rep. (Regina Maris) 36.

Perkins I S, Bryant P I, Nichols G & D R Pattern. 1982a An ecological study of whales, seabirds and the marine environment off West Greenland and eastern Canada. Ocean Research & Education Soc. Cruise Rep. (Regina Maris). 28.
Perrin W F & G P Donovan. 1984 Report of the workshop. Rep. int. Whal. Commn. (Sp. Is. 6):1-24.
Pervushin A S. 1968 Observations on the feeding and behaviour of the baleen whales at the Crozet Isles. Oceanology 8(1): 139-145.
Peters H. 1955 Uber das Vorkommen des Walkrebschens Euphausia superba Dana und seine Bedeutung fur die Ernahrung der sudlichen Bartenwale. Arch. Fishreiwiss. 6(5):288-304.
Piggot W R. 1977 The importance of the Antarctic in atmospheric science. Phil Trans. R. Soc. Lond. B. 279:278-286.
Pike G C. 1951 Lamprey marks on whales. J. Fish. Res. Bd. Canada 8(4):275-280.
Pike G C. 1964 Identification of Whale stocks. Document Presented to IWC, 16th meeting : pp 2.
Pike G C. 1968 Ear plug lamination counts for North Pacific humpback and Antarctic sei whales. IWC doc to Age Determination Meeting, Oslo. (mimeo).
Pike G C & I B MacAskie. 1969 Marine mammals of British Columbia. Fish. Res. Bd. Canada Bull. 171: 1-54.
Pope J G. 1972 An investigation of the accuracy of virtual population analysis using cohort analysis. Res. Bull. int. Commn. N.W. Atl. Fish. 9:65-74.
Price W S. 1985 Whaling in the Caribbean: historical perspective and update. Rep. int. Whal Commn. 35:413-420.
Privalikhin V I & A A Berzin. 1978 Abundance and distribution of Bryde's whale (Balaenoptera edeni) in the Pacific Ocean. Rep. Int. Whal. Comm. 28: 301-302.
Purves P E. 1955 The wax plug of the external auditory meatus of the Mysticeti. Discovery Rep. 27: 293-302.
Purves P E & M D Mountford. 1959 Ear plug laminations in relation to the age composition of a population of fin whales (Balaenoptera physalus). Bull. Br. Mus. nat. Hist. Zool. 5(6):125-161.
Rankin N. 1951 Antarctic Isle. Collins, London. pp383.
Rayner G W. 1940 Whale marking. Progress and results to December 1939. Discovery Rep. 19:245-284.
Reeves R R, Leatherwood S, Karl S A & E R Yohe. 1985 Whaling results at Akutan (1912-1939) and Port Hebron (1926-1937), Alaska. Rep. int. Whal.

Commn.35:441-458.
Rice D W. 1961 Sei whales with rudimentary baleen. Norsk. Hvalfangst Tid. 50:189-193.
Rice D W. 1963 Progress report on biological studies of the large cetacea in the waters off California. Norsk. Hvalfangst Tid. 52(7):181-187.
Rice D W. 1968 Stomach contents and feeding behaviour of killer whales in the eastern North Pacific. Norsk. Hvalfangst Tid. 57:35-38.
Rice D W. 1974 Whales and whale research in the eastern North Pacific. p170-195 In (ed W E Schevill) The Whale Problem. Harvard University Press, Mass. pp420.
Rice D W. 1977 Synopsis of biological data on the sei whale and Bryde's whale in the Eastern North Pacific. Rep. int. Whal. Commn. (Sp. Is. 1):92-97
Rice D W. 1979 Bryde's whales in the equatorial eastern Pacific. Rep. int. Whal. Commn. 29:321-324
Rice D W & A A Wolman. 1971 The Life History of the Gray Whale (Eschrichtius robustus). Sp. Publ. Am. Soc. Mammologists (3).
Rice D W & A A Wolman. 1982 Whale census in the Gulf of Alaska June to August 1980. Rep. int. Whal. Commn. 32:491-497.
Ricker W E. 1975 Computation and interpretation of biological statistics of fish populations. Bull. Fish. Res. Bd. Canada, 191:1-382.
Risting S. 1928 Whales and whale foetuses. Cons. int. Explor. Mer, Rapp. Proc. Verb. Reun. 50:1-122
Robins J P. 1954 Ovulations and pregnancy corpora lutea in the ovaries in the humpback whale. Nature Lond. 173:201-203.
Robson D S. 1963 Maximum liklihood estimation of a sequence of annual survival rates from a capture-recapture series. Spec. Publ. ICNAF 4:330-335.
Robson D S & D G Chapman. 1961 Catch curves and mortality rates. Trans. Am. Fish. Soc. 90:181-189.
Rocha J M da. 1980 Progress report on Brazilian minke whaling. IWC doc. SC/31/43. (mimeo).
Rocha J M da. 1983 Revision of Brazilian whaling data. Rep. int. Whal. Commn. 33:419-427.
Roe H J S. 1967b The rate of lamina formation in the ear plug of the fin whale. Norsk. Hvalfangst Tid. 56(2):41-43.
Roe H S J. 1967a Seasonal formation of laminae in the ear plug of the sei whale. Discovery Rep. 35:1-30.
Roe H S J. 1968 The ear plug of the sei whale with reference to its validity for age determination. IWC doc. to Age Determination Meeting, Oslo. pp 8. (mimeo).

Rorvik C J. 1980 Whales and whaling off Mozambique. Rep. int. Whal. Commn. 30:213-218.
Rorvik C J & A Jonsgard. 1981 Review of balaenopterids in the North Atlantic Ocean. FAO Fisheries Series,5, Vol 3:269-286.
Rubinstein E S & L G Polozova. 1966 Recent Climate Variations. Hydrometerological Publ. Hse. Leningrad. (In Russian). p268.
Rudolphi D K A. 1822 Einige anatomische Bemerkungen uber Balaena rostrata. Abh. Akad. Wiss. Berlin, 1820-1821:27-40.
Ruud J T. 1932 On the biology of Southern Euphausiidae. Hvalradets Skr. 2:1-105.
Ruud J T. 1937 Seihvalen, Balaenoptera borealis (Lesson) og Brydehvalen, Balaenoptera brydei (Olsen). Norsk. Hvalfangst Tid. 26:145-150.
Ruud J T. 1940 The surface structure of the baleen plates as a possible clue to age in whales. Hvaldradets Skr. 23:1-24.
Ruud J T. 1945 Further studies on the structure of the baleen plates and their application to age determination. Hvaldradets Skr. 29:1-69.
Ruud J T. 1952 Do Sword-fish attack the large baleen whales? Norsk. Hvalfangst Tid. 41(4):191-192
Ruud J T. 1959 The use of baleen plates in age determination of whales. Proc. 15th int. Congr. Zool. 302-303.
Ruud J T, Jonsgard A & P Ottestad. 1950 Age studies on blue whales. Hvaldradets Skr. 33:1-72.
Sacher G A. 1980 The constitutional basis for longevity in the Cetacea: do the whales and the terrestrial mammals obey the same laws? Rep. int. Whal. Commn. (Sp. Is. 3):209-213.
Saemundsson B. 1939 Mammalia. The Zoology of Iceland. Vol 4. Pt.76:1-52. Copenhagen and Reykavik, E. Munksgaard.
Sanpera C & A Aguilar. 1984 Historical review of catch statistics in Atlantic waters off the Iberian Peninsula. IWC doc SC/36/O14 pp22. (mimeo)
Sanpera C & L Jover. 1985 Results of the "Ballena 4" fin whale sightings cruise. IWC doc SC/37/Ba2, pp8. (mimeo).
Sanpera C, Aguilar A, Grau E, Jover L & S A Mizroch. 1984 Report of the 'Ballena 2' whale marking and sighting cruise in the Atlantic waters off Spain. Rep. int. Whal. Commn. 34:663-666.
Scheffer V B & D W Rice. 1963 A list of the marine mammals of the world. Fish. Wildl. Serv. U.S. Spec. Sci. Rep. Fish. No.431:1-12.
Schmidly D J. 1981 Marine mammals of the Southeastern United States coast and Gulf of Mexico.

U.S. Fish Wildl. Serv., Biol. Serv. Prog. FWS/OBS-80/41, pp 163.

Schubert K. 1953 Does there exist a sanctuary for baleen whales in the Antarctic? Norsk Hvalfangst Tid. 42:574-577.

Schulte H von W. 1916 Monographs of the Pacific Cetacea, II. The sei whale (Balaenoptera borealis Lesson). 2. Anatomy of a foetus of Balaenoptera borealis. Mem. Amer. Mus. Nat. Hist., New Series Vol.1, Part VI:389-502.

Schwerdtfeger W. 1959 Meterologia descriptiva del sector Antarctica sud Americano. Inst. Antarctica Argentino, Publ 7.

Sciara G N di. 1983 Bryde's whale(Balaenoptera edeni Anderson 1878) off Eastern Venezuela (Cetacea, Balaenopteridae). Hubbs-Sea World Research Inst. Tech. Rep. no 83-153,pp15.

Seber G A F. 1973 The Estimation of Animal Abundance. Griffin, London. pp506.

Sergeant D E. 1966 Populations of large whale species in the western North Atlantic with special reference to the fin whales. Fish. Res. Bd. Canada, Circ. 9:1-30.

Sergeant D E. 1969 Feeding rates of Cetaceans. Fisk Dir. Skr. Hav. 15:246-258.

Shaler N S. 1873 Notes on the right and sperm whales. Amer. Naturalist 7:1-4.

Sharp G D. 1981 Biochemical genetic studies, their value and limitations in stock identification and discrimination of pelagic mammal species. In Mammals in the Sea. FAO Fisheries Series 5(3):131-136.

Sheldrick M C. 1976 Trends in the strandings of Cetacea on the British coasts 1913-1972. Mammal Rev. 6(1):15-23.

Sheldrick M C. 1979 The nature and occurrence of marine mammal strandings. In (ed J R Geraci and D J St Aubin) Biology of Marine Mammals: Insights Through Strandings. U.S. Marine Mammal Commission Rep. MMc-77/13.

Shevchenko V I. 1970 A riddle of white scars on the body of whales. Priroda 6:72-73. (in Russian).

Shevchenko V I. 1977 Application of white scars to the study of the location and migrations of sei whale populations in Area III of the Antarctic. Rep. int. Whal. Commn. (Sp. Is. 1):130-134.

Shimadzu Y & F Kasamatsu. 1983 Operating pattern of Antarctic minke whaling by the Japanese expedition in 1981/82. Rep. int. Whal. Commn. 33:389-391.

Sigurjonsson J. 1983 The cruise of the Ljosfari in the Denmark Strait (June-July 1981) and recent

marking and sightings off Iceland. Rep. int. Whal. Commn. 33:667-682.
Simpson G G. 1945 The principles of clarification and class- ification of mammals. Bull. Am. Mus. Nat. Hist. 85:1-350.
Sladen W J L. 1964 The distribution of the Adelie and chin- strap penguins. p359-365 In (eds R Carrick, M W Holgate and J Prevost) Biologie Antartique. Hermann. Paris. pp651.
Slijper E J. 1962 Whales. Hutchinson & Co. pp 475.
Slijper E J. 1966 Functional morphology of the reproductive system in Cetacea. p277-319 In (ed K S Norris) Whales, Dolphins and Porpoises. Uni. Calif. Press, Los Angeles. p78.
Slijper E J, Van Utrecht W L & C Naaktgeboren. 1964 Remarks on the distribution and migration of whales based on observations from Netherland ships. Brijdragen tot de Dierkunde, 34:1-93.
Small G L. 1971 The Blue Whale. Columbia Uni. Press. pp 248.
Southwell T. 1904 On the whale fishery from Scotland, with some account of the changes in that industry and of the species hunted. Annals of Scottish Natural History: 77-90.
Southwell T. 1905 Some results of the North Atlantic fin-whale fishery. Annals and Magazine of Natural History. 16: 403-421.
Stewart B S, Yochem P K, Karl S A, Leatherwood S & J L Laake. 1985 Aerial surveys of the former Aleutian, Alaska, whaling grounds. IWC doc SC/37/02, pp15. (mimeo).
Stonehouse B. 1965 Birds and Mammals. p153-186 In (ed T Hatherton) Antarctica. Methuen and Co. Lond. pp 511.
Stonehouse B. 1967 Expanding populations of Pygoscelis Antarctica in South Georgia. Ibis. 109: 277-278.
Struthers J. 1893 On the rudimentary hind-limb of a great fin whale (B. musculus). J. Anat. Physiol. 24:291-335.
Swartz S L & M L Jones. 1983 Gray whale (Eschrichtius robustus) calf production and mortality in the winter range. Rep. int. Whal. Commn. 33:503-507.
Tarpy C. 1979 Killer whale attack! Nat. Geog. Mag. 155:542-45.
Taylor R H. 1962 The Adelie penguin (Pygoscelis adelie) at Cape Royde. Ibis. 104:176-204.
Thompson D'Arcy W. 1928 On whales landed at the Scottish whaling stations, during the years 1908-1914 and 1920-1927. Fishery, Scotland, Sci.

Invest., III. pp 39.
Thompson R J. 1940 Analysis of stomach contents of whales taken during the years 1937 and 1938 from the North Pacific. M. Sc. Thesis, University of Washington Seattle. pp 82.
Tillman M F. 1977 Estimates of population size for the North Pacific sei whale. Rep. int. Whal. Commn.(Sp. Is. 1):98-106.
Tillman M F & J M Breiwick. 1977 Estimates of stock size for the Antarctic sei whale. Rep. int. Whal. Commn. (Sp. Is. 1):107-112.
Tillman M F & B Grenfell. 1980 Estimates of abundance for the Western North Pacific stock of Bryde's whales. Rep. int. Whal. Commn. 30:369-373
Tillman M F & S Ohsumi. 1981 Japanese Antarctic pelagic whaling prior to World War 2: review of catch data. Rep. int. Whal. Commn. 31:625-627.
Tomilin A G. 1945 The age of whales as determined from their baleen apparatus. Comptes Rendus (Doklady) Ac. Sci. USSR 59(6):460.
Tomilin A G. 1954 Adaption types among whales. Zoologicheskii Zhurnal 33(3):677-692.
Tomilin A G. 1967 Cetacea, Vol.9. Mammals of the USSR and adjacent countries. Israel Prog. Sci. Transl. Jerusalem. p717.
Tonnessen J N. 1967 Den Moderne Hvalfangsts Historie. Vol 2. Norges Hvalfangstforbund, Sandefjord. pp 620.
Tonnessen J N & A O Johnsen. 1982 The History of Modern Whaling. C. Hursta Co. London, pp 798.
Townsend C H. 1935 The distribution of certain whales as shown by logbook records of American whaleships. Sci. Contr. New York Zool. Soc., Zoologica 19(1):1-50.
True F W. 1903 First record of the pollack whale Balaenoptera borealis in the Western North Atlantic. Science 17(421):150.
True F W. 1904 The whalebone whales of the Western North Atlantic compared with those occurring in European waters with some observations on the species of the North Pacific. Smithsonion Contributions to Knowledge 33:1-332 + 50 plates
Tsuyuki H, Roberts E & G Pike. 1966 Comparative zone electrophorograms of muscle myogens and blood proteins of five species of whales from the coastal waters of British Columbia. IWC doc. N. Pacific working group Hawaii, p8, (mimeo).
Turner W. 1882 A specimen of Rudophi's whale (Balaenoptera borealis or laticeps) captured in the Firth of Forth. J. Anat. and Phys. 16:471.
Turner W. 1912 The Marine Mammals in the Anatomical

Museum of the University of Edinburgh. MacMillan and Co., Lond, p207.
Uda M. 1954 Studies of the relation between the whaling grounds and the hydrographical conditions. Sci. Rep. Whales Res. Inst. 9:179-187.
Uda M. 1956 Studies of the whaling grounds in the northern sea region of the Pacific Ocean in relation to the meteorological and oceanographic conditions. Sci. Rep. Whales Res. Inst. 11:163-180
Uda M & K Nasu. 1956 Studies of the whaling ground in the northern sea-region of the Pacific Ocean in relation to the meterological and oceanographical conditions. Pt 1. Sci. Rep. Whales Res. Inst. 11: 163-179.
Uda M & A Dairokuno. 1957 A study of the relation between the whaling grounds and hydrographical condition.2. A study of the relationship between the whaling grounds off Kinkazan and the boundary water mass. Sci. Rep. Whales Res. Inst. 12:209-224
Uda M & N Suzuki. 1958 Studies of the relation between the whaling grounds and the hydrographical conditions. 3. Sci. Rep. Whales Res. Inst. 13:215-230.
Ulys C T & P B Best. 1966 Pathology of lesions observed in whales flensed at Saldanha Bay, South Africa. J. Comp. Path. 76:407-412.
Utrecht W L van & C N Utrecht-Cock. 1968 Comparison of records of baleen plates and ear plugs of female fin whales. IWC doc. to Age Determination Meeting, Oslo. (mimeo).
Valdivia J & P Ramirez. 1981 Peru progress report on cetacean research 1979 - 1980. Rep. int. Whal. Commn. 31:211-214.
Valdivia J, Franco F & P Ramirez. 1981 The exploitation of Bryde's whales in the Peruvian sea. Rep. int. Whal. Commn. 31:441-448.
Valdivia J, Landa A & P Ramirez. 1984 Peru progress report on cetacean research 1982 to 1983. Rep. int. Whal. Commn. 34: 223-228.
Valdivia J, Ramirez P, Humberto T & F Franco. 1981 Report of a cruise to mark and assess Bryde's whales of the Peruvian stock, February 1980. Rep. int. Whal. Commn. 31:435-440.
Valdivia J, Landa A, Ramirez P & H Tovar. 1982 Peru progress report on cetacean research May 1980-March 1981. Rep. int Whal. Commn. 32:199-203.
Varona L S. 1965 Balaenoptera borealis Lesson (Mammalia, Cetacea) capturada en Cuba. Pocyana (A) 7:1-4.
Viale D. 1977 Big whale populations on the Atlantic coasts of Spain and the western Mediterranean.

Rep. int. Whal. Commn. 27:235.
Viale D. 1981 Ecologie des Cetaces de la Mediterranie occidental. FAO Fisheries Series, 5, Vol 3: 287-300.
Villa-Ramirez B. 1969 La ballena rorcual o ballena de Aleta, Balaenoptera borealis Lesson 1828, en la costa de Veracruz, Mexico. An. Inst. Biol. Nal. Auton. Mexico. 40 Ser. Zool. (1):129-138.
Wada S. 1973 The ninth memorandum on the stock assessment of whales in the North Pacific. Rep. int. Whal. Commn. 23:164-169.
Wada S. 1974 Status of biochemical studies on whale stock identification in Japan. IWC doc. SC/SP 74/31 (mimeo).
Wada S. 1975 Indices of abundance of large sized whales in the North Pacific in 1973 whaling season Rep. int. Whal. Commn. 25:129-165.
Wada S. 1976 Indices of abundance of large sized whales in the North Pacific in the 1974 whaling season. Rep. int. Whal. Commn. 26:382-391.
Wada S. 1977 Indices of abundance of large sized whales in the North Pacific in the 1975 whaling season. Rep. int. Whal. Commn. 27:189-194.
Wada S. 1978 Indices of abundance of large sized whales in the North Pacific in 1976 whaling season Rep. int. Whal. Commn. 28:319-324.
Wada S. 1979 Indices of abundance of large sized whales in the North Pacific in the 1977 whaling season. Rep. int. Whal. Commn. 29:252-264.
Wada S. 1980 Japanese whaling and whale sighting in the North Pacific 1978 season. Rep. int. Whal. Commn. 30:415-424.
Wada S. 1981 Japanese whaling and whale sighting in the North Pacific 1979 season. Rep. int. Whal. Commn. 31:783-792.
Wada S. 1983 Genetic structure and taxonomic status of minke whales in the coastal waters of Japan. Rep. int. Whal. Commn. 33:361-363.
Wada S. 1984 A note on the gene frequency differences between minke whales from Korean and Japanese coastal waters. Rep. int. Whal. Commn. 34:345-347.
Wada S, Okumoto N & S Ohsumi. 1973 Comparative study of genetic polymorphism in four whale species as an approach to stock identification. Document presented to IWC, 25th meeting : pp 6. (mimeo).
Watkins W A & W E Schevill. 1979 Aerial observations of feeding behavoir in four baleen whales: Eubalaena glacialis, Balaenoptera borealis, Megaptera novaeangliae. J. Mammal. 60:155-163.

Went A E. 1968 Whaling from Ireland. J. Roy. Soc. Antiquaries Ireland, 98:31-36.
Whitehead H. 1979 The baleen whales off the northest coast of Newfoundland. IWC doc SC/31/5 pp9.
Williamson G R. 1973 Counting and measuring baleen and ventral grooves of whales. Sci. Rep. Whales Res. Inst 25:279-292.
Williamson G R. 1975 Minke whales off Brazil. Sci. Rep. Whales Res. Inst. 27:37-59.
Winn H E, Thompson T J, Cummings W C, Hain J, Hudnall J, Hays H & W W Steiner. 1981 Song of the humpback whale - population comparisons. Behav. Ecol. Sociobiol. 8:41-46.
Woolner L & J W Horwood. 1980 Length at capture of sei whales and Areal associations. Rep. int. Whal. Commn. 30:537-544.
Yablokov A V, Belkovich V M & V I Borisov. 1972 Whales and Dolphins. Nauka, Moscow.
Yates G W. 1975 Microclimate, climate and breeding success in Antarctic penguins. p321-330 In (ed B Stonehouse) The Biology of Penguins. Macmillan, Lond. pp555.
Yukhov V I. 1969 On the structure of sei whale (Balaenoptera borealis) within Areas V-VI of the Antarctic. Sbornik morskye mlekopitayuschye, Nauka:305-307.
Zahl S. 1983 Correcting the bias of the CPUE due to a varying whale density. Rep. int Whal. Commn. 33: 307-308.
Zahl S. 1985 Revised model for adjustment of Antarctic Japanese minke whale data 1973-1982. Rep. int. Whal. Commn. 35:223-226.
Zemsky V A. 1980 Atlas of the Marine Mammals of the USSR. Pischevaya Promyshlenost, Moscow. pp184.
Zenkovich B A. 1969 Whales and their yielding in the second Antarctic sector. Trudy VNIRO 66:249-66

INDEX

Geographical

Subject